Studies in Fuzziness and Soft Computing

Volume 396

Series Editor

Janusz Kacprzyk, Systems Research Institute, Polish Academy of Sciences, Warsaw, Poland

The series "Studies in Fuzziness and Soft Computing" contains publications on various topics in the area of soft computing, which include fuzzy sets, rough sets, neural networks, evolutionary computation, probabilistic and evidential reasoning, multi-valued logic, and related fields. The publications within "Studies in Fuzziness and Soft Computing" are primarily monographs and edited volumes. They cover significant recent developments in the field, both of a foundational and applicable character. An important feature of the series is its short publication time and world-wide distribution. This permits a rapid and broad dissemination of research results.

Indexed by ISI, DBLP and Ulrichs, SCOPUS, Zentralblatt Math, GeoRef, Current Mathematical Publications, IngentaConnect, MetaPress and Springerlink. The books of the series are submitted for indexing to Web of Science.

More information about this series at http://www.springer.com/series/2941

Xunjie Gou · Zeshui Xu

Double Hierarchy Linguistic Term Set and Its Extensions

Theory and Applications

Springer

Xunjie Gou
Business School
Sichuan University
Chengdu, Sichuan, China

Zeshui Xu
Business School
Sichuan University
Chengdu, Sichuan, China

ISSN 1434-9922 ISSN 1860-0808 (electronic)
Studies in Fuzziness and Soft Computing
ISBN 978-3-030-51322-1 ISBN 978-3-030-51320-7 (eBook)
https://doi.org/10.1007/978-3-030-51320-7

This Springer imprint is published by the registered company Springer Nature Switzerland AG
The registered company address is: Gewerbestrasse 11, 6330 Cham, Switzerland

Preface

The essence of decision-making is to evaluate alternatives and choose the best alternative(s) based on the established goal in the complex decision-making environment. In the actual decision-making problem, due to the limitations of people's cognitive structure, the variability of decision-making environment, and the complexity and fuzziness of objective things, people usually use relative quantifiers like natural languages or extended linguistic terms to describe or evaluate the uncertainty of objective things. To solve this kind of decision-making problem with linguistic evaluation information, the scholars have put forward lots of linguistic expression models, which can be used to transform the linguistic evaluation information into the corresponding linguistic variables, and rank the alternatives using the traditional linguistic decision-making methods. However, with the rapid development of science and technology and the acceleration of information updating, the complexity of decision-making problems becomes increasingly obvious, so the traditional way of single linguistic evaluation or the way of adding linguistic modifier, or the way of only using numbers or the combination of numbers and linguistic variables can no longer be applied to the dynamic and complex decision-making problems that need multiple linguistic terms to describe. Based on this, taking the complex decision-making problems as the research objects, Gou et al. (2017) firstly put forward the concept of double hierarchy linguistic term set (DHLTS). Soon afterwards, some extensions of DHLTS have been developed such as hesitant fuzzy environment named double hierarchy hesitant fuzzy linguistic term set (DHHFLTS), unbalanced DHLTS (UDHLTS), linguistic preference ordering (LPO), double hierarchy hesitant fuzzy linguistic preference relation (DHLPR), double hierarchy hesitant fuzzy linguistic preference relation (DHHFLPR), etc.

Based on the DHLTS and its extensions, this book gives a thorough and systematic introduction to the latest research results on double hierarchy linguistic term set theory and applications, which includes the measure methodologies of double hierarchy hesitant fuzzy linguistic information, the consistency methodologies, the group consensus decision-making methodologies of DHHFLPRs, LPOs and self-confident DHLPR and the large-scale group consensus decision-making methodologies of DHHFLPRs in depth and systematically. We apply these methodologies to some practical decision-making problems under different double hierarchy linguistic environments. The book is constructed into five chapters that deal with different but related issues, which are listed as follows:

Chapter 1 mainly introduces the state of the art of DHLTSs to fully reflect people's true evaluation of objective things in complex decision-making environments from the perspectives of fully analyzing human cognition and parsing complex linguistic information structure. Additionally, this chapter develops some extensions of DHLTS under different decision-making environments including DHHFLTS, UDHLTS, LPO, DHLPR, DHHFLPR, etc. Furthermore, some operational laws of DHLTSs and DHHFLTSs are introduced. Finally, several comparative methods of DHLTSs and DHHFLTSs are proposed based on the expected and variance values, the hesitance degrees, and the envelopes of DHHFLTSs.
Chapter 2 studies some measure methodologies of double hierarchy hesitant fuzzy linguistic information to optimize the deviation between different parameters such as linguistic terms and preference information, etc., Firstly, this chapter studies some traditional distance measures of double hierarchy hesitant fuzzy linguistic information such as Hamming distance, Euclidean measure and Hausdorff distance, introduces the concept of the hesitance degree of DHHFLTS, and discusses some distance and similarity measures of double hierarchy hesitant fuzzy linguistic information with hesitance degrees. Additionally, this chapter investigates the ordered weighted distance and similarity measures and the hybrid ordered weighted distance and similarity measures of double hierarchy hesitant fuzzy linguistic information. Finally, this chapter develops decision-making methods based on the proposed distance measures and applies them to solve the practical problems of Sichuan liquor brands evaluation.
Chapter 3 establishes the consistency theory framework of double hierarchy hesitant fuzzy linguistic preference information, including additive consistency and multiplicative consistency. Firstly, this chapter introduces some consistency checking methods and inconsistency repairing methods from the perspectives of additive consistency and multiplicative consistency, respectively. Then, this chapter proposes group decision-making methods based on the discussed additive consistency and multiplicative consistency and applies them to solve the actual problems of Sichuan province water resources situation assessment and venture capital assessment of real estate market, respectively. These two consistency methods can comprehensively evaluate the consistencies of DHHFLPRs from different angles.
Chapter 4 firstly studies the group consensus decision-making method based on DHHFLPRs. In group decision-making processes, based on the additive consistencies or multiplicative consistencies of DHHFLPRs, this chapter studies the correlation coefficient and correlation measures of DHHFLPRs and constructs the group consensus decision-making method and applies this method to the venture

capital assessment of real estate market. Then, this chapter studies the consensus decision-making methods with LPOs. This chapter develops models to equivalently transform each LPO into the corresponding DHLPR with complete consistency and proposes some consensus models with DHLPRs to obtain the final decision-making result which is equal to the decision-making result with LPO information. Additionally, this chapter defines the concept of self-confident DHLPR, in which the basic element consists of the DHLT and the self-confident degree simultaneously. Finally, this chapter develops a double hierarchy linguistic preference values and self-confident degrees modifying-based consensus model to manage the group decision-making (GDM) problems with self-confident DHLPRs based on the priority ordering theory.

Chapter 5 studies the large-scale group consensus decision-making methods based on DHHFLPRs and applies them to deal with practical large-scale group decision-making (LSGDM) problems. Specially, the concept of a large-scale group consensus decision-making-related problem is not meant here in the sense of computational social choice and related areas (Brandt et al. 2016; Chevalayre et al. 2007). On the one hand, this chapter mainly discusses the large-scale group clustering method and the weight-determining method, proposes the large-scale group consensus decision-making method and applies this method to the assessments of water resources in some cities of Sichuan province. On the other hand, by constructing new clustering method and consensus model, and from the perspective of in-depth analyzing minority opinions and non-cooperative behaviors in LSGDM, this chapter puts forward a novel large-scale group consensus decision-making method based on DHHFLPRs, which is more in line with human cognition, and applies this method to the comprehensive assessments of the reasons of haze formation.

This book is suitable for the engineers, technicians, and researchers in the fields of fuzzy mathematics, operations research, information science, management science and engineering, etc. It can also be used as a textbook for postgraduate and senior-year undergraduate students of the relevant professional institutions of higher learning.

This work was supported in part by the National Natural Science Foundation of China under Grant 71771155 and Grant 71571123 and the Fundamental Research Funds for the Central Universities under Grant YJ202015 and Grant 2020ZY-SX-C01.

Special thanks to Prof. Francisco Herrera at the University of Granada and Prof. Huchang Liao at the Sichuan University for lots of insightful ideas and great suggestions.

Chengdu, China
March 2020

Xunjie Gou
Zeshui Xu

References

Brandt F, Conitzer V, Endriss U, Lang J, Procaccia AD (2016) Handbook of computational social choice. Cambridge University Press

Chevaleyre Y, Endriss U, Lang J, Maudet N (2007) A short introduction to computational social choice. In: van Leeuwen J, Italiano GF, van der Hoek W, Meinel C, Sack H, Plášil F (eds) SOFSEM 2007: theory and practice of computer science. lecture notes in computer science, vol 4362. Springer, Berlin, Heidelberg, pp 51–69

Gou XJ, Liao HC, Xu ZS, Herrera F (2017) Double hierarchy hesitant fuzzy linguistic term set and MULTIMOORA method: a case of study to evaluate the implementation status of haze controlling measures. Inf Fusion 38:22–34

Contents

Chapter 1
Double Hierarchy Linguistic Term Set and Its Extensions

Recently, Artificial Intelligence (AI) has become more and more popular and important in real life and consists of many research fields such as language recognition, image recognition, natural language processing and expert systems. From an AI point of view, one of the most important parts is how to collect and represent natural languages correctly. In order to deal with natural languages, Zadeh (2012) introduced the concept of Computing with Words (CW), and explained it with the following "*Computing with words is a system of computation in which the objects of computation are words, phrases and propositions drawn from a natural language. The carriers of information are propositions. It is important to note that Computing with words is the only system of computation which offers a capability to compute with information described in a natural language.*" Some linguistic models based on CW, such as the hesitant fuzzy linguistic term set (HFLTS) (Rodríguez et al. 2012), the 2-tuple linguistic model (Herrera and Martínez 2000), the virtual linguistic model (Xu 2004a; Xu and Wang 2017) and the linguistic terms with weakened hedges (Wang et al. 2018), etc., have been developed by corresponding syntax and semantic rules. However, the existing linguistic models have some gaps. For example, we cannot use them to express some words such as "*only a little high*" or linguistic sets as "{*only a little good, very good*}", etc. Therefore, it is necessary to consider an important issue: Does it make sense if we split each complex linguistic term into two parts with the form of "adverb + adjective" and express them by different kinds of linguistic terms? Based on this idea, Gou et al. (2017) defined the concept of DHLTS. Soon afterwards, some scholars have developed lots of extensions of DHLTS including double hierarchy hesitant fuzzy linguistic term set (DHHFLTS) (Gou et al. 2017), unbalanced DHLTS (UDHLTS) (Fu and Liao 2019), linguistic preference orderings (LPOs) (Gou et al. 2020b), double hierarchy linguistic preference relation (DHLPR) (Gou et al. 2020a) and double hierarchy hesitant fuzzy linguistic preference relation (DHHFLPR) (Gou et al. 2018a, 2019). In this chapter, we mainly introduce the concepts of DHLTS and its extensions. In addition, the operational laws and the comparative methods of DHLTSs and DHHFLTSs are proposed, respectively.

X. Gou and Z. Xu, *Double Hierarchy Linguistic Term Set and Its Extensions*, Studies in Fuzziness and Soft Computing 396,
https://doi.org/10.1007/978-3-030-51320-7_1

1.1 Double Hierarchy Linguistic Term Set

HFLTS can be used to express the evaluation information for an event or a decision-making problem such as "*fast*", "*more*", "*between high and perfect*", etc. However, when we need to describe some more detailed sentences like "*a little fast*", "*almost 90% perfect*", and "*between much high and very high*", the HFLTS cannot describe them accurately and in detail. Therefore, Gou et al. (2017) defined the concept of DHLTS, which consists of two hierarchy fully independent linguistic term sets (LTSs).

Definition 1.1 (Gou et al. 2017) Let $S = \{s_t | t = -\tau, \ldots, -1, 0, 1, \ldots, \tau\}$ be the first hierarchy LTS, $O^t = \{o_k^t | k = -\varsigma, \ldots, -1, 0, 1, \ldots, \varsigma\}$ be the second hierarchy LTS of s_t. Then a DHLTS can be denoted by

$$S_O = \{s_{t<o_k^t>} | t = -\tau, \ldots, -1, 0, 1, \ldots, \tau;\ k = -\varsigma, \ldots, -1, 0, 1, \ldots, \varsigma\} \quad (1.1)$$

where $s_{t<o_k^t>}$ is called the double hierarchy linguistic term (DHLT), and the $<o_k^t>$ $(k = -\varsigma, \ldots, -1, 0, 1, \ldots, \varsigma)$ express the different degrees of linguistic term s_t. For convenience, the DHLT can be simplified using $s_{t<o_k>}$. Accordingly, the DHLTS can be simplified by $S_O = \{s_{t<o_k>} | t = -\tau, \ldots, -1, 0, 1, \ldots, \tau;$ $k = -\varsigma, \ldots, -1, 0, 1, \ldots, \varsigma\}$.

To understand DHLTS more clearly, Gou et al. (2017) drawn the semantic rule of DHLT in Fig. 1.1.

In Fig. 1.1, the second hierarchy LTS $O^1 = \{o_k^1 | k = -2, -1, 0, 1, 2\}$ of the first hierarchy linguistic term s_1. Meanwhile, four characteristics of DHLTS can be summarized: firstly, all DHLTs are expressed by linguistic labels instead of any numerical scales or something; Secondly, When the linguistic information included in the first hierarchy LTS is large, we only propose the second hierarchy LTS; Thirdly, Each second hierarchy LTS can be regarded as a set of adverbs; Finally,

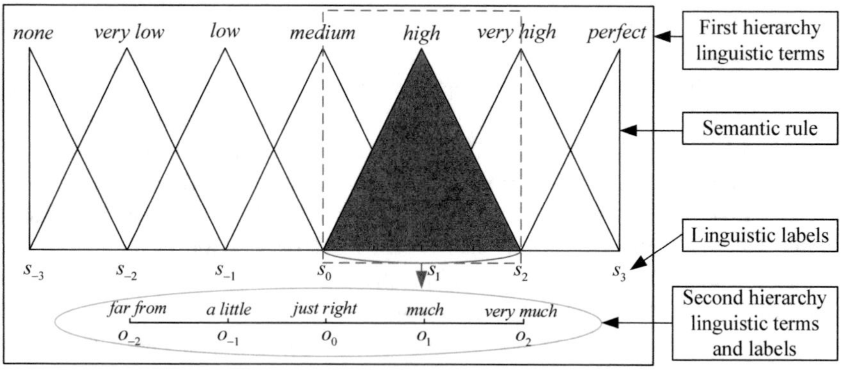

Fig. 1.1 The second hierarchy LTS of a linguistic term in the first hierarchy LTS

Each linguistic term in the first hierarchy LTS has its own second hierarchy LTS, and they are usually different (Gou et al. 2017; Montserrat-Adell et al. 2019).

Remark 1.1 (Gou et al. 2017) Several details about the selections of the second hierarchy LTSs need to be further explained on the basis of the value of t.

(1) For the first hierarchy LTS $S = \{s_t | t = -\tau, \ldots, -1, 0, 1, \ldots, \tau\}$, if $t \geq 0$, then the second hierarchy LTS needs to be described in ascending order just like Fig. 1.1. On the contrary, if $t \leq 0$, then the second hierarchy LTS needs to be described in descending order. Moreover, we only change the orders of linguistic information, and do not change the orders of the linguistic terms $o_k(k = -\varsigma, \ldots, -1, 0, 1, \ldots, \varsigma)$. For example, suppose that $\varsigma = 3$, then we let

$$O^+ = \left\{ \begin{array}{l} o_{-3} = \textit{far from}, o_{-2} = \textit{only a little}, o_{-1} = \textit{a little}, \\ o_0 = \textit{just right}, \\ o_1 = \textit{much}, o_2 = \textit{very much}, o_3 = \textit{entirely} \end{array} \right\},$$
$$\textit{if } t \geq 0$$

and

$$O^- = \left\{ \begin{array}{l} o_{-3} = \textit{entirely}, o_{-2} = \textit{very much}, o_{-1} = \textit{much}, o_0 = \textit{just right}, \\ o_1 = \textit{a little}, o_2 = \textit{only a little}, o_3 = \textit{far from} \end{array} \right\},$$
$$\textit{if } t \leq 0$$

be the second hierarchy LTSs, respectively.

(2) If $t = \tau$, then we only consider the front half of the second hierarchy LTS, i.e., $O = \{o_k | k = -\varsigma, \ldots, -1, 0\}$. On the contrary, if $t = -\tau$, then we only consider the latter half of the second hierarchy LTS, i.e., $O = \{o_k | k = 0, 1, \ldots, \varsigma\}$.

(3) In Fig. 1.1, we can utilize any one linguistic term included in the second hierarchy LTS O to describe the linguistic term s_1 (*high*). For example, we can use "*a little high*" and "*much high*" to express the different meanings of the "*high*". Obviously, the description is more correct and detailed.

1.2 Some Extensions of DHLTS

In recent years, lots of extensions of DHLTS have been developed under different decision-making environments. Firstly, considering that sometimes experts may be hesitant among some different linguistic information when they want to express their assessments more exactly, Gou et al. (2017) extended DHLTS into hesitant fuzzy environment and defined the concept of DHHFLTS. Secondly, because the unbalanced semantics may appear in the first and second hierarchy LTSs, Fu and Liao (2019) developed the concept of UDHLTS. Furthermore, even though preference ordering structures (Chiclana et al. 1998; Dong et al 2008; He and Xu 2018; Tanino 1984) are useful and popular tools to represent experts' preferences in decision-making processes, they usually lack the research on the precise relationship between any two adjacent alternatives. Therefore, Gou et al. (2020b) established the concepts of LPOs in which the ordering of alternatives and the relationships between two adjacent alternatives are fused well. Moreover, considering that the pairwise comparison methods are more accurate than non-pairwise methods because experts only need to focus exclusively on two alternatives at a time (Chiclana et al. 2009; Millet 1997). Then, the concepts of DHLPR (Gou et al. 2020a) and DHHFLPR (Gou et al. 2018a, 2019) were defined.

1.2.1 Double Hierarchy Hesitant Fuzzy Linguistic Term Set

In decision-making processes with double hierarchy linguistic information, considering that sometimes experts may be hesitant among some possible DHLTs. Therefore, by extending DHLTS into hesitant fuzzy environment, Gou et al. (2017) introduced the concept of DHHFLTS.

Definition 1.2 (Gou et al. 2017) Let S_O be a DHLTS. A double hierarchy hesitant fuzzy linguistic term set (DHHFLTS) on X, H_{S_O}, is in mathematical form of

$$H_{S_O} = \{<x_i, h_{S_O}(x_i)> | x_i \in X\} \tag{1.2}$$

where $h_{S_O}(x_i)$ is a set of some values in S_O, denoted as:

$$\begin{aligned} h_{S_O}(x_i) = \{ & s_{\phi_l<o_{\varphi_l}>}(x_i) | s_{\phi_l<o_{\varphi_l}>} \in S_O; l = 1, 2, \ldots, L; \phi_l = -\tau, \ldots, \tau; \\ & \varphi_l = -\varsigma, \ldots, \varsigma\} \end{aligned} \tag{1.3}$$

with L being the number of DHLTs in $h_{S_O}(x_i)$ and $s_{\phi_l<o_{\varphi_l}>}(x_i)$ $(l = 1, \ldots, L)$ in each $h_{S_O}(x_i)$ being the continuous terms in S_O. $h_{S_O}(x_i)$ denotes the possible degree of the linguistic variable x_i to S_O. For convenience, we call $h_{S_O}(x_i)$ the double hierarchy hesitant fuzzy linguistic element (DHHFLE), and $\Phi \times \Psi$ being the set of all DHHFLEs.

To understand the DHHFLTS much more clearly, Gou et al. (2017) established a context-free grammar $\aleph_{DHH}$, which generates some simple but rich linguistic expressions represented by DHHFLTSs.

Definition 1.3 (Gou et al. 2017) Let S_O be a DHLTS, $\aleph_{DHH}$ be a context-free grammar. The element of $\aleph_{DHH} = \{\dot{V}_N, \dot{V}_T, \dot{I}, \dot{P}\}$ can be defined as:

$\dot{V}_N = \{$ <double hierarchy primary term > ,
<double hierarchy composite term > ,
<unary relation > , <binary relation > , <conjunction > $\}$

$\dot{V}_T = \{$less than; more than; between; and; $s_{-\tau}, s_{1-\tau}, \ldots, s_0, \ldots, s_{\tau-1}, s_\tau$;
$o_{-\varsigma}, o_{1-\varsigma}, \ldots, o_0, \ldots, o_{\varsigma-1}, o_\varsigma\}$

$\dot{I} \in \dot{V}_N$.

For the context-free grammar $\aleph_{DHH}$, the production rules $\dot{P}$ can be defined as:

$\dot{P} = \{\dot{I}$ :: = <double hierarchy primary term > | <double hierarchy composite term >
<double hierarchy composite term > ::= <unary relation >
<double hierarchy primary term > | <binary relation >
<double hierarchy primary term > <conjunction >
<double hierarchy primary term >
<double hierarchy primary term > ::= $s_{-\tau<o_{-\varsigma}>} | s_{-\tau<o_{-\varsigma+1}>} | \ldots | s_{\tau<o_{\varsigma-1}>} | s_{\tau<o_\varsigma>}$
<unary relation > ::= less than | more than
<binary relation > ::= between
<conjunction > ::= and$\}$.

Remark 1.2 (Gou et al. 2017) (1) There exist some limitations about the "*unary relation*". The "*double hierarchy primary term*" cannot be $s_{-\tau<o_0>}$ if the nonterminal symbol is "*less than*". Similarly, the "*double hierarchy primary term*" cannot be $s_{\tau<o_0>}$ if the nonterminal symbol is "*more than*".

(2) For the "*binary relation*", the "*double hierarchy primary term*" on the left-hand side must be less than the "*double hierarchy primary term*" on the right-hand side.

Additionally, to understand the DHHFLTS much better, Gou et al. (2017) introduced the concept of the envelope of a DHHFLE:

Definition 1.4 (Gou et al. 2017) The envelope of a DHHFLE, $env(h_{S_O})$, is a double hierarchy linguistic interval whose limits are obtained by means of the upper bound (max) and the lower bound (min). That is

$$env(h_{S_O}) = [h_{S_O}^-, h_{S_O}^+] \tag{1.4}$$

which is just an uncertain linguistic variable (Xu 2004b). Clearly, the DHHFLE h_{S_O} contains all the elements from the lower bound $h_{S_O}^-$ to the upper bound $h_{S_O}^+$.

Example 1.1 (Gou et al. 2017) Let S_O be a DHLTS. Suppose that $\tau = \varsigma = 3$ and the linguistic labels are the same as those in Fig. 1.1. Three linguistic expressions are listed as:

(1) "*a little high*";
(2) "*between much medium and just right very high*";
(3) "*just right perfect*".

Then we can utilize the DHHFLEs $\{s_{1<o_{-1}>}\}$, $\{s_{0<o_1>}, s_1, s_{2<o_0>}\}$, and $\{s_{3<o_0>}\}$ to transform the above sentences. Besides, $env(\{s_{0<o_1>}, s_1, s_{2<o_0>}\}) = [s_{0<o_1>}, s_{2<o_0>}]$. Specially, for the second linguistic expression "*between much medium and just right very high*", it contains all the linguistic terms from "*much medium*" to "*just right very high*". Therefore, we can utilize s_1 to represent the middle linguistic term without using the form of DHHFLE.

1.2.2 Unbalanced Double Hierarchy Linguistic Term Set

Considering that the DHLTS consists of two hierarchy LTSs which are fully independent. In other words, these two hierarchy LTSs have their own unique syntax and semantics. Therefore, the second hierarchy linguistic terms are of great uncertainty and fuzziness and the uneven distribution is more likely to happen in the second hierarchy (Fu and Liao 2019). Therefore, Fu and Liao (2019) proposed some new linguistic scale functions for the second hierarchy LTS by considering unbalanced cases appearing in the DHLTS. Accordingly, the concept of UDHLTS was constructed by developing five distributions of the second hierarchy linguistic terms, and all of them are shown in Fig. 1.2.

Let $O = \{o_k | k = -\varsigma, \ldots, -1, 0, 1, \ldots, \varsigma\}$ be the second hierarchy LTS, for the first situation shown as Fig. 1.2a, Gou et al. (2017) has research it and the linguistic scale function of the second hierarchy LTS is defined as:

$$v_1(o_k) = \frac{k + \varsigma}{2\varsigma} \tag{1.5}$$

Then, Fu and Liao (2019) defined two linguistic scale functions of the second hierarchy LTS for the remaining situations shown in Fig. 1.2b–e. Firstly, if the semantics of the second hierarchy linguistic terms are unbalanced, shown as Fig. 1.2b and c, then the linguistic scale function of the second hierarchy LTS is obtained as:

$$v_2(o_k) = \left[1 + (2\varsigma + 1)^{-\sigma_1 k}\right]^{-1} \times \frac{\max(k, 0)}{|k|} + \left[1 + (2\varsigma + 1)^{-\sigma_2 k}\right]^{-1} \times \frac{\min(k, 0)}{k} \tag{1.6}$$

where $\sigma_i (\sigma_i \geq 0)$ $(i = 1, 2)$ are the cognitive bias parameters of expert with the "positive" and "negative" preferences, respectively, and can be determined by the model provided by Fu and Liao (2019).

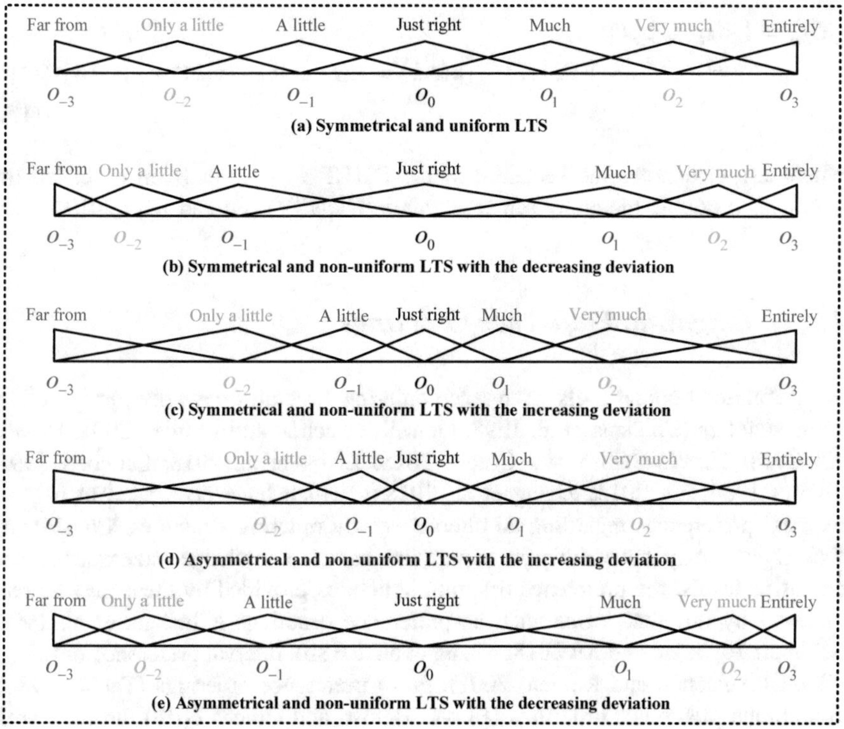

Fig. 1.2 Five distributions of the second LTS

In addition, if the semantics of the second hierarchy linguistic terms are unbalanced, shown as Fig. 1.2d and e, then the linguistic scale function of the second hierarchy LTS is obtained as:

$$
\begin{aligned}
v_3(o_k) = & \left\{ \frac{4\varsigma(k+\varsigma)}{4\varsigma^2 - 1} - \left[1 + (2\varsigma+1)^{-\sigma_3 \frac{(4\varsigma^2+1)k+2\varsigma}{4\varsigma^2-1}} \right]^{-1} \right\} \times \frac{\max(k,0)}{|k|} \\
& + \left\{ \frac{4\varsigma(k+\varsigma)}{4\varsigma^2 - 1} - \left[1 + (2\varsigma+1)^{-\sigma_4 \frac{(4\varsigma^2+1)k+2\varsigma}{4\varsigma^2-1}} \right]^{-1} \right\} \times \frac{\min(k,0)}{k}
\end{aligned}
\tag{1.7}
$$

Similarly, $\sigma_i(\sigma_i \geq 0)(i = 3, 4)$ are the cognitive bias parameters of expert.

Then, the concept of UDHLTS was developed:

Definition 1.5 (Fu and Liao 2019) Let $S = \{s_t | t = -\tau, \ldots, -1, 0, 1, \ldots, \tau\}$ and $O = \{o_k | k = -\varsigma, \ldots, -1, 0, 1, \ldots, \varsigma\}$ be the first hierarchy LTS and the second hierarchy LTS, respectively, and $\mu : s_t \rightarrow [0, 1\}$ and $v : o_k \rightarrow [0, 1]$ be their linguistic scale functions, respectively. If any one of S and O is non-uniformly distributed, then, the UDHLTS can be defined as:

$$U_{S_O} = \{ <s_{t<o_k>}, u(t,k)> | t = -\varsigma, \ldots, -1, 0, 1, \ldots, \varsigma; u(t,k) = \mu(t-1)(1 - v(k)) + \mu(t+1)v(k) \} \tag{1.8}$$

where $u(t,k)$ denotes the semantic of the DHLT $s_{t<o_k>}$, and can be called the unbalanced double hierarchy semantic value (UDHSV) (Fu and Liao 2019).

1.2.3 Linguistic Preference Orderings

As useful and popular tools, preference ordering structures have been proposed by some scholars (Chiclana et al. 1998; González-Pachón and Romero 2001; He and Xu 2018; Hervés-Beloso and Cruces 2018; Liang et al. 2018; Schubert 1995; Tanino 1984; Xu 2013; Zhang et al. 2018a), which have been used to express experts' preferences regarding all alternatives by ordering structures. Considering that experts may have different expression habits, knowledge backgrounds and cognitive levels, the preference ordering structures provided by them may be represented by different forms such as preference orderings (Chiclana et al. 1998; Schubert 1995; He and Xu 2018; Zhang et al. 2018b), interval preference orderings (González-Pachón and Romero 2001), fuzzy preference orderings (Tanino 1984), continuous preference orderings (Hervés-Beloso and Cruces 2018), hesitant preference ordering sets (He and Xu 2018), etc. For instance, suppose that $\{A_1, A_2, A_3, A_4\}$ is a set of alternatives, then $\{2, 1, 4, 3\}$ is a preference ordering in which the positive integers are used by the experts to show the order positions of alternatives. Similarly, other preference orderings can be represented by different forms of preference information such as interval positive integers, hesitant fuzzy numbers, etc.

However, two critical problems about the existing preference orderings have not yet been fully addressed:

Firstly, the existing preference orderings can only reflect the ranking order of alternatives, but lack the research on the precise relationship between any two adjacent alternatives in the preference orderings. In other words, the existing preference orderings default to the same relationship between two adjacent alternatives in the preference orderings (Chiclana et al. 1998). For example, when evaluating the comprehensive quality of three cars $\{A_1, A_2, A_3\}$, expert may say "*A_2 is very better than A_1, and A_1 is slightly better than A_3*". However, we can only obtain the preference ordering $\{2, 1, 3\}$, but ignore the words that involve the relationships between any two adjacent alternatives such as " *very better*" and " *slightly better*". Therefore, to represent preferences more comprehensively and correctly, the preference ordering should include not only the ordering of alternatives, but also the relationships between any two adjacent alternatives in preference ordering.

Secondly, the existing methods mainly aggregate the preference orderings and then obtain the final ordering of all alternatives (He and Xu 2018). However, because the relationships between any two adjacent alternatives are usually unbalanced, so there is an urgent need to deal with the preference orderings that contain both the ordering of alternatives and the relationships between two adjacent alternatives.

To overcome the first problem, it is necessary to establish a novel preference ordering structure in which the order of alternatives and the relationships between two adjacent alternatives should be fused well. Considering that natural languages are more in line with the real thoughts of people, and DHLTS can be used to handle complex linguistic information well. Therefore, Gou et al. (2020b) developed the concepts of LPOs by combining the preference ordering and DHLTs. Additionally, depending on the interests of the experts, there exist two situations: One is that all alternatives are in a preference ordering, which is called the LPO in continuous form; The other one is that a set contains several preference orderings, and each of them only consists of the relationship of two alternatives, which is called the LPO in decentralized form. The descriptions of the above two LPOs can be given in detail below:

(a) **The LPO in continuous form**

Definition 1.6 (Gou et al. 2020b) Let S_O be a DHLTS, and $A = \{A_1, A_2, \ldots, A_m\}$ be a set of alternatives. Then the LPO in continuous form can be defined as follows:

$$LPO' = \left\{ A_{\sigma(1)} \overset{s_{t<o_k>}^{(\sigma(1),\sigma(2))}}{>} A_{\sigma(2)} \overset{s_{t<o_k>}^{(\sigma(2),\sigma(3))}}{>} \cdots \overset{s_{t<o_k>}^{(\sigma(m-1),\sigma(m))}}{>} A_{\sigma(m)} \right\} \tag{1.9}$$

where $\bigoplus_{i=1}^{m-1} s_{t<o_k>}^{(\sigma(i),\sigma(i+1))} \le s_{\tau<o_\varsigma>}$, $A_{\sigma(i)}$ denotes the i-th largest alternative, and the linguistic preference, denoted as a DHLT $s_{t<o_k>}^{(\sigma(i),\sigma(i+1))}$, means that the degree of the i-th largest alternative is better than the $(i+1)$-th largest alternative.

(b) **The LPO in decentralized form**

Considering that sometimes some experts like to provide several pairwise comparisons between any two alternatives. Then this kind of preference information can be expressed as follows:

$$LPO'' = \left\{ A_i \overset{s_{t<o_k>}^{ij}}{>} A_j \middle| s_{t<o_k>}^{ij} \in S_O, \quad i,j = 1,2,\ldots,m; i \ne j \right\} \tag{1.10}$$

where $s_{t<o_k>}^{ij}$ expresses the relationship between the alternatives A_i and A_j $(i,j = 1,2,\ldots,m; i \ne j)$.

1.2.4 Double Hierarchy Linguistic Preference Relation

In recent years, the pairwise comparison methods are more accurate than non-pairwise methods (Millet 1997), and the main advantage of pairwise comparison is that experts only need to focus exclusively on two alternatives at a time when expressing their preferences (Chiclana et al. 2009). In decision-making processes, more and more experts prefer to provide their preferences by making pairwise comparisons between any two alternatives, meanwhile, this kind of preference reflects the relationships between different alternatives intuitively. Therefore, preference relation becomes one of the popular and effective tools. Motivated by the advantages of preference relation and DHLTS, Gou and Xu (2020a) defined the concept of DHLPR and used it to express the evaluation information of all experts more reasonably.

Before giving the definition of DHLPR, Gou and Xu (2020a) developed the operational laws, addition and multiplication, for DHLTs under some special conditions:

Definition 1.7 (Gou and Xu, 2020a) Let S_O be a DHLTS, $s_{t^1<o_{k^1}>}$, $s_{t^2<o_{k^2}>}$ and $s_{t<o_k>}$ be three DHLTs, $\lambda(0<\lambda<1)$ be a real number. Then

(1) $s_{t^1<o_{k^1}>} \oplus s_{t^2<o_{k^2}>} = s_{t^1+t^2<o_{k^1+k^2}>}$, if $t^1+t^2 \leq \tau$ and $k^1+k^2 \leq \varsigma$;
(2) $\lambda s_{t<o_k>} = s_{\lambda t<o_{\lambda k}>}$, $0<\lambda<1$.

In decision-making processes, let $A = \{A_1, A_2, \ldots, A_m\}$ be a fixed set of alternatives, then the DHLPR can be developed:

Definition 1.8 (Gou and Xu, 2020a) Let S_O be a DHLTS. Then a DHLPR $\mathbb{R}$ is presented by a matrix $\mathbb{R} = (r_{ij})_{m\times m} \subset A \times A$, where $r_{ij} \in S_O$ $(i,j = 1,2,\ldots,m)$ is a DHLT, indicating that the degree of A_i is superior to A_j. For all $i,j = 1,2,\ldots,m$, $r_{ij}(i<j)$ satisfies the conditions $r_{ij} + r_{ji} = s_{0<o_0>}$ and $r_{ii} = s_{0<o_0>}$.

1.2.5 Double Hierarchy Hesitant Fuzzy Linguistic Preference Relation

Similar to Definition 1.7, Gou et al. (2019) defined the concept of DHHFLPR under some special conditions:

Definition 1.9 (Gou et al. 2019) Let S_O be a DHLTS. Suppose that $h_{S_O} = \{s_{\phi_l<o_{\varphi_l}>} | s_{\phi_l<o_{\varphi_l}>} \in S_O;\ l = 1,2,\ldots,\#h_{S_O}\}$ and $h_{S_{O_i}} = \{s_{\phi_l^i<o_{\varphi_l^i}>} | s_{\phi_l^i<o_{\varphi_l^i}>} \in S_O;\ l = 1,2,\ldots,\#h_{S_O}^i\}(i = 1,2;\ \#h_{S_O}^1 = \#h_{S_O}^2)$ are three DHHFLEs, and $\lambda(0 \leq \lambda \leq 1)$ is a real number. Some operational laws of DHHFLEs were developed as follows:

(1) Addition: $h_{S_{O_1}} \oplus h_{S_{O_2}} = \bigcup_{s_{\phi_l^1 < o_{\varphi_l^1} >} \in h_{S_{O_1}}, s_{\phi_l^2 < o_{\varphi_l^2} >} \in h_{S_{O_2}}} \left\{ s_{\phi_l^1 + \phi_l^2 < o_{\varphi_l^1 + \varphi_l^2} >} \right\}$, if $\phi_l^1 + \phi_l^2 \leq \tau, \varphi_l^1 + \varphi_l^2 \leq \varsigma$;

(2) Multiplication: $\lambda h_{S_O} = \bigcup_{s_{\phi_l < o_{\varphi_l}>} \in h_{S_O}} \{ s_{\lambda \phi_l < o_{\lambda \varphi_l}>} \}, 0 \leq \lambda \leq 1$.

Definition 1.10 (Gou et al. 2019) A DHHFLPR $\tilde{H}_{S_O}$ is presented by a matrix $\tilde{H}_{S_O} = (h_{S_{O_{ij}}})_{m \times m} \subset A \times A$, where $h_{S_{O_{ij}}} = \{ h_{S_{O_{ij}}}^{\sigma(l)} | l = 1, 2, \ldots, \# h_{S_{O_{ij}}} \}$ ($\# h_{S_{O_{ij}}}$ is the number of DHLTs in $h_{S_{O_{ij}}}$, $h_{S_{O_{ij}}}^{\sigma(l)}$ is the l - th DHLT in $h_{S_{O_{ij}}}$) is a DHHFLE, indicating hesitant degrees to which A_i is preferred to A_j. For all $i, j = 1, 2, \ldots, m$, $h_{S_{O_{ij}}} (i < j)$ satisfies the conditions:

(1) $h_{S_{O_{ij}}}^{\sigma(l)} + h_{S_{O_{ji}}}^{\sigma(l)} = s_{0 < o_0 >}$, $h_{S_{O_{ii}}} = s_{0 < O_0 >}$, and $\# h_{S_{O_{ij}}} = \# h_{S_{O_{ji}}}$;

(2) $h_{S_{O ij}}^{\sigma(l)} < h_{S_{O ij}}^{\sigma(l+1)}$ and $h_{S_{O ji}}^{\sigma(l)} > h_{S_{O ji}}^{\sigma(l+1)}$.

Remark 1.3 (Gou et al. 2019) Based on the operations of DHHFLEs, we can utilize $h_{S_{O_{ij}}}^{\sigma(l)} + h_{S_{O_{ji}}}^{\sigma(l)} = s_{0 < o_0 >}$ to check the first condition of the DHHFLPR. Furthermore, considering that a DHHFLE is an ordered finite subset of the consecutive linguistic terms of a DHLTS, then we can also define that the DHLTs in the upper triangle are arranged in an ascending order, while in the lower triangle are arranged in a descending order. That is to say, $h_{S_{O ij}}^{\sigma(l)} < h_{S_{O ij}}^{\sigma(l+1)}$ and $h_{S_{O ji}}^{\sigma(l)} > h_{S_{O ji}}^{\sigma(l+1)}$. For example, one DHHFLPR can be established as:

$$\tilde{H}_{S_O} = \begin{pmatrix} \{ s_{0 < o_0 >} \} & \{ s_{-1 < o_1 >}, s_0, s_{1 < o_2 >} \} & \{ s_{-1 < o_{-2} >}, s_{0 < o_1 >} \} \\ \{ s_{1 < o_{-1} >}, s_0, s_{-1 < o_{-2} >} \} & \{ s_{0 < o_0 >} \} & \{ s_{2 < o_1 >} \} \\ \{ s_{1 < o_2 >}, s_{0 < o_{-1} >} \} & \{ s_{-2 < o_{-1} >} \} & \{ s_{0 < o_0 >} \} \end{pmatrix}$$

1.3 Operational Laws

Note that, in Definition 1.1, the DHLTs are chosen in discrete form from S_O and the value range of subscripts of $s_{\phi_l < o_{\varphi_l} >}(x_i)$ is $\{ \phi_l = -\tau, \ldots, -1, 0, 1, \ldots, \tau;$ $\varphi_l = -\varsigma, \ldots, -1, 0, 1, \ldots, \varsigma \}$. Similar to the continuous LTS (Xu 2005), Gou et al. (2017) extended it to continuous form, i.e., $\bar{S}_O = \{ s_{t < o_k^t >} | t = [-\tau, \tau]; k = [-\varsigma, \varsigma] \}$. Then, based on the discussion of monotonic function of Dubois (2011), Gou et al. (2017) defined two monotone functions for making the mutual transformations between the DHLT and the numerical scale before defining the operational laws of the DHHFLEs.

Definition 1.10 (Gou et al. 2017) Let $\bar{S}_O$ be a DHLTS. $h_{S_O} = \{s_{\phi_l<o_{\varphi l}>} | s_{\phi_l<o_{\varphi l}>} \in S_O; \quad l = 1,2,\ldots,L; \phi_l = -\tau,\ldots,-1,0,1,\ldots,\tau; \varphi_l = -\varsigma,\ldots,-1,0,1,\ldots,\varsigma\}$ be a DHHFLE with L being the number of DHLTs in h_{S_O}, and $h_\gamma = \{\gamma_l | \gamma_l \in [0,1]; l = 1,\ldots,L\}$ be a hesitant fuzzy element (HFE) (Xia and Xu 2011). Then the membership degree γ_l and the subscript $\phi_l<\varphi_l>$ of the DHLT $s_{\phi_l<o_{\varphi_l}>}$ that expresses the equivalent information to the membership degree γ_l can be transformed to each other by the following functions f and f^{-1}, respectively:

$$f : [-\tau,\tau] \times [-\varsigma,\varsigma] \to [0,1], f(\phi_l,\varphi_l) = \frac{\varphi_l + (\tau + \phi_l)\varsigma}{2\varsigma\tau} = \gamma_l \tag{1.11}$$

$$\begin{aligned} f^{-1} &: [0,1] \to [-\tau,\tau] \times [-\varsigma,\varsigma], f^{-1}(\gamma_l) = [2\tau\gamma_l - \tau] < o_{\varsigma(2\tau\gamma_l - \tau - [2\tau\gamma_l - \tau])} > \\ &= [2\tau\gamma_l - \tau] + 1 < o_{\varsigma((2\tau\gamma_l - \tau - [2\tau\gamma_l - \tau]) - 1)} > \end{aligned} \tag{1.12}$$

Based on Definition 1.10, Gou et al. (2017) introduced the transformation functions F and F^{-1} between the DHHFLE h_{S_O} and the HFE h_γ:

$$\begin{aligned} &F : \Phi \times \Psi \to \Theta, \\ &F(h_{S_O}) = F(\{s_{\phi_l<o_{\varphi_l}>} | s_{\phi_l<o_{\varphi_l}>} \in S_O; l = 1,\ldots,L; \phi_l \in [-\tau,\tau]; \varphi_l \in [-\varsigma,\varsigma]\}) = \{\gamma_l | \gamma_l \\ &\quad = f(\phi_l,\varphi_l)\} = h_\gamma \end{aligned} \tag{1.13}$$

$$\begin{aligned} F^{-1} &: \Theta \to \Phi \times \Psi, F^{-1}(h_\gamma) = F^{-1}(\{\gamma_l | \gamma_l \in [0,1]; l = 1,\ldots,L\}) \\ &= \{s_{\phi_l<o_{\varphi_l}>} | \phi_l < o_{\varphi_l} > = f^{-1}(\gamma_l)\} = h_{S_O} \end{aligned} \tag{1.14}$$

Remark 1.4 (Gou et al. 2017) It is noted that the second hierarchy linguistic term is a linguistic feature or detailed supplementary of each linguistic term included in the first LTS, and the second hierarchy LTSs are different when describing the upper bound, the lower bound or the median term of the first hierarchy LTS. Therefore, the function f can be divided into three parts according to the different values of ϕ_l. Suppose that $\tau = \varsigma = 3$, then we can utilize the functions f to transform the three DHLTs $s_{-3<o_1>}$, $s_{0<o_1>}$ and $s_{3<o_{-1}>}$ into $1/18$, $5/9$, $17/18$, respectively. This can be illustrated in Fig. 1.3.

In Fig. 1.3, firstly we need to use different second hierarchy LTSs considering that the values of ϕ_l included in $s_{-3<o_1>}$, $s_{0<o_1>}$ and $s_{3<o_{-1}>}$ are different. Then we utilize the function f to calculate the equivalent real numbers: $f(s_{-3<o_1>}) = 1/18$, $f(s_{0<o_1>}) = 5/9$ and $f(s_{3<o_{-1}>}) = 17/18$.

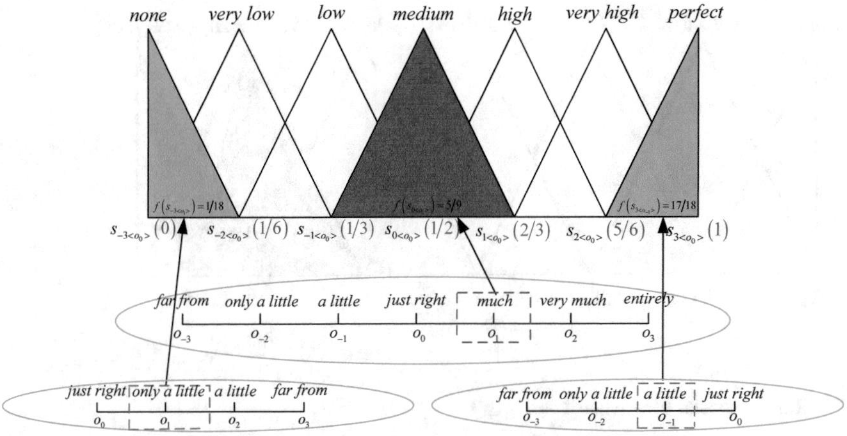

Fig. 1.3 Some operation results based on the equivalent transformation function f

Remark 1.5 (Gou et al. 2017) For the real number $\gamma \in [0, 1]$, let $\tau = \varsigma = 3$. The function f^{-1} can be described in three different cases:

(1) Let $\gamma = 3/4$, then $-2 < 2\tau\gamma - \tau = 1.5 < 2$. It follows that $s_1 < s_{1.5} < s_2$. Thus, there is $f^{-1}(3/4) = s_{1<o_{1.5}>} = s_{2<o_{-1.5}>}$. This can be shown in Fig. 1.4.
(2) Let $\gamma = 11/12$, then $2 < 2\tau\gamma - \tau = 2.5 < 3$. It follows that $s_2 < s_{2.5} < s_3$. Thus, there is $f^{-1}(11/12) = s_{2<o_{1.5}>} = s_{3<o_{-1.5}>}$. This can be shown in Fig. 1.5.
(3) Let $\gamma = 1/12$, then $-3 < 2\tau\gamma - \tau = -2.5 < -2$. Thus, $s_{-3} < s_{-2.5} < s_{-2}$, and so there is $f^{-1}(1/12) = s_{-2<o_{-1.5}>} = s_{-3<o_0>}$. This can be shown in Fig. 1.6.

Remark 1.6 (Gou et al. 2017) It is noted that, based on the equivalent transformation function f^{-1}, there are two equivalent DHLTs in each situation as discussed

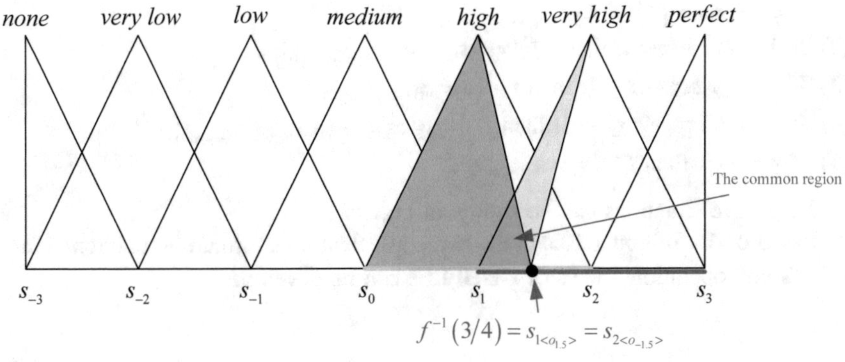

Fig. 1.4 A special case when $1 - \tau \leq 2\tau\gamma - \tau \leq \tau - 1$

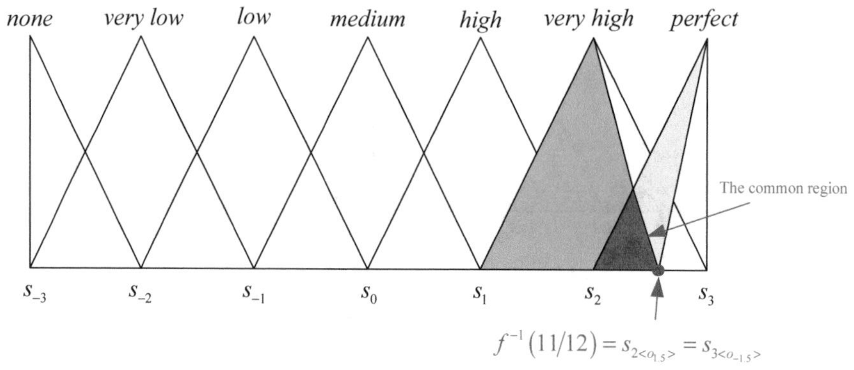

Fig. 1.5 A special case when $1-\tau \le 2\tau\gamma-\tau \le \tau$

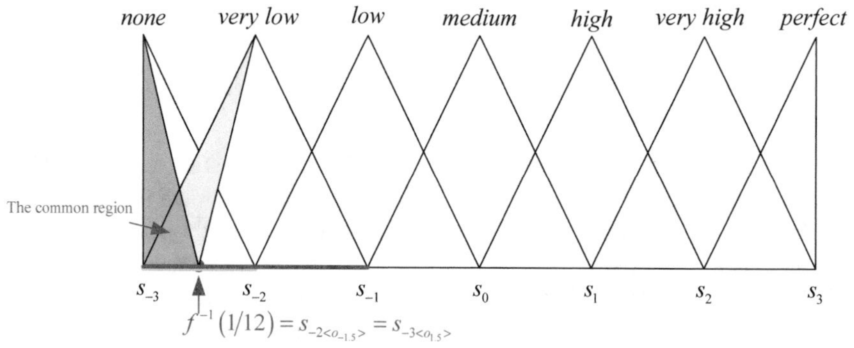

Fig. 1.6 A special case when $-\tau \le 2\tau\gamma-\tau \le 1-\tau$

in Remark 1.5. To make the calculations more convenient, we can introduce some rules regarding to f^{-1}.

(1) If $\gamma=1$, then $f^{-1}(\gamma)=s_{\tau<o_0>}$;
(2) If $1 \le 2\tau\gamma-\tau<\tau$, then $f^{-1}(\gamma)=s_{[2\tau\gamma-\tau]<o_{\varsigma(2\tau\gamma-\tau-[2\tau\gamma-\tau])}>}$;
(3) If $-1 \le 2\tau\gamma-\tau \le 1$, then $f^{-1}(\gamma)=s_{0<o_{\varsigma(2\tau\gamma-\tau)}>}$;
(4) If $-\tau<2\tau\gamma-\tau \le -1$, then $f^{-1}(\gamma)=s_{[2\tau\gamma-\tau]+1<o_{\varsigma(2\tau\gamma-\tau-[2\tau\gamma-\tau]-1)}>}$;
(5) If $\gamma=-1$, then $f^{-1}(\gamma)=s_{-\tau<o_0>}$.

These five situations can be shown in Fig. 1.7.

Based on Definition 1.10 and the two equivalent transformation functions F and F^{-1}. Some operational laws of DHHFLEs can be developed:

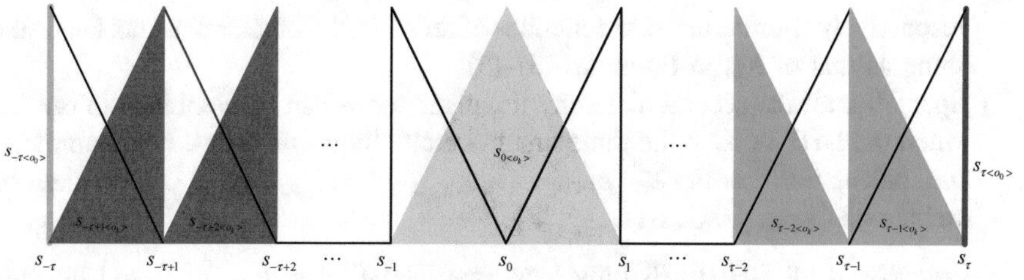

Fig. 1.7 The regions of each situation regarding to the function f^{-1}

Definition 1.11 (Gou et al. 2017) Let S_O be a DHLTS, $h_{S_O} = \{s_{\phi_l<o_{\varphi_l}>} | s_{\phi_l<o_{\varphi_l}>} \in S_O;\ l=1,2,\ldots,L; \phi_l = -\tau,\ldots,-1,0,1,\ldots,\tau; \varphi_l = -\varsigma,\ldots,-1,0,1,\ldots,\varsigma\}$, $h_{S_{O_i}} = \{s^i_{\phi_l<o_{\varphi_l}>} | s^i_{\phi_l<o_{\varphi_l}>} \in S_O;\ l=1,2,\ldots,L_i; \phi_l = -\tau,\ldots,-1,0,1,\ldots,\tau; \varphi_l = -\varsigma,\ldots,-1,0,1,\ldots,\varsigma\}$ $(i=1,2)$ be three DHHFLEs, and λ be a real number. Then

(1) (Addition) $h_{S_{O_1}} \oplus h_{S_{O_2}} = F^{-1}\left(\bigcup_{\eta_1 \in F(h_{S_{O_1}}),\eta_2 \in F(h_{S_{O_2}})} \{\eta_1+\eta_2-\eta_1\eta_2\}\right)$;

(2) (Multiplication) $h_{S_{O_1}} \otimes h_{S_{O_2}} = F^{-1}\left(\bigcup_{\eta_1 \in F(h_{S_{O_1}}),\eta_2 \in F(h_{S_{O_2}})} \{\eta_1\eta_2\}\right)$;

(3) (Multiplication) $\lambda h_{S_O} = F^{-1}\left(\bigcup_{\eta \in F(h_{S_O})} \{1-(1-\eta)^\lambda\}\right)$;

(4) (Power) $(h_{S_O})^\lambda = F^{-1}\left(\bigcup_{\eta \in F(h_{S_O})} \{\eta^\lambda\}\right)$;

(5) (Complementary) $\overline{h_{S_O}} = F^{-1}\left(\bigcup_{\eta \in F(h_{S_O})} \{1-\eta\}\right)$;

(6) (Union) $h_{S_{O_1}} \cup h_{S_{O_2}} = \{s_{t<o_{k_t}>} | s_{t<o_{k_t}>} \subset h_{S_{O_1}}\ or\ s_{t<o_{k_t}>} \subset h_{S_{O_2}}\}$;

(7) (Intersection) $h_{S_{O_1}} \cap h_{S_{O_2}} = \{s_{t<o_{k_t}>} | s_{t<o_{k_t}>} \subset h_{S_{O_1}}\ and\ s_{t<o_{k_t}>} \subset h_{S_{O_2}}\}$.

Remark 1.7 (Gou et al. 2017) For Definition 1.11, the following points are remarkable:

(1) Based on the equivalent transformation function F, the DHHFLEs can be transformed to the HFEs. Therefore, the operational laws of DHHFLEs can be developed based on the operational laws of HFEs. Then the results of DHHFLEs can be obtained by transforming the HFEs to the DHHFLEs equivalently according to the other transformation function F^{-1}.
(2) For Formulas (1) and (2), the number of terms in the obtained results must be $L_1 \times L_2$, where L_1 and L_2 are the number of DHLTs included in $h_{S_{O_1}}$ and $h_{S_{O_2}}$,

respectively. Furthermore, the number of terms in the obtained results keeps the same as that of h_{S_O} in Formulas (3)–(5).

(3) Specially, all the second hierarchy linguistic terms can be combined to one set when the DHLTs have the same first hierarchy linguistic terms. For example, if the calculation result is $\{s_{0<o_1>}, s_{0<o_{1.5}>}, s_{1<o_{-0.5}>}, s_{1<o_0>}\}$, it can be written as $\{s_{0<o_1,o_{1.5}>}, s_{1<o_{-0.5},o_0>}\}$.

Specially, if all DHHFLEs only have one DHLT, then h_{S_O}, $h_{S_{O_1}}$, and $h_{S_{O_2}}$ are reduced to $s_{t<o_k>}$, $s_{t_1<o_{k_1}>}$ and $s_{t_2<o_{k_2}>}$, respectively. Accordingly, the operational laws can be reduced to the operational laws of DHLTs:

(1) (Addition) $s_{t_1<o_{k_1}>} \oplus s_{t_2<o_{k_2}>} = f^{-1}(f(s_{t_1<o_{k_1}>}) + f(s_{t_2<o_{k_2}>}) - f(s_{t_1<o_{k_1}>}) f(s_{t_2<o_{k_2}>}))$;
(2) (Multiplication) $s_{t_1<o_{k_1}>} \otimes s_{t_2<o_{k_2}>} = f^{-1}(f(s_{t_1<o_{k_1}>}) f(s_{t_2<o_{k_2}>}))$;
(3) (Multiplication) $\lambda s_{t<o_k>} = f^{-1}(1 - (1 - f(s_{t<o_k>}))^{\lambda})$;
(4) (Power) $(s_{t<o_k>})^{\lambda} = f^{-1}((f(s_{t<o_k>}))^{\lambda})$;
(5) (Complementary) $\overline{s_{t<o_k>}} = f^{-1}(1 - f(s_{t<o_k>}))$;
(6) (Union) $s_{t_1<o_{k_1}>} \cup s_{t_2<o_{k_2}>} = \{s_{t_1<o_{k_1}>}, s_{t_2<o_{k_2}>}\}$.

Remark 1.8 (Gou et al. 2017) As we know, there is a one-to-one correspondence between the double hierarchy linguistic information and the fuzzy information (real numbers). Therefore, based on Yager's CW scheme (Yager 2004), we can transform the double hierarchy linguistic information into fuzzy information, and make computations between real numbers based on the operational laws of HFEs. Finally, the result can be transformed into double hierarchy linguistic information based on the anti-function. Similarly, the operational laws of DHHFLEs also can be obtained. The flow of the operations is shown in Fig. 1.8.

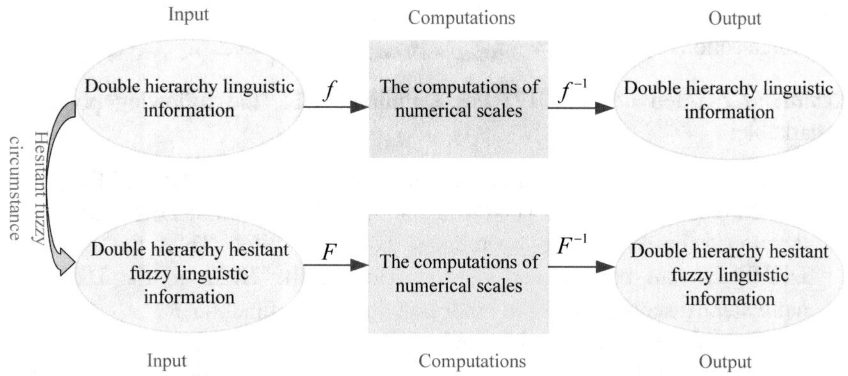

Fig. 1.8 The flow of the operations of DHLTs and DHHFLEs

Additionally, for the UDHLTS, considering that both the first and second LTSs have three linguistic scale functions, respectively, so there are nine cases when combining these two hierarchy LTSs. Therefore, Fu and Liao (2019) discussed these cases and developed the corresponding semantics, and then defined some operational laws for UDHLTSs. For more information about the operational laws of UDHLTSs, please refer to (Fu and Liao 2019).

Example 1.2 (Gou et al. 2017) Let $S_O = \{s_{t<o_k>} | t = -3, \ldots, 3,\ k = -3, \ldots, 3\}$ be a DHLTS. $h_{S_{O_1}} = \{s_{1<o_2>}, s_{2<o_0>}\}$ and $h_{S_{O_2}} = \{s_{-1<o_{-2}>}, s_0, s_{1<o_0>}\}$ be two DHHFLEs, and $\lambda = \frac{1}{2}$. Then

(1) $h_{S_{O_1}} \oplus h_{S_{O_2}} = F^{-1}\left(\bigcup_{\eta_1 \in F(h_{S_{O_1}}), \eta_2 \in F(h_{S_{O_2}})} \{\eta_1 + \eta_2 - \eta_1\eta_2\}\right)$
$= F^{-1}\{\frac{7}{9} + \frac{5}{18} - \frac{7}{9} \times \frac{5}{18}, \frac{7}{9} + \frac{5}{9} - \frac{7}{9} \times \frac{5}{9}, \frac{7}{9} \quad + \frac{2}{3} - \frac{7}{9} \times \frac{2}{3}, \frac{5}{6} +$
$\frac{5}{18} - \frac{5}{6} \times \frac{5}{18}, \frac{5}{6} + \frac{5}{9} - \frac{5}{6} \times \frac{5}{9}, \frac{5}{6} + \frac{2}{3} - \frac{5}{6} \times \frac{2}{3}\} = \{s_{2<o_{0.11}, o_{0.83}, o_{1.22}, o_{1.67}, o_2>}\}$

(2) $h_{S_{O_1}} \otimes h_{S_{O_2}} = F^{-1}\left(\bigcup_{\eta_1 \in F(h_{S_{O_1}}), \eta_2 \in F(h_{S_{O_2}})} \{\eta_1\eta_2\}\right)$
$= F^{-1}\{\frac{7}{9} \times \frac{5}{18}, \frac{7}{9} \times \frac{5}{9}, \frac{7}{9} \times \frac{2}{3}, \frac{5}{6} \times \frac{5}{18}, \frac{5}{6} \times \frac{5}{9}, \frac{5}{6}$
$\times \frac{2}{3}\} = \{s_{-1<o_{-2.11}, o_{-1.83}>}, s_{0<o_{-1.22}, o_{-0.66}, o_{0.33}, o_1>}\}$

(3) $\lambda h_{S_{O_1}} = F^{-1}\left(\bigcup_{\eta \in F(h_{S_{O_1}})} \{1 - (1 - \eta)^\lambda\}\right) =$
$F^{-1}\left\{1 - \left(1 - \frac{7}{9}\right)^{\frac{1}{2}}, 1 - \left(1 - \frac{5}{6}\right)^{\frac{1}{2}}\right\} = \{s_{0<o_{0.51}, o_{1.65}>}\}$

(4) $(h_{S_{O_1}})^\lambda = F^{-1}\left(\bigcup_{\eta \in F(h_{S_O})} \{\eta^\lambda\}\right) = F^{-1}\left\{\left(\frac{7}{9}\right)^{\frac{1}{2}}, \left(\frac{5}{6}\right)^{\frac{1}{2}}\right\} = \{s_{2<o_{0.87}, o_{1.43}>}\}$;

(5) $\overline{h_{S_{O_1}}} = F^{-1}\left(\bigcup_{\eta \in F(h_{S_{O_1}})} \{1 - \eta\}\right) = F^{-1}\{1 - \frac{7}{9}, 1 - \frac{5}{6}\} =$
$\{s_{-2<o_0>}, s_{-1<o_{-2}>}\}$.

(6) $h_{S_{O_1}} \cup h_{S_{O_2}} = \{s_{t<o_{k_t}>} | s_{t<o_{k_t}>} \subset h_{S_{O_1}}\ or\ s_{t<o_{k_t}>} \subset h_{S_{O_2}}\} =$
$\{s_{-1<o_{-2}>}, s_0, s_{1<o_0>}, s_{1<o_2>}, s_{2<o_0>}\}$;

(7) $h_{S_{O_1}} \cap h_{S_{O_2}} = \{s_{t<o_{k_t}>} | s_{t<o_{k_t}>} \subset h_{S_{O_1}}\ and\ s_{t<o_{k_t}>} \subset h_{S_{O_2}}\} = \{\varnothing\}$.

1.4 Comparative Methods

For any two DHHFLEs, it is necessary to develop comparative methods to compare the sizes of them. Therefore, lots of scholars have studied some comparative methods for DHHFLEs.

(a) **The comparative method based on expected and variance values**

Firstly, Gou et al. (2017) defined the concepts of expected and variance values of a DHHFLE. Let S_O be a DHLTS, $h_{S_O} = \{s_{\phi_l<o_{\varphi_l}>} | s_{\phi_l<o_{\varphi_l}>} \in S_O; l = 1, 2, \ldots, \#h_{S_O}; \phi_l = -\tau, \ldots, -1, 0, 1, \ldots, \tau; \varphi_l = -\varsigma, \ldots, -1, 0, 1, \ldots, \varsigma\}$ be a DHHFLE. Then the expected value and the variance of h_{S_O} can be defined respectively as follows:

$$E(h_{S_O}) = \frac{1}{\#h_{S_O}} \sum_{l=1}^{\#h_{S_O}} f(s_{\phi_l<o_{\varphi_l}>}) \tag{1.15}$$

$$v(h_{S_O}) = \frac{1}{\#h_{S_O}} \sum_{l=1}^{\#h_{S_O}} (f(s_{\phi_l<o_{\varphi_l}>}) - E(h_{S_O}))^2 \tag{1.16}$$

Based on Eqs. (1.15) and (1.16), for two DHHFLEs $h_{S_{O_1}}$ and $h_{S_{O_2}}$,

(1) If $E(h_{S_{O_1}}) > E(h_{S_{O_2}})$, then $h_{S_{O_1}}$ is bigger than $h_{S_{O_2}}$, denoted by $h_{S_{O_1}} \succ h_{S_{O_2}}$.
(2) If $E(h_{S_{O_1}}) = E(h_{S_{O_2}})$, then

  i. If $v(h_{S_{O_1}}) < v(h_{S_{O_2}})$, then $h_{S_{O_1}}$ is bigger than $h_{S_{O_2}}$, denoted by $h_{S_{O_2}} \prec h_{S_{O_1}}$;
  ii. If $v(h_{S_{O_1}}) = v(h_{S_{O_2}})$, then $h_{S_{O_1}}$ is equivalent with $h_{S_{O_2}}$, denoted by $h_{S_{O_1}} \sim h_{S_{O_2}}$.

(b) **The comparative method based on hesitant degrees**

As we know, the DHHFLEs have different hesitant degrees when two DHHFLEs have different number of DHLTs. Therefore, the accuracy of the comparison results can be improved by introducing the hesitant degrees of DHHFLEs. The definition of the hesitant degree of DHHFLE was proposed by Gou et al. (2018b): Let S_O be a DHLTS, h_{S_O} be a DHHFLE. Then we call

$$u(h_{S_O}) = 1 - \frac{1}{\#h_{S_O}} \tag{1.17}$$

the hesitant degree of h_{S_O}, where $\#h_{S_O}$ is the number of DHLTs included in h_{S_O}.

Then, Wang et al. (2020) developed another score value for DHHFLE, denoted as Score-DHHeLiSF, shown as follows

$$E'(h_{S_O}) = (1 - u(h_{S_O})) \times \frac{1}{\#h_{S_O}} \sum_{l=1}^{\#h_{S_O}} f(s_{\phi_l<o_{\varphi_l}>}) \tag{1.18}$$

For two DHHFLEs $h_{S_{O_1}}$ and $h_{S_{O_2}}$, there are:

(1) If $E'(h_{S_{O_1}}) > E'(h_{S_{O_2}})$, then $h_{S_{O_1}}$ is bigger than $h_{S_{O_2}}$, denoted by $h_{S_{O_1}} \succ h_{S_{O_2}}$;
(2) If $E'(h_{S_{O_1}}) = E'(h_{S_{O_2}})$, then $h_{S_{O_1}}$ is equivalent with $h_{S_{O_2}}$, denoted by $h_{S_{O_1}} \sim h_{S_{O_2}}$.

(c) **The comparative method based on the envelopes of DHHFLTSs**

Gou et al. (2017) proposed the concept of the envelope of DHHFLE, which is consisted by the upper bound and the lower bound of a DHHFLE. Obviously, the envelope of a DHHFLE is equal to the interval-valued linguistic term (Xu 2004a, b). Therefore, Gou and Xu (2019) proposed the possible degree of DHHFLE and used it to compare the size of DHHFLEs. Let S_O be a DHLTS, $h_{S_{O_1}}$ and $h_{S_{O_2}}$ be two DHHFLEs, then

$$p(env(h_{S_{O_1}}) \geq env(h_{S_{O_2}})) = min\left\{ \max\left(\frac{f(h^+_{S_{O_1}}) - f(h^-_{S_{O_2}})}{(f(h^+_{S_{O_1}}) - f(h^-_{S_{O_1}})) + f(h^+_{S_{O_2}}) - f(h^-_{S_{O_2}})}, 0 \right), 0 \right\} \tag{1.19}$$

can be called the possible degree of that $h_{S_{O_1}}$ is bigger than $h_{S_{O_2}}$.

Therefore, there are

(1) If $p(env(h_{S_{O_1}}) \geq env(h_{S_{O_2}})) > 0.5$, then $h_{S_{O_1}}$ is bigger than $h_{S_{O_2}}$, denoted by $h_{S_{O_1}} \succ h_{S_{O_2}}$;
(2) If $p(env(h_{S_{O_1}}) \geq env(h_{S_{O_2}})) = 0.5$, then $h_{S_{O_1}}$ is equivalent with $h_{S_{O_2}}$, denoted by $h_{S_{O_1}} \sim h_{S_{O_2}}$.

References

Chiclana F, Herrera F, Herrera-Viedma E (1998) Integrating three representation models in fuzzy multipurpose decision making based on fuzzy preference relations. Fuzzy Sets Syst 97:33–48

Chiclana F, Herrera-Viedma E, Alonso S, Herrera F (2009) Cardinal consistency of reciprocal preference relations: a characterization of multiplicative transitivity. IEEE Trans Fuzzy Syst 17(1):14–23

Dong YC, Xu YF, Li HY (2008) On consistency measures of linguistic preference relations. Eur J Oper Res 189(2):430–444

Dubois D (2011) The role of fuzzy sets indecision sciences: old techniques and new directions. Fuzzy Sets Syst 184:3–28

Fu ZG, Liao HC (2019) Unbalanced double hierarchy linguistic term set: the TOPSIS method for multi-expert qualitative decision making involving green mine selection. Inf Fusion 51:271–286

González-Pachón J, Romero C (2001) Aggregation of partial ordinal rankings: an interval goal programming approach. Comput Oper Res 28:827–834

Gou XJ, Liao HC, Xu ZS, Herrera F (2017) Double hierarchy hesitant fuzzy linguistic term set and MULTIMOORA method: a case of study to evaluate the implementation status of haze controlling measures. Inf Fusion 38:22–34

Gou XJ, Liao HC, Xu ZS, Min R, Herrera F (2019) Group decision making with double hierarchy hesitant fuzzy linguistic preference relations: consistency based measures, index and repairing algorithms and decision model. Inf Sci 489:93–112
Gou XJ, Xu ZS, Herrera F (2018a) Consensus reaching process for large-scale group decision making with double hierarchy hesitant fuzzy linguistic preference relations. Knowl-Based Syst 157:20–33
Gou XJ, Xu ZS, Liao HC, Herrera F (2018b) Multiple criteria decision making based on distance and similarity measures with double hierarchy hesitant fuzzy linguistic environment. Comput Ind Eng 126:516–530
Gou XJ, Xu ZS, Liao HC, Herrera F (2020a) A consensus model to manage minority opinions and noncooperative behaviors in large-scale GDM with double hierarchy linguistic preference relations. IEEE Trans Cybern. https://doi.org/10.1109/TCYB.2020.2985069
Gou XJ, Xu ZS, Zhou W (2020b) Managing consensus by multiple stages optimization models with linguistic preference orderings and double hierarchy linguistic preferences. Technol Econ Dev Econ 26(3):642–674
He Y, Xu ZS (2018) A consensus framework with different preference ordering structures and its applications in human resource selection. Comput Ind Eng 118:80–88
Herrera F, Martínez L (2000) A 2-tuple fuzzy linguistic representation model for computing with words. IEEE Trans Fuzzy Syst 8(6):746–752
Hervés-Beloso C, Cruces HV (2018) Continuous preference orderings representable by utility functions. J Econ Surv 33(1):179–194
Liang HM, Xiong W, Dong YC (2018) A prospect theory-based method for fusing the individual preference-approval structures in group decision making. Comput Ind Eng 117:237–248
Millet I (1997) The effectiveness of alternative preference elicitation methods in the analytic hierarchy process. J Multi-Criteria Decis Anal 6:41–51
Montserrat-Adell J, Xu ZS, Gou XJ, Agell N (2019) Free double hierarchy hesitant fuzzy linguistic term sets: an application on raking alternatives in GDM. Inf Fusion 47:45–59
Rodríguez RM, Martínez L, Herrera F (2012) Hesitant fuzzy linguistic terms sets for decision making. IEEE Trans Fuzzy Syst 20:109–119
Schubert J (1995) On p in a decision-theoretic apparatus of Dempster-Shafer theory. Int J Approx Reason 13:185–200
Tanino T (1984) Fuzzy preference orderings in group decision making. Fuzzy Sets Syst 12: 117–131
Wang XD, Gou XJ, Xu ZS (2020) Assessment of Traffic congestion with ORESTE method under double hierarchy hesitant fuzzy linguistic term set. Appl Soft Comput 86:105864
Wang H, Xu ZS, Zeng XJ (2018) Linguistic terms with weakened hedges: a model for qualitative decision making under uncertainty. Inf Sci 433:37–45
Xia MM, Xu ZS (2011) Hesitant fuzzy information aggregation in decision making. Int J Approx Reason 52:395–407
Xu ZS (2004a) A method based on linguistic aggregation operators for group decision making with linguistic preference relations. Inf Sci 166(1–4):19–30
Xu ZS (2004b) Uncertain linguistic aggregation operators based approach to multiple attribute group decision making under uncertain linguistic environment. Inf Sci 168(1–4):171–184
Xu ZS (2005) Deviation measures of linguistic preference relations in group decision making. Omega 33(3):249–254
Xu ZS (2013) Group decision making model and approach based on interval preference orderings. Comput Ind Eng 64:797–803
Xu ZS, Wang H (2017) On the syntax and semantics of virtual linguistic terms for information fusion in decision making. Inf Fusion 34:43–48
Yager RR (2004) On the retranslation process in Zadeh's paradigm of computing with words. IEEE Trans Syst Man Cybern Part B Cybern 34(2):1184–1195
Zadeh LA (2012) Computing with words: what is computing with words (CWW)?. Springer, Berlin, Heidelberg

Zhang BW, Liang HM, Zhang GQ, Xu YF (2018a) Minimum deviation ordinal consensus reaching in GDM with heterogeneous preference structures. Appl Soft Comput 67:658–676

Zhang BW, Liang HM, Gao Y, Zhang GQ (2018b) The optimization-based aggregation and consensus with minimum-cost in group decision making under incomplete linguistic distribution context. Knowl-Based Syst 162:92–102

Chapter 2
Measure Methods for DHHFLTSs

In decision-making processes, measure methods, such as distance and similarity measures, correlation measure, entropy and cross entropy measure, etc., play an important role in many research fields including decision-making (Liao et al. 2015b; Xu and Wang 2011; Xu and Xia 2011), pattern recognition (Arevalillo-Herráez et al. 2013; Li et al. 1993), intelligent computing (Chen et al. 2010), recommended system (Liao et al. 2014), distance learning techniques (Gao et al. 2017), electricity markets (Gao et al. 2018), and ontological sparse vector learning (Gao et al. 2015), etc. In addition, measure methods are also the theoretical basis of some well-known decision-making methods such as TOPSIS (Biswas and Sarkar 2019; Hussian and Yang 2019; Liu et al. 2019; Wu et al. 2019), VIKOR (Wu et al. 2019; Wang and Pang 2019; Ren et al. 2016, 2017), MULTIMOORA (Hafezalkotob et al. 2019; Xian et al. 2020; Zhang et al. 2019), ORESTE (Liao et al. 2018; Wang et al. 2020; Wu and Liao 2018), TODIM (Liu et al. 2019; Ren et al. 2016), LINMAP (Xue et al. 2018), etc. Meanwhile, under double hierarchy hesitant fuzzy linguistic environment, considering that measure methods are very important in consistency index, consensus reaching process, clustering method, etc. Therefore, in this chapter, we will introduce some distance and similarity measures for DHHFLTS (Gou et al. 2018).

2.1 Distance and Similarity Measures for DHHFLEs

Distance and similarity measures can be utilized to measure the deviation and closeness degrees between different arguments (Liao et al. 2014). Up to now, amounts of scholars have developed a lot of distance and similarity measures including some traditional distance measures (Zavadskas et al. 2016) as the Hamming distance (Hamming 1950), the Euclidean distance (Danielsson 1980), and the Hausdorff metric (Hausdorff 1957), and some ordered weighted distance measures (Hung and Yang 2004; Liao et al. 2014; Xu 2005; Xu and Chen 2008a; Yager 1988). Additionally, these distance and similarity measures have been

X. Gou and Z. Xu, *Double Hierarchy Linguistic Term Set and Its Extensions*, Studies in Fuzziness and Soft Computing 396,
https://doi.org/10.1007/978-3-030-51320-7_2

extended into different uncertain circumstances, such as fuzzy sets (Xu 2012), intuitionistic fuzzy sets (Grzegorzewski 2004; Hung and Yang 2004; Xu and Chen 2008b), hesitant fuzzy sets (Farhadinia 2014; Xu and Xia 2011), LTSs (Xu 2005; Xu and Wang 2011) and HFLTSs (Liao and Xu 2015; Liao et al. 2014), etc.

In this subsection, we will develop some distance and similarity measures between the DHHFLEs by those previous distance and similarity measures. Firstly, the axioms of distance and similarity measures between any two single DHHFLEs are discussed; then some specific distance measures are defined including three basic distances, the hybrid distances, and some distances with preference information.

2.1.1 Axioms of Distance and Similarity Measures Between the DHHFLEs

Gou et al. (2018) defined the axioms of distance and similarity measures between any two DHHFLEs with four properties such as Boundary, Symmetry, Complementarity and Reflexivity:

Definition 2.1 (Gou et al. 2018) Let S_O be a DHLTS, $h_{S_{O_i}}(i = 1, 2)$ be two DHHFLEs. Then $d(h_{S_{O_1}}, h_{S_{O_2}})$ is called the distance measure between $h_{S_{O_1}}$ and $h_{S_{O_2}}$ if it satisfies the following properties:

(I) **Boundary**: $0 \leq d(h_{S_{O_1}}, h_{S_{O_2}}) \leq 1$;
(II) **Symmetry**: $d(h_{S_{O_1}}, h_{S_{O_2}}) = d(h_{S_{O_2}}, h_{S_{O_1}})$;
(III) **Complementarity**: $d(h_{S_{O_1}}, \overline{h_{S_{O_1}}}) = 1$ iff $F(h_{S_{O_1}}) = \{0\}$ or $F(h_{S_{O_1}}) = \{1\}$;
(IV) **Reflexivity**: $d(h_{S_{O_1}}, h_{S_{O_2}}) = 0$ iff $h_{S_{O_1}} = h_{S_{O_2}}$.

Where $\overline{h_{S_{O_1}}}$ is the complement set of $h_{S_{O_1}}$, and F is the equivalent transformation function shown in Eq. (1.13).

Definition 2.2 (Gou et al. 2018) Let S_O be a DHLTS, $h_{S_{O_i}}(i = 1, 2)$ be two DHFLEs. Then $s(h_{S_{O_1}}, h_{S_{O_2}})$ is called the similarity measure between $h_{S_{O_1}}$ and $h_{S_{O_2}}$ if it satisfies the following properties:

(I) **Boundary**: $0 \leq s(h_{S_{O_1}}, h_{S_{O_2}}) \leq 1$;
(II) **Symmetry**: $s(h_{S_{O_1}}, h_{S_{O_2}}) = s(h_{S_{O_2}}, h_{S_{O_1}})$;
(III) **Complementarity**: $s(h_{S_{O_1}}, \overline{h_{S_{O_1}}}) = 0$ iff $F(h_{S_{O_1}}) = \{0\}$ or $F(h_{S_{O_1}}) = \{1\}$;
(IV) **Reflexivity**: $s(h_{S_{O_1}}, h_{S_{O_2}}) = 1$ iff $h_{S_{O_1}} = h_{S_{O_2}}$,

where $\overline{h_{S_{O_1}}}$ is the complement set of $h_{S_{O_1}}$, and F is the equivalent transformation function.

As we know, there usually exist some relationships between the distance measure $d(h_{S_{O_1}}, h_{S_{O_2}})$ and the similarity measure $s(h_{S_{O_1}}, h_{S_{O_2}})$, and the most simple and common relationship is $d(hs_{O_1}, hs_{O_2}) = 1 - s(hs_{O_1}, hs_{O_2})$. Additionally, Gou et al. (2018) developed a more suitable and comprehensive formula to show the relationship between the distance measure and the similarity measure of the DHHFLEs.

Theorem 2.1 (Gou et al. 2018) *Let* $\Im : [0,1] \to [0,1]$ *be a strictly monotonically decreasing real function, and* $d(hs_{O_1}, hs_{O_2})$ *be the distance measure between any two DHHFLEs* hs_{O_1} *and* hs_{O_2}. *Then*

$$s(hs_{O_1}, hs_{O_2}) = \frac{\Im(d(hs_{O_1}, hs_{O_2})) - \Im(1)}{\Im(0) - \Im(1)} \tag{2.1}$$

can be called the similarity measure between hs_{O_1} *and* hs_{O_2} *based on the corresponding distance measure* $d(hs_{O_1}, hs_{O_2})$.

Obviously, Eq. (2.1) satisfies all conditions of similarity measures and its proof can be omitted.

Remark 2.1 (Gou et al. 2018) For Theorem 2.1, we can establish different formulas to calculate the similarity measures between any two DHHFLEs using different strictly monotonically decreasing real function such as (1) $\Im(\upsilon) = 1 - \upsilon$, (2) $\Im(\upsilon) = \frac{1-\upsilon}{1+\upsilon}$, (3) $\Im(\upsilon) = 1 - \upsilon e^{\upsilon - 1}$, and (4) $\Im(\upsilon) = 1 - \upsilon^2$.

Generally, different DHHFLEs mainly have different numbers of DHLTs. Therefore, it is necessary to add DHLTs to the shorter DHHFLE for calculating the distance and similarity measures between two DHHFLEs. Let $hs_O = \{s_{\phi_l < o_{\varphi_l} >} | s_{\phi_l < o_{\varphi_l} >} \in S_O; l = 1, 2, \ldots, \#hs_O\}$ be a DHHFLE, and $\varepsilon (0 \le \varepsilon \le 1)$ be an optimized parameter. Because all DHLTs included in DHHFLE are ranked in ascending order, $s_{\phi_1 < o_{\varphi_1} >}$ and $s_{\phi_{\#hs_O} < o_{\varphi_{\#hs_O}} >}$ are the minimum and maximum DHLTs in hs_O, respectively. Then we can add the DHLT

$$\tilde{s}_{\phi < o_\varphi >} = s_{(1-\varepsilon)\phi_1 + \varepsilon\phi_{\#hs_O} < o_{(1-\varepsilon)\varphi_1 + \varepsilon\varphi_{\#hs_O}} >} \tag{2.2}$$

to the shorter DHHFLE. The optimized parameter ε mainly reflects the risk preferences of experts with $\tilde{s}_{\phi < o_\varphi >} = s_{\phi_{\#hs_O} < o_{\varphi_{\#hs_O}} >}$ and $\tilde{s}_{\phi < o_\varphi >} = s_{\phi_1 < o_{\varphi_1} >}$ with respect to the optimism rule $\varepsilon = 1$ and the pessimism rule $\varepsilon = 0$, respectively. In this chapter, we let $\varepsilon = 0.5$ and $\tilde{s}_{\phi < o_\varphi >} = s_{\frac{(\phi_1 + \phi_{\#hs_O})}{2} < o_{\frac{(\varphi_1 + \varphi_{\#hs_O})}{2}} >}$.

2.1.2 *Some Basic Distance and Similarity Measures Between DHHFLEs*

This subsection mainly discusses some basic distance measures between DHHFLEs. Then the corresponding similarity measures can be obtained by Eq. (2.1) and thus we omit them.

Definition 2.3 (Gou et al. 2018) Let S_O be a DHLTS, $h_{S_{O_i}} = \{h_{S_{O_i}}^{(l)} | h_{S_{O_i}}^{(l)} \in S_O; l = 1, 2, \ldots, \# h_{S_{O_i}}\}$ $(i = 1, 2)$ be two DHHFLEs ($\# h_{S_{O_1}}$ and $\# h_{S_{O_2}}$ being the number of DHLTs in $h_{S_{O_1}}$ and $h_{S_{O_2}}$ respectively and $\# h_{S_{O_1}} = \# h_{S_{O_2}} = L$. If not, the shorter one can be extended by adding DHLTs obtained by Eq. (2.2). Based on the well-known Hamming distance and the Euclidean distance, the Hamming distance and the Euclidean distance between $h_{S_{O_1}}$ and $h_{S_{O_2}}$ can be developed respectively:

$$d_{hd}(h_{S_{O_1}}, h_{S_{O_2}}) = \frac{1}{L}\sum_{l=1}^{L} \left| f(h_{S_{O_1}}^{(l)}) - f(h_{S_{O_2}}^{(l)}) \right| \tag{2.3}$$

$$d_{ed}(h_{S_{O_1}}, h_{S_{O_2}}) = \left(\frac{1}{L}\sum_{l=1}^{L} \left| f(h_{S_{O_1}}^{(l)}) - f(h_{S_{O_2}}^{(l)}) \right|^2 \right)^{1/2} \tag{2.4}$$

where f is the equivalent transformation function.

Based on the generalized idea provided by Yager (2004), let $\lambda > 0$, the Hamming distance and the Euclidean distance can be extended into the generalized distance between $h_{S_{O_1}}$ and $h_{S_{O_2}}$:

$$d_{gd}(h_{S_{O_1}}, h_{S_{O_2}}) = \left(\frac{1}{L}\sum_{l=1}^{L} \left| f(h_{S_{O_1}}^{(l)}) - f(h_{S_{O_2}}^{(l)}) \right|^\lambda \right)^{1/\lambda} \tag{2.5}$$

Additionally, the generalized Hausdorff distance between $h_{S_{O_1}}$ and $h_{S_{O_2}}$ can be given as:

$$d_{ghd}(h_{S_{O_1}}, h_{S_{O_2}}) = \left(\max_{l=1,2,\ldots,L} \left| f(h_{S_{O_1}}^{(l)}) - f(h_{S_{O_2}}^{(l)}) \right|^\lambda \right)^{1/\lambda} \tag{2.6}$$

where $\lambda > 0$, and f is the equivalent transformation function.

If $\lambda = 1$ and $\lambda = 2$, then Eq. (2.6) reduces to the Hamming-Hausdorff distance and Euclidean-Hausdorff distance between $h_{S_{O_1}}$ and $h_{S_{O_2}}$, respectively:

$$d_{hhd}(h_{S_{O_1}}, h_{S_{O_2}}) = \max_{l=1,2,\ldots,L} \left| f(h_{S_{O_1}}^{(l)}) - f(h_{S_{O_2}}^{(l)}) \right| \tag{2.7}$$

$$d_{ehd}(h_{S_{O_1}}, h_{S_{O_2}}) = \left(\max_{l=1,2,\ldots,L} \left| f(h_{S_{O_1}}^{(l)}) - f(h_{S_{O_2}}^{(l)}) \right|^2 \right)^{1/2} \tag{2.8}$$

Furthermore, considering that the hesitance degree (Li et al. 2015) is an important factor in the calculations about hesitant fuzzy environment, based on the hesitance degree of the DHHFLE shown in Eq. (1.18), Gou et al. (2018) defined some distance and similarity measures between the DHHFLEs with hesitance degrees.

Remark 2.2 (Gou et al. 2018) The hesitance degrees $u(h_{S_O}(x_i))$ reflects the degree of hesitance of an expert. Therefore, the larger the values are, the more hesitant the expert should be.

Based on the hesitance degrees of the DHHFLEs, the generalized hesitance degree-based distance between two DHHFLEs $h_{S_{O_1}}$ and $h_{S_{O_2}}$ can be defined as follows:

$$d_{ghdd}(h_{S_{O_1}}, h_{S_{O_2}}) = \left(\left| u(h_{S_{O_1}}) - u(h_{S_{O_2}}) \right|^{\lambda} \right)^{1/\lambda} \tag{2.9}$$

Specially, if $\lambda = 1$ and $\lambda = 2$, then Eq. (2.9) reduces to the Hamming-hesitance degree-based distance and the Euclidean-hesitance degree-based distance between $h_{S_{O_1}}$ and $h_{S_{O_2}}$, respectively:

$$d_{hhdd}(h_{S_{O_1}}, h_{S_{O_2}}) = \left| u(h_{S_{O_1}}) - u(h_{S_{O_2}}) \right| \tag{2.10}$$

$$d_{ehdd}(h_{S_{O_1}}, h_{S_{O_2}}) = \left(\left| u(h_{S_{O_1}}) - u(h_{S_{O_2}}) \right|^2 \right)^{1/2} \tag{2.11}$$

Based on the three basic distance measures shown as Eqs. (2.5), (2.6) and (2.9), Gou et al. (2018) developed some generalized hybrid distance measures between the DHHFLEs, including the generalized hybrid Hausdorff distance, the generalized hybrid hesitance degree-based distance, and the generalized hybrid Hausdorff-hesitance degree-based distance between $h_{S_{O_1}}$ and $h_{S_{O_2}}$, respectively:

$$d_{ghhd}(h_{S_{O_1}}, h_{S_{O_2}}) = \left(\frac{1}{2} \left(\frac{1}{L} \sum_{l=1}^{L} \left| f(h_{S_{O_1}}^{(l)}) - f(h_{S_{O_2}}^{(l)}) \right|^{\lambda} + \max_{l=1,2,\ldots,L} \left| f(h_{S_{O_1}}^{(l)}) - f(h_{S_{O_2}}^{(l)}) \right|^{\lambda} \right) \right)^{1/\lambda} \tag{2.12}$$

$$d_{ghhdd}(h_{S_{O_1}}, h_{S_{O_2}}) = \left(\frac{1}{2}\left(\frac{1}{L}\sum_{l=1}^{L}\left|f(h_{S_{O_1}}^{(l)}) - f(h_{S_{O_2}}^{(l)})\right|^{\lambda} + \left|u(h_{S_{O_1}}) - u(h_{S_{O_2}})\right|^{\lambda}\right)\right)^{1/\lambda} \quad (2.13)$$

$$d_{ghhhdd}(h_{S_{O_1}}, h_{S_{O_2}}) = \left(\frac{1}{2}\left(\max_{l=1,2,\ldots,L}\left|f(h_{S_{O_1}}^{(l)}) - f(h_{S_{O_2}}^{(l)})\right|^{\lambda} + \left|u(h_{S_{O_1}}) - u(h_{S_{O_2}})\right|^{\lambda}\right)\right)^{1/\lambda} \quad (2.14)$$

Specially, if $\lambda = 1$, then Eqs. (2.12)–(2.14) reduce to the hybrid Hamming-Hausdorff distance, the hybrid Hamming-hesitance degree-based distance, and the hybrid Hamming-Hausdorff-hesitance degree-based distance between $h_{S_{O_1}}$ and $h_{S_{O_2}}$, respectively:

$$d_{hhhd}(h_{S_{O_1}}, h_{S_{O_2}}) = \frac{1}{2}\left(\frac{1}{L}\sum_{l=1}^{L}\left|f(h_{S_{O_1}}^{(l)}) - f(h_{S_{O_2}}^{(l)})\right| + \max_{l=1,2,\ldots,L}\left|f(h_{S_{O_1}}^{(l)}) - f(h_{S_{O_2}}^{(l)})\right|\right) \quad (2.15)$$

$$d_{hhhdd}(h_{S_{O_1}}, h_{S_{O_2}}) = \frac{1}{2}\left(\frac{1}{L}\sum_{l=1}^{L}\left|f(h_{S_{O_1}}^{(l)}) - f(h_{S_{O_2}}^{(l)})\right| + \left|u(h_{S_{O_1}}) - u(h_{S_{O_2}})\right|\right) \quad (2.16)$$

$$d_{hhhhdd}(h_{S_{O_1}}, h_{S_{O_2}}) = \frac{1}{2}\left(\max_{l=1,2,\ldots,L}\left|f(h_{S_{O_1}}^{(l)}) - f(h_{S_{O_2}}^{(l)})\right| + \left|u(h_{S_{O_1}}) - u(h_{S_{O_2}})\right|\right) \quad (2.17)$$

If $\lambda = 2$, then Eqs. (2.12)–(2.14) reduce to the hybrid Euclidean-Hausdorff distance, the hybrid Euclidean-hesitance degree-based distance, and the hybrid Euclidean-Hausdorff-hesitance degree-based distance between $h_{S_{O_1}}$ and $h_{S_{O_2}}$, respectively:

$$d_{hehd}(h_{S_{O_1}}, h_{S_{O_2}}) = \left(\frac{1}{2}\left(\frac{1}{L}\sum_{l=1}^{L}\left|f(h_{S_{O_1}}^{(l)}) - f(h_{S_{O_2}}^{(l)})\right|^{2} + \max_{l=1,2,\ldots,L}\left|f(h_{S_{O_1}}^{(l)}) - f(h_{S_{O_2}}^{(l)})\right|^{2}\right)\right)^{1/2} \quad (2.18)$$

$$d_{hehdd}(h_{S_{O_1}}, h_{S_{O_2}}) = \left(\frac{1}{2}\left(\frac{1}{L}\sum_{l=1}^{L}\left|f(h_{S_{O_1}}^{(l)}) - f(h_{S_{O_2}}^{(l)})\right|^{2} + \left|u(h_{S_{O_1}}) - u(h_{S_{O_2}})\right|^{2}\right)\right)^{1/2} \quad (2.19)$$

$$d_{hehhdd}(h_{S_{O_1}}, h_{S_{O_2}}) = \left(\frac{1}{2}\left(\max_{l=1,2,\ldots,L}\left|f(h_{S_{O_1}}^{(l)}) - f(h_{S_{O_2}}^{(l)})\right|^2 + \left|u(h_{S_{O_1}}) - u(h_{S_{O_2}})\right|^2\right)\right)^{1/2} \tag{2.20}$$

Moreover, combining all these three basic distance measures together, the generalized completely hybrid Hausdorff-hesitance degree-based distance between $h_{S_{O_1}}$ and $h_{S_{O_2}}$ can be defined as:

$$d_{chehhdd}(h_{S_{O_1}}, h_{S_{O_2}}) = \left(\frac{1}{3}\left(\frac{1}{L}\sum_{l=1}^{L}\left|f(h_{S_{O_1}}^{(l)}) - f(h_{S_{O_2}}^{(l)})\right|^\lambda + \max_{l=1,2,\ldots,L}\left|f(h_{S_{O_1}}^{(l)}) - f(h_{S_{O_2}}^{(l)})\right|^\lambda + \left|u(h_{S_{O_1}}) - u(h_{S_{O_2}})\right|^\lambda\right)\right)^{1/\lambda} \tag{2.21}$$

Similarly, if $\lambda = 1$ and $\lambda = 2$, then Eq. (2.21) reduces to the completely hybrid Hamming-Hausdorff-hesitance degree-based distance and the completely hybrid Euclidean-Hausdorff-hesitance degree-based distance between $h_{S_{O_1}}$ and $h_{S_{O_2}}$, respectively:

$$d_{chhhhdd}(h_{S_{O_1}}, h_{S_{O_2}}) = \frac{1}{3}\left(\frac{1}{L}\sum_{l=1}^{L}\left|f(h_{S_{O_1}}^{(l)}) - f(h_{S_{O_2}}^{(l)})\right| + \max_{l=1,2,\ldots,L}\left|f(h_{S_{O_1}}^{(l)}) - f(h_{S_{O_2}}^{(l)})\right| + \left|u(h_{S_{O_1}}) - u(h_{S_{O_2}})\right|\right) \tag{2.22}$$

$$d_{chehhdd}(h_{S_{O_1}}, h_{S_{O_2}}) = \left(\frac{1}{3}\left(\frac{1}{L}\sum_{l=1}^{L}\left|f(h_{S_{O_1}}^{(l)}) - f(h_{S_{O_2}}^{(l)})\right|^\lambda + \max_{l=1,2,\ldots,L}\left|f(h_{S_{O_1}}^{(l)}) - f(h_{S_{O_2}}^{(l)})\right|^\lambda + \left|u(h_{S_{O_1}}) - u(h_{S_{O_2}})\right|^\lambda\right)\right)^{1/\lambda} \tag{2.23}$$

In the following, we can set up an example to show these distance measures:

Example 2.1 (Gou et al. 2018) Let $S_O = \{s_{t<o_k>} | t = -3, \ldots, 3, k = -3, \ldots, 3\}$ be a DHLTS, $h_{S_O}^2 = \{s_{-1<o_{-2}>}, s_0, s_{1<o_{-1}>}\}$ and $h_{S_O}^1 = \{s_{-2<o_{-1}>}, s_{-1}, s_0, s_{1<o_2>}\}$ be two DHHFLEs. We can calculate these basic distance measures between $h_{S_{O_1}}$ and $h_{S_{O_2}}$ discussed above by different values of λ. The results are shown in Fig. 2.1 and Table 2.1.

Remark 2.3 (Gou et al. 2018) In Fig. 2.1 and Table 2.1, firstly, for any distance measure, we can find that the bigger the value of λ is, the greater (or at least the same) the distance measures would be. Secondly, no matter what the value of λ is, the values of the three hybrid distance measures are between two corresponding basic distance measures. Similarly, the value of the completely hybrid distance measure is among three basic distance measures.

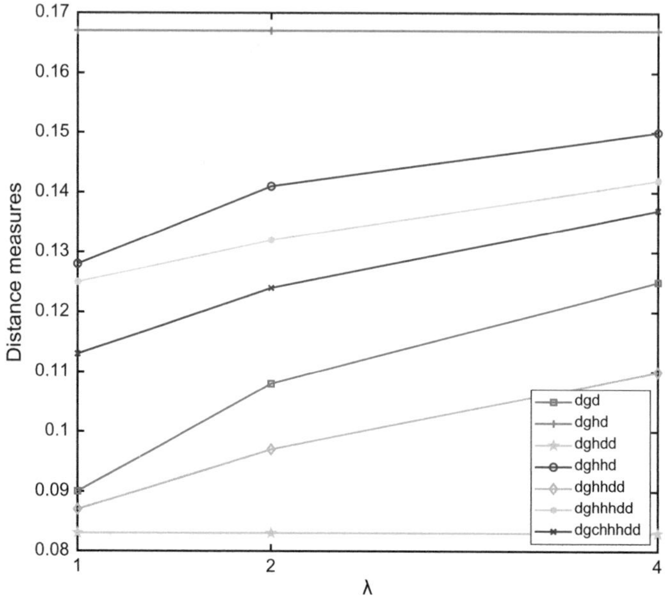

Fig. 2.1 The distributions of some basic distance measures based on different values of λ

2.1.3 Some Distance and Similarity Measures with Preference Information

In Sect. 2.1.2, the same preference is given to membership values, Hausdorff distances and hesitance degrees. However, the expert usually owns different preferences for different kinds of distance measures in actual situations. Therefore, Gou et al. (2018) defined some distance measures with preference information between any two DHHFLEs. Similarly, the corresponding similarity measures can be omitted by Eq. (2.1).

For any two DHHFLEs $h_{S_{O_1}}$ and $h_{S_{O_2}}$, and combining all the three basic distance measures discussed in Sect. 2.1.2, the generalized completely hybrid Hausdorff-hesitance degree-preference distance between $h_{S_{O_1}}$ and $h_{S_{O_2}}$ can be defined as:

$$d_{gchhhdpd}(h_{S_{O_1}}, h_{S_{O_2}}) = \left(\frac{a}{L} \sum_{l=1}^{L} \left| f(h_{S_{O_1}}^{(l)}) - f(h_{S_{O_2}}^{(l)}) \right|^{\lambda} + b \max_{l=1,2,\ldots,L} \left| f(h_{S_{O_1}}^{(l)}) - f(h_{S_{O_2}}^{(l)}) \right|^{\lambda} + c \left| u(h_{S_{O_1}}) - u(h_{S_{O_2}}) \right|^{\lambda} \right)^{1/\lambda} \tag{2.24}$$

where $\lambda > 0$, $0 \leq a, b, c \leq 1$, $a + b + c = 1$.

Next, some distance measures can be obtained by taking the values of λ, a, b and c:

Table 2.1 The results of some basic distance measures with different values of λ

	$d_{gd}(h^1_{S_O}, h^2_{S_O})$	$d_{ghd}(h^1_{S_O}, h^2_{S_O})$	$d_{ghdd}(h^1_{S_O}, h^2_{S_O})$	$d_{ghhd}(h^1_{S_O}, h^2_{S_O})$	$d_{ghhdd}(h^1_{S_O}, h^2_{S_O})$	$d_{ghhhdd}(h^1_{S_O}, h^2_{S_O})$	$d_{gchhdd}(h^1_{S_O}, h^2_{S_O})$
$\lambda = 1$	0.090	0.167	0.083	0.129	0.087	0.125	0.113
$\lambda = 2$	0.109	0.167	0.083	0.141	0.097	0.132	0.125
$\lambda = 4$	0.125	0.167	0.083	0.150	0.110	0.142	0.137

(1) If $\lambda = 1$ and $\lambda = 2$, then Eq. (2.24) reduces to the completely hybrid Hamming-Hausdorff-hesitance degree-preference distance and the completely hybrid Euclidean-Hausdorff-hesitance degree-preference distance between $h_{S_{O_1}}$ and $h_{S_{O_2}}$, respectively:

$$d_{chhhhdpd}(h_{S_{O_1}}, h_{S_{O_2}}) = \frac{a}{L}\sum_{l=1}^{L}\left|f(h_{S_{O_1}}^{(l)}) - f(h_{S_{O_2}}^{(l)})\right| + b\max_{l=1,2,\ldots,L}\left|f(h_{S_{O_1}}^{(l)}) - f(h_{S_{O_2}}^{(l)})\right| + c\left|u(h_{S_{O_1}}) - u(h_{S_{O_2}})\right| \tag{2.25}$$

$$d_{chehhdpd}(h_{S_{O_1}}, h_{S_{O_2}}) = \left(\frac{a}{L}\sum_{l=1}^{L}\left|f(h_{S_{O_1}}^{(l)}) - f(h_{S_{O_2}}^{(l)})\right|^2 + b\max_{l=1,2,\ldots,L}\left|f(h_{S_{O_1}}^{(l)}) - f(h_{S_{O_2}}^{(l)})\right|^2 + c\left|u(h_{S_{O_1}}) - u(h_{S_{O_2}})\right|^2\right)^{1/2} \tag{2.26}$$

(2) If $a + b = 1$ and $c = 0$, then Eq. (2.24) reduces to the generalized hybrid Hausdorff-preference distance between $h_{S_{O_1}}$ and $h_{S_{O_2}}$:

$$d_{ghhpd}(h_{S_{O_1}}, h_{S_{O_2}}) = \left(\frac{a}{L}\sum_{l=1}^{L}\left|f(h_{S_{O_1}}^{(l)}) - f(h_{S_{O_2}}^{(l)})\right|^{\lambda} + b\max_{l=1,2,\ldots,L}\left|f(h_{S_{O_1}}^{(l)}) - f(h_{S_{O_2}}^{(l)})\right|^{\lambda}\right)^{1/\lambda} \tag{2.27}$$

If $\lambda = 1$ and $\lambda = 2$, then Eq. (2.27) reduces to the hybrid Hamming-Hausdorff-preference distance and the hybrid Euclidean-Hausdorff-preference distance between $h_{S_{O_1}}$ and $h_{S_{O_2}}$, respectively:

$$d_{hhhpd}(h_{S_{O_1}}, h_{S_{O_2}}) = \frac{a}{L}\sum_{l=1}^{L}\left|f(h_{S_{O_1}}^{(l)}) - f(h_{S_{O_2}}^{(l)})\right| + b\max_{l=1,2,\ldots,L}\left|f(h_{S_{O_1}}^{(l)}) - f(h_{S_{O_2}}^{(l)})\right| \tag{2.28}$$

$$d_{hehpd}(h_{S_{O_1}}, h_{S_{O_2}}) = \left(\frac{a}{L}\sum_{l=1}^{L}\left|f(h_{S_{O_1}}^{(l)}) - f(h_{S_{O_2}}^{(l)})\right|^2 + b\max_{l=1,2,\ldots,L}\left|f(h_{S_{O_1}}^{(l)}) - f(h_{S_{O_2}}^{(l)})\right|^2\right)^{1/2} \tag{2.29}$$

(3) If $a + c = 1$ and $b = 0$, then Eq. (2.24) reduces to the generalized hybrid hesitance degree-preference distance between $h_{S_{O_1}}$ and $h_{S_{O_2}}$:

$$d_{ghhdpd}(h_{S_{O_1}}, h_{S_{O_2}}) = \left(\frac{a}{L} \sum_{l=1}^{L} \left| f(h_{S_{O_1}}^{(l)}) - f(h_{S_{O_2}}^{(l)}) \right|^{\lambda} + c \left| u(h_{S_{O_1}}) - u(h_{S_{O_2}}) \right|^{\lambda} \right)^{1/\lambda} \quad (2.30)$$

Furthermore, if $\lambda = 1$ and $\lambda = 2$, then Eq. (2.30) reduces to the hybrid Hamming-hesitance degree-preference distance and the hybrid Euclidean-hesitance degree-preference distance between $h_{S_{O_1}}$ and $h_{S_{O_2}}$, respectively:

$$d_{hhhdpd}(h_{S_{O_1}}, h_{S_{O_2}}) = \frac{a}{L} \sum_{l=1}^{L} \left| f(h_{S_{O_1}}^{(l)}) - f(h_{S_{O_2}}^{(l)}) \right| + c \left| u(h_{S_{O_1}}) - u(h_{S_{O_2}}) \right| \quad (2.31)$$

$$d_{hehdpd}(h_{S_{O_1}}, h_{S_{O_2}}) = \left(\frac{a}{L} \sum_{l=1}^{L} \left| f(h_{S_{O_1}}^{(l)}) - f(h_{S_{O_2}}^{(l)}) \right|^{2} + c \left| u(h_{S_{O_1}}) - u(h_{S_{O_2}}) \right|^{2} \right)^{1/2} \quad (2.32)$$

(4) If $b + c = 1$ and $a = 0$, then Eq. (2.24) reduces to the generalized hybrid Hausdorff-hesitance degree-preference distance between $h_{S_{O_1}}$ and $h_{S_{O_2}}$:

$$d_{ghhhdpd}(h_{S_{O_1}}, h_{S_{O_2}}) = \left(b \max_{l=1,2,\ldots,L} \left| f(h_{S_{O_1}}^{(l)}) - f(h_{S_{O_2}}^{(l)}) \right|^{\lambda} + c \left| u(h_{S_{O_1}}) - u(h_{S_{O_2}}) \right|^{\lambda} \right)^{1/\lambda} \quad (2.33)$$

Furthermore, if $\lambda = 1$ and $\lambda = 2$, then Eq. (2.33) reduces to the hybrid Hamming-Hausdorff-hesitance degree-preference distance and the hybrid Euclidean-Hausdorff-hesitance degree-preference distance between $h_{S_{O_1}}$ and $h_{S_{O_2}}$, respectively:

$$d_{hhhhdpd}(h_{S_{O_1}}, h_{S_{O_2}}) = b \max_{l=1,2,\ldots,L} \left| f(h_{S_{O_1}}^{(l)}) - f(h_{S_{O_2}}^{(l)}) \right| + c \left| u(h_{S_{O_1}}) - u(h_{S_{O_2}}) \right| \quad (2.34)$$

$$d_{hehhdpd}(h_{S_{O_1}}, h_{S_{O_2}}) = \left(b \max_{l=1,2,\ldots,L} \left| f(h_{S_{O_1}}^{(l)}) - f(h_{S_{O_2}}^{(l)}) \right|^{2} + c \left| u(h_{S_{O_1}}) - u(h_{S_{O_2}}) \right|^{2} \right)^{1/2} \quad (2.35)$$

2.2 Distance and Similarity Measures for DHHFLTSs

In Sect. 2.1, we have developed some distance and similarity measures between two DHHFLEs over only one double hierarchy linguistic variable. However, in some practical problems especially in the multiple criteria decision-making (MCDM) problems, the experts usually use a set to express their evaluation information when evaluating each alternative (or object) with respect to all attributes (or criteria). Therefore, the DHHFLTS is a perfect expression to take into account all aspects. Additionally, the weights of criteria are very important in the MCDM problems, and we need to consider them. When the evaluation information of each alternative (or object) with respect to all criteria is expressed by the DHHFLTS, the distance and similarity measures are very important to deal with the MCDM problems. Therefore, this subsection mainly establishes some weighted distance and similarity measures between the DHHFLTSs. Firstly, the axioms of the distance and similarity measures between the DHHFLTSs are shown. Then, we develop some weighted distance and similarity measures between the DHHFLTSs in discrete case and continuous case, respectively. Finally, we propose some ordered weighted distance and similarity measures between the DHHFLTSs.

2.2.1 Axioms of Distance and Similarity Measures for DHHFLTSs

Definition 2.4 (Gou et al. 2018) Let S_O be a DHLTS, $H_{S_{O_1}} = \{h_{S_{O_{11}}}, h_{S_{O_{12}}}, \ldots, h_{S_{O_{1n}}}\}$ and $H_{S_{O_2}} = \{h_{S_{O_{21}}}, h_{S_{O_{22}}}, \ldots, h_{S_{O_{2n}}}\}$ be two DHHFLTSs. Then $d(H_{S_{O_1}}, H_{S_{O_2}})$ is called the distance measure between $H_{S_{O_1}}$ and $H_{S_{O_2}}$ if it satisfies the following properties:

(I) **Boundary**: $0 \leq d(H_{S_{O_1}}, H_{S_{O_2}}) \leq 1$;
(II) **Symmetry**: $d(H_{S_{O_1}}, H_{S_{O_2}}) = d(H_{S_{O_2}}, H_{S_{O_1}})$;
(III) **Complementarity**: $d(H_{S_{O_1}}, \overline{H_{S_{O_1}}}) = 1$ iff $H_{S_{O_1}} = \{s_{\tau<o_\varsigma>}\}$ or $H_{S_{O_1}} = \{s_{-\tau<o_{-\varsigma}>}\}$;
(IV) **Reflexivity**: $d(H_{S_{O_1}}, H_{S_{O_2}}) = 0$ iff $H_{S_{O_1}} = H_{S_{O_2}}$.

where $\overline{H_{S_{O_1}}} = \{\overline{h_{S_{O_{11}}}}, \overline{h_{S_{O_{12}}}}, \ldots, \overline{h_{S_{O_{1n}}}}\}$ is the complement set of $H_{S_{O_1}}$.

Definition 2.5 (Gou et al. 2018) Let S_O be a DHLTS, $H_{S_{O_1}}$ and $H_{S_{O_2}}$ be two DHHFLTSs. Then $s(H_{S_{O_1}}, H_{S_{O_2}})$ is called the similarity measure between $H_{S_{O_1}}$ and $H_{S_{O_2}}$ if it satisfies the following properties:

(I) **Boundary**: $0 \leq s(H_{S_{O_1}}, H_{S_{O_2}}) \leq 1$;
(II) **Symmetry**: $s(H_{S_{O_1}}, H_{S_{O_2}}) = s(H_{S_{O_2}}, H_{S_{O_1}})$;

(III) **Complementarity**: $s(H_{S_{O_1}}, \overline{H_{S_{O_1}}}) = 0$ iff $H_{S_{O_1}} = \{s_{\tau<o_\varsigma>}\}$ or $H_{S_{O_1}} = \{s_{-\tau<o_{-\varsigma}>}\}$;

(IV) **Reflexivity**: $s(H_{S_{O_1}}, H_{S_{O_2}}) = 1$ iff $H_{S_{O_1}} = H_{S_{O_2}}$.

where $\overline{H_{S_{O_1}}} = \{\overline{h_{S_{O_{11}}}}, \overline{h_{S_{O_{12}}}}, \ldots, \overline{h_{S_{O_{1n}}}}\}$ is the complement set of $H_{S_{O_1}}$.

Similar to Eq. (2.1), Gou et al. (2018) established the relationship between the distance measure and the similarity measure of the DHHFLTSs by the following formula:

$$s(H_{S_{O_1}}, H_{S_{O_2}}) = \frac{\Im(d(H_{S_{O_1}}, H_{S_{O_2}})) - \Im(1)}{\Im(0) - \Im(1)} \tag{2.36}$$

Similarly, the strictly monotonically decreasing real function can be (1) $\Im(\upsilon) = 1 - \upsilon$, (2) $\Im(\upsilon) = \frac{1-\upsilon}{1+\upsilon}$, (3) $\Im(\upsilon) = 1 - \upsilon e^{\upsilon-1}$, and (4) $\Im(\upsilon) = 1 - \upsilon^2$.

2.2.2 *Weighted Distance and Similarity Measures for DHHFLTSs in Discrete Case*

Definition 2.6 (Gou et al. 2018) Let S_O be a DHLTS, $H_{S_{O_i}} = \{h_{S_{O_{i1}}}, h_{S_{O_{i2}}}, \ldots, h_{S_{O_{in}}}\}(i = 1, 2)$ be two DHHFLTSs, where $h_{S_{O_{ij}}} = \{h^{(l)}_{S_{O_{ij}}} | h^{(l)}_{S_{O_{ij}}} \in S_O; l = 1, 2, \ldots, \#h_{S_{O_{ij}}}\}(i = 1, 2; j = 1, 2, \ldots, n)$ ($h^{(l)}_{S_{O_{ij}}}$ being the l - th DHLT in $h_{S_{O_{ij}}}$, $\#h_{S_{O_{ij}}}$ being the number of DHLTs in $h_{S_{O_{ij}}}$, and let $\#h_{S_{O_{1j}}} = \#h_{S_{O_{2j}}} = L$). For $H_{S_{O_i}} (i = 1, 2)$ with the associated weighting vector $w = (w_1, w_2, \ldots, w_n)^T$, where $0 \leq w_j \leq 1$ and $\sum_{j=1}^n w_j = 1$, the generalized weighted distance, the generalized weighted Hausdorff distance, and the generalized weighted hesitance degree-based distance between $H_{S_{O_1}}$ and $H_{S_{O_2}}$ can be defined, respectively:

$$d_{gwd}(H_{S_{O_1}}, H_{S_{O_2}}) = \left(\sum_{j=1}^{n} \frac{w_j}{L} \sum_{l=1}^{L} \left| f(h^{(l)}_{S_{O_{1j}}}) - f(h^{(l)}_{S_{O_{2j}}}) \right|^{\lambda} \right)^{1/\lambda} \tag{2.37}$$

$$d_{gwhd}(H_{S_{O_1}}, H_{S_{O_2}}) = \left(\sum_{j=1}^{n} w_j \max_{l=1,2,\ldots,L} \left| f(h^{(l)}_{S_{O_{1j}}}) - f(h^{(l)}_{S_{O_{2j}}}) \right|^{\lambda} \right)^{1/\lambda} \tag{2.38}$$

$$d_{gwhdd}(H_{S_{O_1}}, H_{S_{O_2}}) = \left(\sum_{j=1}^{n} w_j \left| u(h^{(l)}_{S_{O_{1j}}}) - u(h^{(l)}_{S_{O_{2j}}}) \right|^{\lambda} \right)^{1/\lambda} \tag{2.39}$$

where $\lambda > 0$, and f is the equivalent transformation function. Specially, if $\lambda = 1$ and $\lambda = 2$, then Eqs. (2.37)–(2.39) can be reduced to the corresponding Hamming and Euclidean distances, here we omit them.

Additionally, Gou et al. (2018) defined some generalized hybrid weighted distance measures as the generalized hybrid weighted Hausdorff distance, the generalized hybrid weighted hesitance degree-based distance, the generalized hybrid weighted Hausdorff-hesitance degree-based distance, and the generalized completely hybrid weighted Hausdorff-hesitance degree-based distance between $H_{S_{O_1}}$ and $H_{S_{O_2}}$, respectively:

$$d_{ghwhd}(H_{S_{O_1}}, H_{S_{O_2}}) = \left(\sum_{j=1}^{n}\frac{w_j}{2}\left(\frac{1}{L}\sum_{l=1}^{L}\left|f(h_{S_{O_{1j}}}^{(l)}) - f(h_{S_{O_{2j}}}^{(l)})\right|^{\lambda} + \max_{l=1,2,\ldots,L}\left|f(h_{S_{O_{1j}}}^{(l)}) - f(h_{S_{O_{2j}}}^{(l)})\right|^{\lambda}\right)\right)^{1/\lambda} \tag{2.40}$$

$$d_{ghwhdd}(H_{S_{O_1}}, H_{S_{O_2}}) = \left(\sum_{j=1}^{n}\frac{w_j}{2}\left(\frac{1}{L}\sum_{l=1}^{L}\left|f(h_{S_{O_{1j}}}^{(l)}) - f(h_{S_{O_{2j}}}^{(l)})\right|^{\lambda} + \left|u(h_{S_{O_{1j}}}^{(l)}) - u(h_{S_{O_{2j}}}^{(l)})\right|^{\lambda}\right)\right)^{1/\lambda} \tag{2.41}$$

$$d_{ghwhhdd}(H_{S_{O_1}}, H_{S_{O_2}}) = \left(\sum_{j=1}^{n}\frac{w_j}{2}\left(\max_{l=1,2,\ldots,L}\left|f(h_{S_{O_{1j}}}^{(l)}) - f(h_{S_{O_{2j}}}^{(l)})\right|^{\lambda} + \left|u(h_{S_{O_{1j}}}^{(l)}) - u(h_{S_{O_{2j}}}^{(l)})\right|^{\lambda}\right)\right)^{1/\lambda} \tag{2.42}$$

$$d_{gchwhhdd}(H_{S_{O_1}}, H_{S_{O_2}}) = \left(\sum_{j=1}^{n}\frac{w_j}{3}\left(\frac{1}{L}\sum_{l=1}^{L}\left|f(h_{S_{O_{1j}}}^{(l)}) - f(h_{S_{O_{2j}}}^{(l)})\right|^{\lambda} + \max_{l=1,2,\ldots,L}\left|f(h_{S_{O_{1j}}}^{(l)}) - f(h_{S_{O_{2j}}}^{(l)})\right|^{\lambda} + \left|u(h_{S_{O_{1j}}}^{(l)}) - u(h_{S_{O_{2j}}}^{(l)})\right|^{\lambda}\right)\right)^{1/\lambda} \tag{2.43}$$

where $\lambda > 0$, and f is the equivalent transformation function. Similarly, if $\lambda = 1$ and $\lambda = 2$, then Eqs. (2.41)–(2.43) can be reduced to the corresponding Hamming and Euclidean distances, we can also omit them.

In addition, if we consider the preference information about the Hausdorff distances, the hesitance degrees and the membership values, then the generalized completely hybrid weighted Hausdorff-hesitance degree-preference distance between $H_{S_{O_1}}$ and $H_{S_{O_2}}$ can be defined as (Gou et al. 2018):

$$d_{gchwhhdpd}(H_{S_{O_1}}, H_{S_{O_2}}) = \left(\sum_{j=1}^{n} w_j\left(\frac{a}{L}\sum_{l=1}^{L}\left|f(h_{S_{O_{1j}}}^{(l)}) - f(h_{S_{O_{2j}}}^{(l)})\right|^{\lambda} + b\max_{l=1,2,\ldots,L}\left|f(h_{S_{O_{1j}}}^{(l)}) - f(h_{S_{O_{2j}}}^{(l)})\right|^{\lambda} + c\left|u(h_{S_{O_{1j}}}) - u(h_{S_{O_{2j}}})\right|^{\lambda}\right)\right)^{1/\lambda} \tag{2.44}$$

where $\lambda > 0$, $0 \le a, b, c \le 1$, $a + b + c = 1$, and F' is the equivalent transformation function.

Next, we can obtain different distance measures based on the values of λ, a, b, and c (Gou et al. 2018):

(1) If $a + b = 1$ and $c = 0$, then Eq. (2.44) reduces to the generalized hybrid weighted Hausdorff-preference distance $H_{S_{O_1}}$ and $H_{S_{O_2}}$:

$$d_{ghwhpd}(H_{S_{O_1}}, H_{S_{O_2}}) = \left(\sum_{j=1}^{n} w_j \left(\frac{a}{L} \sum_{l=1}^{L} \left| f(h_{S_{O_{1j}}}^{(l)}) - f(h_{S_{O_{2j}}}^{(l)}) \right|^{\lambda} + b \max_{l=1,2,\ldots,L} \left| f(h_{S_{O_{1j}}}^{(l)}) - f(h_{S_{O_{2j}}}^{(l)}) \right|^{\lambda} \right) \right)^{1/\lambda} \tag{2.45}$$

(2) If $a + c = 1$ and $b = 0$, then Eq. (2.44) reduces to the generalized hybrid weighted hesitance degree-preference distance between $H_{S_{O_1}}$ and $H_{S_{O_2}}$:

$$d_{ghwhdpd}(H_{S_{O_1}}, H_{S_{O_2}}) = \left(\sum_{j=1}^{n} w_j \left(\frac{a}{L} \sum_{l=1}^{L} \left| f(h_{S_{O_{1j}}}^{(l)}) - f(h_{S_{O_{2j}}}^{(l)}) \right|^{\lambda} + c \left| u(h_{S_{O_{1j}}}) - u(h_{S_{O_{2j}}}) \right|^{\lambda} \right) \right)^{1/\lambda} \tag{2.46}$$

(3) If $b + c = 1$ and $a = 0$, then Eq. (2.44) reduces to the generalized hybrid weighted Hausdorff-hesitance degree-preference distance between $H_{S_{O_1}}$ and $H_{S_{O_2}}$:

$$d_{ghwhhdpd}(H_{S_{O_1}}, H_{S_{O_2}}) = \left(\sum_{j=1}^{n} w_j \left(b \max_{l=1,2,\ldots,L} \left| f(h_{S_{O_{1j}}}^{(l)}) - f(h_{S_{O_{2j}}}^{(l)}) \right|^{\lambda} + c \left| u(h_{S_{O_{1j}}}) - u(h_{S_{O_{2j}}}) \right|^{\lambda} \right) \right)^{1/\lambda} \tag{2.47}$$

Similarly, if $\lambda = 1$ and $\lambda = 2$, then Eqs. (2.44)–(2.47) reduce to the corresponding Hamming and Euclidean distances, here we omit them.

2.2.3 Weighted Distance and Similarity Measures for DHHFLTSs in Continuous Case

In Sect. 2.2.2, all the distance and similarity measures discussed above are in discrete case. If both the universe of discourse and the weights of elements are continuous, Gou et al. (2018) defined some distance and similarity measures between the DHHFLTSs in continuous case.

Let $x \in [\alpha, \beta]$, and $w(x)$ be the weight of x, where $0 \le w(x) \le 1$ and $\int_{\alpha}^{\beta} w(x)dx = 1$. Let $H_{S_{O_1}}$ and $H_{S_{O_2}}$ be two DHHFLTSs over the element x. Then Gou et al. (2018) defined the generalized continuous weighted distance, the generalized continuous weighted Hausdorff distance, and the generalized continuous weighted hesitance degree-based distance between $H_{S_{O_1}}$ and $H_{S_{O_2}}$, respectively:

$$d_{gcwd}(H_{S_{O_1}}, H_{S_{O_2}}) = \left(\int_{\alpha}^{\beta} \frac{w(x)}{L} \sum_{l=1}^{L} \left| f(h_{S_{O_{1j}}}^{(l)}(x)) - f(h_{S_{O_{2j}}}^{(l)}(x)) \right|^{\lambda} dx \right)^{1/\lambda} \quad (2.48)$$

$$d_{gcwhd}(H_{S_{O_1}}, H_{S_{O_2}}) = \left(\int_{\alpha}^{\beta} w(x) \max_{l=1,2,\ldots,L} \left| f(h_{S_{O_{1j}}}^{(l)}(x)) - f(h_{S_{O_{2j}}}^{(l)}(x)) \right|^{\lambda} dx \right)^{1/\lambda} \quad (2.49)$$

$$d_{gcwhdd}(H_{S_{O_1}}, H_{S_{O_2}}) = \left(\int_{\alpha}^{\beta} w(x) \left| u(h_{S_{O_{1j}}}^{(l)}) - u(h_{S_{O_{2j}}}^{(l)}) \right|^{\lambda} dx \right)^{1/\lambda} \quad (2.50)$$

Specially, if $\lambda = 1$, then Eqs. (2.48)–(2.50) reduce to the continuous weighted Hamming distance, the continuous weighted Hamming-Hausdorff distance, and the continuous weighted Hamming-hesitance degree-based distance between $H_{S_{O_1}}$ and $H_{S_{O_2}}$, respectively:

$$d_{cwhd}(H_{S_{O_1}}, H_{S_{O_2}}) = \int_{\alpha}^{\beta} \frac{w(x)}{L} \sum_{l=1}^{L} \left| f(h_{S_{O_{1j}}}^{(l)}(x)) - f(h_{S_{O_{2j}}}^{(l)}(x)) \right| dx \quad (2.51)$$

$$d_{gwhhd}(H_{S_{O_1}}, H_{S_{O_2}}) = \int_{\alpha}^{\beta} w(x) \max_{l=1,2,\ldots,L} \left| f(h_{S_{O_{1j}}}^{(l)}(x)) - f(h_{S_{O_{2j}}}^{(l)}(x)) \right| dx \quad (2.52)$$

$$d_{cwhhdd}(H_{S_{O_1}}, H_{S_{O_2}}) = \int_{\alpha}^{\beta} w(x) \left| u(h_{S_{O_{1j}}}^{(l)}) - u(h_{S_{O_{2j}}}^{(l)}) \right| dx \quad (2.53)$$

If $\lambda = 2$, then Eqs. (2.48)–(2.50) reduce to the continuous weighted Euclidean distance, the continuous weighted Euclidean-Hausdorff distance, and the continuous weighted Euclidean-hesitance degree-based distance between $H_{S_{O_1}}$ and $H_{S_{O_2}}$, respectively:

$$d_{cwed}(H_{S_{O_1}}, H_{S_{O_2}}) = \left(\int_{\alpha}^{\beta} \frac{w(x)}{L} \sum_{l=1}^{L} \left| f(h_{S_{O_{1j}}}^{(l)}(x)) - f(h_{S_{O_{2j}}}^{(l)}(x)) \right|^{2} dx \right)^{1/2} \quad (2.54)$$

$$d_{cwehd}(H_{S_{O_1}}, H_{S_{O_2}}) = \left(\int_{\alpha}^{\beta} w(x) \max_{l=1,2,\ldots,L} \left| f(h_{S_{O_{1j}}}^{(l)}(x)) - f(h_{S_{O_{2j}}}^{(l)}(x)) \right|^{2} dx \right)^{1/2} \quad (2.55)$$

$$d_{cwehdd}(H_{S_{O_1}}, H_{S_{O_2}}) = \left(\int_{\alpha}^{\beta} w(x) \left| u(h_{S_{O_{1j}}}^{(l)}) - u(h_{S_{O_{2j}}}^{(l)}) \right|^2 dx \right)^{1/2} \quad (2.56)$$

Additionally, Gou et al. (2018) defined some hybrid continuous weighted distance measures, such as the generalized hybrid continuous weighted Hausdorff distance, the generalized hybrid continuous weighted hesitance degree-based distance, the generalized hybrid continuous weighted Hausdorff-hesitance degree-based distance, the generalized completely hybrid continuous weighted distance, and the generalized completely hybrid continuous weighted distance between $H_{S_{O_1}}$ and $H_{S_{O_2}}$, respectively:

$$d_{ghcwhd}(H_{S_{O_1}}, H_{S_{O_2}}) = \left(\int_{\alpha}^{\beta} \frac{w(x)}{2} \left(\frac{1}{L} \sum_{l=1}^{L} \left| f(h_{S_{O_{1j}}}^{(l)}(x)) - f(h_{S_{O_{2j}}}^{(l)}(x)) \right|^{\lambda} + \max_{l=1,2,\ldots,L} \left| f(h_{S_{O_{1j}}}^{(l)}(x)) - f(h_{S_{O_{2j}}}^{(l)}(x)) \right|^{\lambda} \right) dx \right)^{1/\lambda} \quad (2.57)$$

$$d_{ghcwhdd}(H_{S_{O_1}}, H_{S_{O_2}}) = \left(\int_{\alpha}^{\beta} \frac{w(x)}{2} \left(\frac{1}{L} \sum_{l=1}^{L} \left| f(h_{S_{O_{1j}}}^{(l)}(x)) - f(h_{S_{O_{2j}}}^{(l)}(x)) \right|^{\lambda} + \left| u(h_{S_{O_{1j}}}^{(l)}) - u(h_{S_{O_{2j}}}^{(l)}) \right|^{\lambda} \right) dx \right)^{1/\lambda} \quad (2.58)$$

$$d_{ghcwhhdd}(H_{S_{O_1}}, H_{S_{O_2}}) = \left(\int_{\alpha}^{\beta} \frac{w(x)}{2} \left(\max_{l=1,2,\ldots,L} \left| f(h_{S_{O_{1j}}}^{(l)}(x)) - f(h_{S_{O_{2j}}}^{(l)}(x)) \right|^{\lambda} + \left| u(h_{S_{O_{1j}}}^{(l)}) - u(h_{S_{O_{2j}}}^{(l)}) \right|^{\lambda} \right) dx \right)^{1/\lambda} \quad (2.59)$$

$$d_{gchcwhhdd}(H_{S_{O_1}}, H_{S_{O_2}}) = \left(\int_{\alpha}^{\beta} \frac{w(x)}{3} \left(\frac{1}{L} \sum_{l=1}^{L} \left| f(h_{S_{O_{1j}}}^{(l)}(x)) - f(h_{S_{O_{2j}}}^{(l)}(x)) \right|^{\lambda} + \max_{l=1,2,\ldots,L} \left| f(h_{S_{O_{1j}}}^{(l)}(x)) - f(h_{S_{O_{2j}}}^{(l)}(x)) \right|^{\lambda} + \left| u(h_{S_{O_{1j}}}^{(l)}) - u(h_{S_{O_{2j}}}^{(l)}) \right|^{\lambda} \right) dx \right)^{1/\lambda} \quad (2.60)$$

Specially, if $\lambda = 1$ and $\lambda = 2$, then Eqs. (2.57)–(2.60) reduce to the corresponding Hamming and Euclidean distances, here we omit them.

Then, by adding the preference information into Eq. (2.60), Gou et al. (2018) defined the generalized completely hybrid continuous weighted Hausdorff-hesitance degree-preference distance between $H_{S_{O_1}}$ and $H_{S_{O_2}}$:

$$d_{gchcwhhdpd}(H_{S_{O_1}}, H_{S_{O_2}}) = \left(\int_{\alpha}^{\beta} w(x) \left(\frac{a}{L} \sum_{l=1}^{L} \left| f(h_{S_{O_{1j}}}^{(l)}(x)) - f(h_{S_{O_{2j}}}^{(l)}(x)) \right|^{\lambda} + b \max_{l=1,2,\ldots,L} \left| f(h_{S_{O_{1j}}}^{(l)}(x)) - f(h_{S_{O_{2j}}}^{(l)}(x)) \right|^{\lambda} + c \left| u(h_{S_{O_{1j}}}^{(l)}) - u(h_{S_{O_{2j}}}^{(l)}) \right|^{\lambda} \right) dx \right)^{1/\lambda} \tag{2.61}$$

where $\lambda > 0$, $0 \le a, b, c \le 1$, $a + b + c = 1$.

Next, some distance measures can be obtained by taking the values of λ, a, b and c:

(1) If $a + b = 1$ and $c = 0$, then Eq. (2.61) reduces to the generalized hybrid continuous weighted Hausdorff-preference distance between $H_{S_{O_1}}$ and $H_{S_{O_2}}$:

$$d_{ghcwhpd}(H_{S_{O_1}}, H_{S_{O_2}}) = \left(\int_{\alpha}^{\beta} w(x) \left(b \max_{l=1,2,\ldots,L} \left| f(h_{S_{O_{1j}}}^{(l)}(x)) - f(h_{S_{O_{2j}}}^{(l)}(x)) \right|^{\lambda} + c \left| u(h_{S_{O_{1j}}}^{(l)}) - u(h_{S_{O_{2j}}}^{(l)}) \right|^{\lambda} \right) dx \right)^{1/\lambda} \tag{2.62}$$

(2) If $a + c = 1$ and $b = 0$, then Eq. (2.61) reduces to the generalized hybrid continuous weighted hesitance degree-preference distance between $H_{S_{O_1}}$ and $H_{S_{O_2}}$:

$$d_{ghcwhdpd}(H_{S_{O_1}}, H_{S_{O_2}}) = \left(\int_{\alpha}^{\beta} w(x) \left(\frac{a}{L} \sum_{l=1}^{L} \left| f(h_{S_{O_{1j}}}^{(l)}(x)) - f(h_{S_{O_{2j}}}^{(l)}(x)) \right|^{\lambda} + c \left| u(h_{S_{O_{1j}}}^{(l)}) - u(h_{S_{O_{2j}}}^{(l)}) \right|^{\lambda} \right) dx \right)^{1/\lambda} \tag{2.63}$$

(3) If $b + c = 1$ and $a = 0$, then Eq. (2.61) reduces to the generalized hybrid continuous weighted Hausdorff-hesitance degree-preference distance between $H_{S_{O_1}}$ and $H_{S_{O_2}}$:

$$d_{ghcwhhdpd}(H_{S_{O_1}}, H_{S_{O_2}}) = \left(\int_{\alpha}^{\beta} w(x) \left(\max_{l=1,2,\ldots,L} \left| f(h_{S_{O_{1j}}}^{(l)}(x)) - f(h_{S_{O_{2j}}}^{(l)}(x)) \right|^{\lambda} + c \left| u(h_{S_{O_{1j}}}^{(l)}) - u(h_{S_{O_{2j}}}^{(l)}) \right|^{\lambda} \right) dx \right)^{1/\lambda} \tag{2.64}$$

2.2.4 *Ordered Weighted Distance and Similarity Measures for DHHFLTSs*

In recent years, lots of scholars have researched the ordered weighted distance and similarity measures under different uncertain environments. Xu (2012) defined several ordered weighted distance measures, which are suitable to be used in many actual fields, including medical diagnosis, data mining, and pattern recognition. Based on Xu and Chen' distance measures, Yager (2010) generalized and provided a variety of ordered weighted averaging norms and similarity measures. Merigó and Gil-Lafuente (2010) introduced an ordered weighted averaging distance operator. Furthermore, on the basis of hesitant fuzzy information, Xu and Xia (2011) developed a variety of distance measures and the corresponding similarity measures for hesitant fuzzy sets. Liao et al. (2014) and Liao and Xu (2015) proposed a family of distance and similarity measures between two HFLTSs. Under double hierarchy hesitant fuzzy linguistic environment, Gou et al. (2018) developed some ordered weighted distance measures between DHHFLTSs.

Firstly, the generalized ordered weighted distance between $H_{S_{O_1}}$ and $H_{S_{O_2}}$ is defined as:

$$d_{gowd}(H_{S_{O_1}}, H_{S_{O_2}}) = \left(\sum_{j=1}^{n} \frac{w_j}{L} \sum_{l=1}^{L} \left| f(h^{(l)}_{S_{O_{1\sigma(j)}}}(x)) - f(h^{(l)}_{S_{O_{2\sigma(j)}}}(x)) \right|^{\lambda} \right)^{1/\lambda} \tag{2.65}$$

where $\lambda > 0$ and $\sigma(j) : (1, 2, \ldots, n) \to (1, 2, \ldots, n)$ is a permutation satisfying

$$\left| f(h^{(l)}_{S_{O_{1\sigma(j+1)}}}(x)) - f(h^{(l)}_{S_{O_{2\sigma(j+1)}}}(x)) \right| \geq \left| f(h^{(l)}_{S_{O_{1\sigma(j)}}}(x)) - f(h^{(l)}_{S_{O_{2\sigma(j)}}}(x)) \right|, \quad j = 1, 2, \ldots, n-1.$$

Similarly, the generalized ordered weighted Hausdorff distance between $H_{S_{O_1}}$ and $H_{S_{O_2}}$ is defined as:

$$d_{gowhd}(H_{S_{O_1}}, H_{S_{O_2}}) = \left(\sum_{j=1}^{n} w_j \max_{l=1,2,\ldots,L} \left| f(h^{(l)}_{S_{O_{1\sigma(j)}}}(x)) - f(h^{(l)}_{S_{O_{2\sigma(j)}}}(x)) \right|^{\lambda} \right)^{1/\lambda} \tag{2.66}$$

where $\lambda > 0$ and $\sigma'(j) : (1, 2, \ldots, n) \to (1, 2, \ldots, n)$ is a permutation satisfying

$$\max_{l=1,2,\ldots,L} \left| f(h^{(l)}_{S_{O_{1\sigma(j+1)}}}(x)) - f(h^{(l)}_{S_{O_{2\sigma(j+1)}}}(x)) \right|^{\lambda} \geq \max_{l=1,2,\ldots,L} \left| f(h^{(l)}_{S_{O_{1\sigma(j)}}}(x)) - f(h^{(l)}_{S_{O_{2\sigma(j)}}}(x)) \right|^{\lambda}, j = 1, 2, \ldots, n-1.$$

and the generalized ordered weighted hesitance degree-based distance between $H_{S_{O_1}}$ and $H_{S_{O_2}}$ is defined as:

$$d_{gowhdd}(H_{S_{O_1}}, H_{S_{O_2}}) = \left(\sum_{j=1}^{n} w_j \left| u(h^{(l)}_{S_{O_{1\sigma(j)}}}(x)) - u(h^{(l)}_{S_{O_{2\sigma(j)}}}(x)) \right|^{\lambda} \right)^{1/\lambda} \tag{2.67}$$

where $\lambda > 0$ and $\sigma''(j) : (1, 2, \ldots, n) \to (1, 2, \ldots, n)$ is a permutation satisfying

$$\left|u(h^{(l)}_{S_{O_{1\sigma(j+1)}}}(x)) - u(h^{(l)}_{S_{O_{2\sigma(j+1)}}}(x))\right|^{\lambda} \geq \left|u(h^{(l)}_{S_{O_{1\sigma(j)}}}(x)) - u(h^{(l)}_{S_{O_{2\sigma(j)}}}(x))\right|^{\lambda},$$
$$j = 1, 2, \ldots, n-1.$$

Specially, if $\lambda = 1$, then Eqs. (2.65)–(2.67) reduce to the ordered weighted Hamming distance, the ordered weighted Hamming-Hausdorff distance, and the ordered weighted Hamming-hesitance degree-based distance between $H_{S_{O_1}}$ and $H_{S_{O_2}}$, respectively:

$$d_{owhd}(H_{S_{O_1}}, H_{S_{O_2}}) = \sum_{j=1}^{n} \frac{w_j}{L} \sum_{l=1}^{L} \left|f(h^{(l)}_{S_{O_{1\sigma(j)}}}(x)) - f(h^{(l)}_{S_{O_{2\sigma(j)}}}(x))\right| \tag{2.68}$$

$$d_{owhhd}(H_{S_{O_1}}, H_{S_{O_2}}) = \sum_{j=1}^{n} w_j \max_{l=1,2,\ldots,L} \left|f(h^{(l)}_{S_{O_{1\sigma(j)}}}(x)) - f(h^{(l)}_{S_{O_{2\sigma(j)}}}(x))\right| \tag{2.69}$$

$$d_{owhhdd}(H_{S_{O_1}}, H_{S_{O_2}}) = \sum_{j=1}^{n} w_j \left|u(h^{(l)}_{S_{O_{1\sigma(j)}}}(x)) - u(h^{(l)}_{S_{O_{2\sigma(j)}}}(x))\right| \tag{2.70}$$

If $\lambda = 2$, then Eqs. (2.65)–(2.67) reduce to the ordered weighted Euclidean distance, the ordered weighted Euclidean-Hausdorff distance, and the ordered weighted Euclidean-hesitance degree-based distance between $H_{S_{O_1}}$ and $H_{S_{O_2}}$, respectively:

$$d_{owed}(H_{S_{O_1}}, H_{S_{O_2}}) = \left(\sum_{j=1}^{n} \frac{w_j}{L} \sum_{l=1}^{L} \left|f(h^{(l)}_{S_{O_{1\sigma(j)}}}(x)) - f(h^{(l)}_{S_{O_{2\sigma(j)}}}(x))\right|^2\right)^{1/2} \tag{2.71}$$

$$d_{owehd}(H_{S_{O_1}}, H_{S_{O_2}}) = \left(\sum_{j=1}^{n} w_j \max_{l=1,2,\ldots,L} \left|f(h^{(l)}_{S_{O_{1\sigma(j)}}}(x)) - f(h^{(l)}_{S_{O_{2\sigma(j)}}}(x))\right|^2\right)^{1/2} \tag{2.72}$$

$$d_{owehdd}(H_{S_{O_1}}, H_{S_{O_2}}) = \left(\sum_{j=1}^{n} w_j \left|u(h^{(l)}_{S_{O_{1\sigma(j)}}}(x)) - u(h^{(l)}_{S_{O_{2\sigma(j)}}}(x))\right|^2\right)^{1/2} \tag{2.73}$$

Additionally, Gou et al. (2018) defined three generalized hybrid distance measures:

(1) The generalized hybrid ordered weighted Hausdorff distance between $H_{S_{O_1}}$ and $H_{S_{O_2}}$:

$$d_{ghowhd}(H_{S_{O_1}}, H_{S_{O_2}}) = \left(\sum_{j=1}^{n} \frac{w_j}{2} \left(\frac{1}{L} \sum_{l=1}^{L} \left|f(h^{(l)}_{S_{O_{1\sigma(j)}}}(x)) - f(h^{(l)}_{S_{O_{2\sigma(j)}}}(x))\right|^{\lambda} + \max_{l=1,2,\ldots,L} \left|f(h^{(l)}_{S_{O_{1\sigma(j)}}}(x)) - f(h^{(l)}_{S_{O_{2\sigma(j)}}}(x))\right|^{\lambda}\right)\right)^{1/\lambda} \tag{2.74}$$

where $\lambda > 0$ and $\dot{\sigma} : (1, 2, \ldots, n) \rightarrow (1, 2, \ldots, n)$ is a permutation satisfying

$$\frac{1}{L}\sum_{l=1}^{L}\left|f(h_{S_{O_{1\sigma(j+1)}}}^{(l)}(x)) - f(h_{S_{O_{2\sigma(j+1)}}}^{(l)}(x))\right|^{\lambda} + \max_{l=1,2,\ldots,L}\left|f(h_{S_{O_{1\sigma(j+1)}}}^{(l)}(x)) - f(h_{S_{O_{2\sigma(j+1)}}}^{(l)}(x))\right|^{\lambda} \geq$$
$$\frac{1}{L}\sum_{l=1}^{L}\left|f(h_{S_{O_{1\sigma(j)}}}^{(l)}(x)) - f(h_{S_{O_{2\sigma(j)}}}^{(l)}(x))\right|^{\lambda} + \max_{l=1,2,\ldots,L}\left|f(h_{S_{O_{1\sigma(j)}}}^{(l)}(x)) - f(h_{S_{O_{2\sigma(j)}}}^{(l)}(x))\right|^{\lambda}, \; j = 1, 2, \ldots, n-1.$$

(2) The generalized hybrid ordered weighted hesitance degrees distance between $H_{S_{O_1}}$ and $H_{S_{O_2}}$:

$$d_{ghowhhd}(H_{S_{O_1}}, H_{S_{O_2}}) = \left(\sum_{j=1}^{n}\frac{w_j}{2}\left(\frac{1}{L}\sum_{l=1}^{L}\left|f(h_{S_{O_{1\sigma(j)}}}^{(l)}(x)) - f(h_{S_{O_{2\sigma(j)}}}^{(l)}(x))\right|^{\lambda} + \left|u(h_{S_{O_{1\sigma(j)}}}^{(l)}(x)) - u(h_{S_{O_{2\sigma(j)}}}^{(l)}(x))\right|^{\lambda}\right)\right)^{1/\lambda} \tag{2.75}$$

where $\lambda > 0$ and $\ddot{\sigma} : (1, 2, \ldots, n) \rightarrow (1, 2, \ldots, n)$ is a permutation satisfying

$$\frac{1}{L}\sum_{l=1}^{L}\left|f(h_{S_{O_{1\sigma(j+1)}}}^{(l)}(x)) - f(h_{S_{O_{2\sigma(j+1)}}}^{(l)}(x))\right|^{\lambda} + \left|u(h_{S_{O_{1\sigma(j+1)}}}^{(l)}(x)) - u(h_{S_{O_{2\sigma(j+1)}}}^{(l)}(x))\right|^{\lambda} \geq$$
$$\frac{1}{L}\sum_{l=1}^{L}\left|f(h_{S_{O_{1\sigma(j)}}}^{(l)}(x)) - f(h_{S_{O_{2\sigma(j)}}}^{(l)}(x))\right|^{\lambda} + \left|u(h_{S_{O_{1\sigma(j)}}}^{(l)}(x)) - u(h_{S_{O_{2\sigma(j)}}}^{(l)}(x))\right|^{\lambda}, \; j = 1, 2, \ldots, n-1.$$

(3) The generalized hybrid ordered weighted Hausdorff-hesitance degree-based distance between $H_{S_{O_1}}$ and $H_{S_{O_2}}$:

$$d_{ghowhhdd}(H_{S_{O_1}}, H_{S_{O_2}}) = \left(\sum_{j=1}^{n}\frac{w_j}{2}\left(\max_{l=1,2,\ldots,L}\left|f(h_{S_{O_{1\sigma(j)}}}^{(l)}(x)) - f(h_{S_{O_{2\sigma(j)}}}^{(l)}(x))\right|^{\lambda} + \left|u(h_{S_{O_{1\sigma(j)}}}^{(l)}(x)) - u(h_{S_{O_{2\sigma(j)}}}^{(l)}(x))\right|^{\lambda}\right)\right)^{1/\lambda} \tag{2.76}$$

where $\lambda > 0$ and $\dddot{\sigma} : (1, 2, \ldots, n) \rightarrow (1, 2, \ldots, n)$ is a permutation satisfying

$$\max_{l=1,2,\ldots,L}\left|f(h_{S_{O_{1\sigma(j+1)}}}^{(l)}(x)) - f(h_{S_{O_{2\sigma(j+1)}}}^{(l)}(x))\right|^{\lambda} + \left|u(h_{S_{O_{1\sigma(j+1)}}}^{(l)}(x)) - u(h_{S_{O_{2\sigma(j+1)}}}^{(l)}(x))\right|^{\lambda}$$
$$\geq \max_{l=1,2,\ldots,L}\left|f(h_{S_{O_{1\sigma(j)}}}^{(l)}(x)) - f(h_{S_{O_{2\sigma(j)}}}^{(l)}(x))\right|^{\lambda} + \left|u(h_{S_{O_{1\sigma(j)}}}^{(l)}(x)) - u(h_{S_{O_{2\sigma(j)}}}^{(l)}(x))\right|^{\lambda},$$
$$j = 1, 2, \ldots, n-1.$$

Specially, if $\lambda = 1$ and $\lambda = 2$, then it is obvious that Eqs. (2.74)–(2.76) reduce to their Hamming and Euclidean distance measures respectively. Here we omit them.

Furthermore, by combining all these three distance measures together, the generalized completely hybrid ordered weighted distance can be defined as:

$$d_{gchowhhdd}(H_{S_{O_1}}, H_{S_{O_2}}) = \left(\sum_{j=1}^{n} \frac{w_j}{2} \left(\frac{1}{L} \sum_{l=1}^{L} \left| f(h_{S_{O_{1\sigma(j)}}}^{(l)}(x)) - f(h_{S_{O_{2\sigma(j)}}}^{(l)}(x)) \right|^{\lambda} + \max_{l=1,2,\ldots,L} \left| f(h_{S_{O_{1\sigma(j)}}}^{(l)}(x)) - f(h_{S_{O_{2\sigma(j)}}}^{(l)}(x)) \right|^{\lambda} + \left| u(h_{S_{O_{1\sigma(j)}}}^{(l)}(x)) - u(h_{S_{O_{2\sigma(j)}}}^{(l)}(x)) \right|^{\lambda} \right) \right)^{1/\lambda} \tag{2.77}$$

where $\lambda > 0$ and $\tilde{\sigma} : (1, 2, \ldots, n) \to (1, 2, \ldots, n)$ is a permutation satisfying

$$\begin{aligned} &\frac{1}{L} \sum_{l=1}^{L} \left| f(h_{S_{O_{1\sigma(j+1)}}}^{(l)}(x)) - f(h_{S_{O_{2\sigma(j+1)}}}^{(l)}(x)) \right|^{\lambda} + \max_{l=1,2,\ldots,L} \left| f(h_{S_{O_{1\sigma(j+1)}}}^{(l)}(x)) \right. \\ &\left. - f(h_{S_{O_{2\sigma(j+1)}}}^{(l)}(x)) \right|^{\lambda} + \left| u(h_{S_{O_{1\sigma(j+1)}}}^{(l)}(x)) - u(h_{S_{O_{2\sigma(j+1)}}}^{(l)}(x)) \right|^{\lambda} \\ &\geq \frac{1}{L} \sum_{l=1}^{L} \left| f(h_{S_{O_{1\sigma(j)}}}^{(l)}(x)) - f(h_{S_{O_{2\sigma(j)}}}^{(l)}(x)) \right|^{\lambda} \\ &+ \max_{l=1,2,\ldots,L} \left| f(h_{S_{O_{1\sigma(j)}}}^{(l)}(x)) - f(h_{S_{O_{2\sigma(j)}}}^{(l)}(x)) \right|^{\lambda} + \left| u(h_{S_{O_{1\sigma(j)}}}^{(l)}(x)) - u(h_{S_{O_{2\sigma(j)}}}^{(l)}(x)) \right|^{\lambda}, \\ &j = 1, 2, \ldots, n-1. \end{aligned}$$

Similarly, when $\lambda = 1$ and $\lambda = 2$, the corresponding Hamming and Euclidean distances can be omitted.

2.3 Distance Measures-Based MCDM Method and Application

Based on the proposed distance and similarity measures, Gou et al. (2018) developed a corresponding decision-making method and applied it to deal with a practical MCDM problem.

2.3.1 Distance Measures-Based MCDM Method

In recent years, lots of MCDM methods are developed such as TOPSIS (Tan et al. 2016), TODIM (Wei et al. 2015), VIKOR (Liao et al. 2015a) and MULTIMOORA (Gou et al. 2017). TOPSIS is attractive as limited subjective input is needed from decision-makers. Many authors argue that TOPSIS is an easy and useful method that helps decision-maker to select the best choice according to both the minimal distance from the positive-ideal solution and the maximal distance from the negative-ideal solution (Zavadskas et al. 2013). Therefore, Gou et al. (2018) proposed a MCDM method with double hierarchy hesitant fuzzy linguistic information based on TOPSIS model and the proposed distance measures.

A MCDM problem with double hierarchy hesitant fuzzy linguistic information can be described as follows: Let $A = \{A_1, A_2, \ldots, A_m\}$ be a set of alternatives, $C = \{C_1, C_2, \ldots, C_n\}$ be a set of criteria, and $w = (w_1, w_2, \ldots, w_n)^T$ be the weight vector of all criteria with $w_j \geq 0$, $j = 1, 2, \ldots, n$, and $\sum_{j=1}^{n} w_j = 1$. Let $S_O = \{s_{t<o_k>} | t = -\tau, \ldots, -1, 0, 1, \ldots, \tau;\ k = -\varsigma, \ldots, -1, 0, 1, \ldots, \varsigma\}$ be a DHLTS. The invited experts can give their linguistic evaluation information about each alternative with respect to each criterion. We can gather the evaluation information and establish a decision-making matrix $DM = (h_{S_{O_{ij}}})_{m \times n}$ $(i = 1, 2, \ldots, m;\ j = 1, 2, \ldots, n)$ shown as:

$$DM = (h_{S_{O_{ij}}})_{m\times n} = \begin{pmatrix} h_{S_{O_{11}}} & h_{S_{O_{12}}} & \cdots & h_{S_{O_{1n}}} \\ h_{S_{O_{21}}} & h_{S_{O_{22}}} & \cdots & h_{S_{O_{2n}}} \\ \vdots & \vdots & \ddots & \vdots \\ h_{S_{O_{m1}}} & h_{S_{O_{m2}}} & \cdots & h_{S_{O_{mn}}} \end{pmatrix} = \begin{pmatrix} H_{S_{O_1}} \\ H_{S_{O_2}} \\ \vdots \\ H_{S_{O_m}} \end{pmatrix}$$

Obviously, the DHHFLTSs $H_{S_{O_i}} = (h_{S_{O_{i1}}}, h_{S_{O_{i2}}}, \ldots, h_{S_{O_{in}}})(i = 1, 2, \ldots, m)$ can be used to express all evaluation information on the alternatives $A_i (i = 1, 2, \ldots, m)$. Then, Gou et al. (2018) developed a MCDM method shown as follows:

Algorithm 2.1 (Gou et al. 2018) Distance measures-based MCDM method

Step 1. For each criterion C_j, we can obtain the smallest DHHFLE $h^-_{S_{O_j}}$ and the largest DHHFLE $h^+_{S_{O_j}}$, respectively:

$$h^-_{S_{O_j}} = \begin{cases} \min\limits_{i=1,2,\ldots,m} \{h_{S_{O_{ij}}}\}, & \text{for benefit criterion } C_j \\ \max\limits_{i=1,2,\ldots,m} \{h_{S_{O_{ij}}}\}, & \text{for cost criterion } C_j \end{cases} \tag{2.78}$$

$$h^+_{S_{O_j}} = \begin{cases} \max\limits_{i=1,2,\ldots,m} \{h_{S_{O_{ij}}}\}, & \text{for benefit criterion } C_j \\ \min\limits_{i=1,2,\ldots,m} \{h_{S_{O_{ij}}}\}, & \text{for cost criterion } C_j \end{cases} \tag{2.79}$$

Combining all the smallest DHHFLEs and the largest DHHFLEs, respectively, we can obtain the double hierarchy hesitant fuzzy linguistic negative ideal solution $H^-_{S_O} = \{h^-_{S_{O_1}}, h^-_{S_{O_2}}, \ldots, h^-_{S_{O_n}}\}$ and the double hierarchy hesitant fuzzy linguistic positive ideal solution $H^+_{S_O} = \{h^+_{S_{O_1}}, h^+_{S_{O_2}}, \ldots, h^+_{S_{O_n}}\}$, respectively.

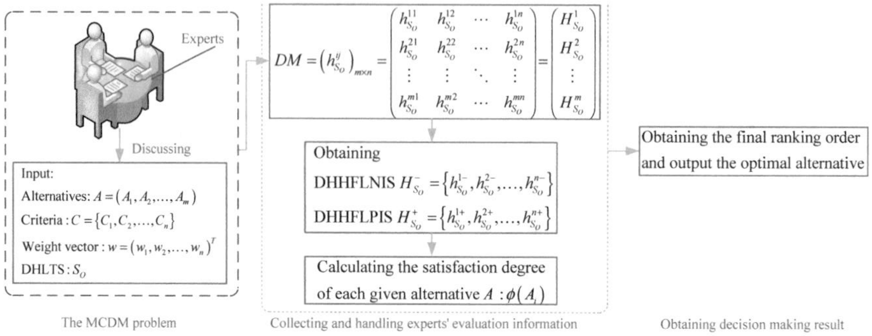

Fig. 2.2 The flowchart of the MCDM method

Step 2. Calculate the distance $d(H_{S_{O_i}}, H_{S_O}^-)$ between each alternative $H_{S_{O_i}}$ and the double hierarchy hesitant fuzzy linguistic negative ideal solution $H_{S_O}^-$, and the distance $d(H_{S_{O_i}}, H_{S_O}^+)$ between each alternative $H_{S_{O_i}}$ and the double hierarchy hesitant fuzzy linguistic positive ideal solution $H_{S_O}^+$, respectively. Clearly, the larger the distance $d(H_{S_{O_i}}, H_{S_O}^-)$ is, the better the alternative would be, while the smaller the value of $d(H_{S_{O_i}}, H_{S_O}^+)$ is, the better the alternative would be.

Step 3. Calculate the satisfaction degree of each given alternative A_i based on the following formula:

$$\partial(A_i) = \frac{(1-\theta)d(H_{S_{O_i}}, H_{S_O}^-)}{\theta d(H_{S_{O_i}}, H_{S_O}^+) + (1-\theta)d(H_{S_{O_i}}, H_{S_O}^-)} \tag{2.80}$$

where the parameter θ expresses the risk preferences of the expert and $0 \le \theta \le 1$. If $\theta > 0.5$, then the expert is pessimist; if $\theta < 0.5$, then the expert is optimist.

Step 4. Obviously, the bigger the satisfaction degree is, the better the alternative should be. Therefore, we can obtain the final ranking order of all alternatives.

Step 5. End.

The flowchart of this MCDM method can be drawn in Fig. 2.2.

2.3.2 *Application of the Distance Measures-Based MCDM Method*

Gou et al. (2018) applied the proposed distance measures-based MCDM method to deal with a practical MCDM problem about Sichuan liquor brand assessment.

Example 2.7 (Gou et al. 2018) Chinese liquor has a thousand years of history, which also carries Chinese culture. Meanwhile, the liquor industry has very high rates of return and profitability. In China, both Sichuan and Guizhou provinces are the largest scale and the optimal production quality white liquor producing regions, and support the development of the entire Chinese liquor industry. At present, the whole liquor market has the following characteristics:

(1) The brand competition will be the main theme of the next stage liquor competition because of young consumers' rational consumption.
(2) The work of government will further affect the development direction of the whole liquor industry, such as forbidding driving after drinking, tax adjustment, etc.
(3) The living spaces of middle and small-sized and low side competition enterprises are more and more small.

Nowadays, according to the development of economy and the constantly improvement of consuming stratums, liquors of middle and top grades will be the theme of Chinese liquor industry development in the future, as well as the main battlefield of Chinese liquor competition. However, Sichuan liquor lacks the hard core in the true sense. Therefore, according to the awkward situation of Sichuan liquor industry, it is necessary to analyze and research the development strategy of Sichuan liquor industry, and then analyze the preference relations and consuming behaviors of consumers from their cognitive perspectives about each Sichuan liquor brand. Thus, the above work can provide a series of adjustment strategy to Sichuan liquor enterprises and promote the development of Sichuan liquor enterprises much better.

In order to investigate the consumers' cognitions about Sichuan liquor, we choose five Sichuan liquor brands, namely, Wuliangye Yibin (A_1), Luzhou Old Cellar (A_2), Ichiro liquor (A_3), Tuopai liquor (A_4) and Jian Nan Chun (A_5). Then we investigate the cognitions of consumers based on four criteria such as product price (C_1), product classification (C_2), consumer group (C_3) and distribution channel (C_4). Based on the following two LTSs:

$$\begin{aligned} S = \{ & s_{-3} = none, s_{-2} = very\ bad, s_{-1} = bad, s_0 = medium, \\ & s_1 = good, s_2 = very\ good, s_3 = perfect\} \end{aligned}$$

$$O = \begin{cases} \{o_{-3} = far\ from, o_{-2} = only\ a\ little, o_{-1} = a\ little, o_0 = just\ right, \\ \quad o_1 = much, o_2 = very\ much, o_3 = extremely\}, if\ s_t \geq s_0. \\ \{o_{-3} = extremely, o_{-2} = very\ much, o_{-1} = much, o_0 = just\ right, \\ \quad o_1 = a\ little, o_2 = only\ a\ little, o_3 = far\ from\}, if\ s_t < s_0. \end{cases}$$

we summarize the survey results and the evaluation information for each alternative with respect to each criterion and express the information by DHHFLEs. All evaluation information establishes the decision-making matrix (Table 2.2). Furthermore, the weight vector of these criteria is $w = (0.1, 0.3, 0.2, 0.4)^T$.

Gou et al. (2018) utilized Algorithm 2.1 to deal with this MCDM problem, which can be given as follows:

Table 2.2 Decision-making matrix with DHHFLEs

	C_1	C_2	C_3	C_4
A_1	$\{s_{0<o_{-2}>}, s_1, s_{2<o_1>}\}$	$\{s_{2<o_1>}\}$	$\{s_{-1<o_2>}, s_{0<o_1>}\}$	$\{s_{1<o_1>}, s_{2<o_2>}\}$
A_2	$\{s_{2<o_0>}, s_{3<o_{-1}>}\}$	$\{s_{-2<o_{-2}>}, s_{-1}, s_0, s_{1<o_2>}\}$	$\{s_{-2<o_{-2}>}, s_{-1}, s_{0<o_2>}\}$	$\{s_{1<o_{-2}>}, s_2, s_{3<o_{-2}>}\}$
A_3	$\{s_{1<o_0>}\}$	$\{s_{1<o_0>}, s_{2<o_0>}\}$	$\{s_{-1<o_{-2}>}, s_{0<o_{-2}>}\}$	$\{s_{0<o_2>}, s_{1<o_2>}\}$
A_4	$\{s_{2<o_1>}, s_{3<o_0>}\}$	$\{s_{0<o_0>}\}$	$\{s_{1<o_1>}\}$	$\{s_{0<o_0>}\}$
A_5	$\{s_{0<o_{-2}>}, s_{1<o_1>}\}$	$\{s_{1<o_{-1}>}, s_2, s_{3<o_{-1}>}\}$	$\{s_{-1<o_{-2}>}, s_0, s_{1<o_{-1}>}\}$	$\{s_{2<o_{-2}>}, s_{3<o_{-2}>}\}$

Firstly, we need to obtain the double hierarchy hesitant fuzzy linguistic negative ideal solution:

$$H_{S_O}^- = \{\{s_{0<o_{-2}>}, s_{1<o_1>}\}, \{s_{-2<o_{-2}>}, s_{-1}, s_0, s_{1<o_2>}\}, \{s_{-1<o_{-2}>}, s_{0<o_{-2}>}\}, \{s_{0<o_0>}\}\}$$

and the double hierarchy hesitant fuzzy linguistic positive ideal solution:

$$H_{S_O}^+ = \{\{s_{2<o_1>}, s_{3<o_0>}\}, \{s_{2<o_1>}\}, \{s_{1<o_1>}\}, \{s_{1<o_1>}, s_{2<o_2>}\}\}$$

Additionally, we utilize the generalized completely hybrid weighted Hausdorff-hesitance degree-based distance to calculate the distance between each alternative and the double hierarchy hesitant fuzzy linguistic negative ideal solution $d_{gchwhhdd}(H_{S_{O_i}}, H_{S_O}^-)$, and the distance between each alternative and the double hierarchy hesitant fuzzy linguistic positive ideal solution $d_{gchwhhdd}(H_{S_{O_i}}, H_{S_O}^+)$, respectively. In this process, we let λ be 1, 2 and 5, respectively.

Furthermore, based on Eq. (2.80), we can calculate the satisfaction degree of each alternative. And then, the ranking orders of all alternatives on the basis of the satisfaction degrees can be obtained and shown in Table 2.3 and Fig. 2.3.

On the other hand, utilizing the generalized completely hybrid ordered weighted distance, we can also obtain the ranking orders of all alternatives in Table 2.4 and Fig. 2.4.

For the generalized completely hybrid weighted Hausdorff-hesitance degree-based distance, as we have seen in Table 2.3 and Fig. 2.3, the changes of the ranking orders are very small when we utilize the different values of λ. Specifically, for the alternatives A_1 and A_4, the satisfaction degrees of them are gradually decreased with the increase of the value of λ; For the alternatives A_2 and A_3, the satisfaction degrees of them are gradually increased with the increase of the

Table 2.3 The satisfaction degrees and the ranking orders based on the generalized completely hybrid weighted Hausdorff-hesitance degree-based distance

	A_1	A_2	A_3	A_4	A_5	Ranking order
$\lambda = 1$	0.681	0.286	0.394	0.689	0.411	$A_4 \succ A_1 \succ A_5 \succ A_3 \succ A_2$
$\lambda = 2$	0.642	0.324	0.452	0.624	0.398	$A_1 \succ A_4 \succ A_3 \succ A_5 \succ A_2$
$\lambda = 5$	0.627	0.383	0.498	0.595	0.428	$A_1 \succ A_4 \succ A_3 \succ A_5 \succ A_2$

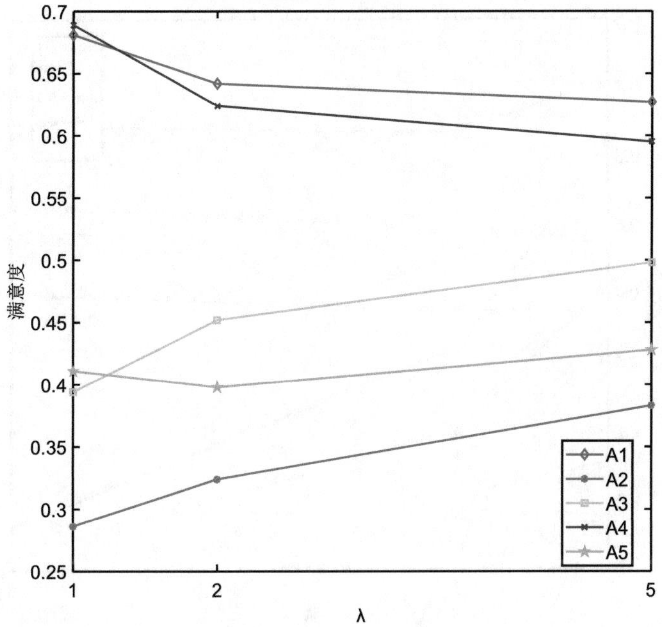

Fig. 2.3 The satisfaction degrees and ranking orders based on the generalized completely hybrid weighted Hausdorff-hesitance degree-based distance

Table 2.4 The satisfaction degrees and ranking orders based on the generalized completely hybrid ordered weighted distance

	A_1	A_2	A_3	A_4	A_5	Ranking order
$\lambda = 1$	0.713	0.374	0.475	0.623	0.415	$A_1 \succ A_4 \succ A_3 \succ A_5 \succ A_2$
$\lambda = 2$	0.635	0.378	0.494	0.324	0.202	$A_1 \succ A_3 \succ A_2 \succ A_4 \succ A_5$
$\lambda = 5$	0.646	0.401	0.528	0.062	0.018	$A_1 \succ A_3 \succ A_2 \succ A_4 \succ A_5$

value of λ; For the alternative A_5, its satisfaction degrees have three different stages of change. Additionally, for the generalized completely hybrid ordered weighted distance, as we have seen in Table 2.4 and Fig. 2.4, the changes of the ranking orders of A_1 and A_2 are small and the changes of the ranking orders of the rest alternatives are very apparent when we utilize different values of λ. Finally, by considering that we change the orders of all DHHFLEs included in DHHFLTS when we utilize the generalized completely hybrid weighted Hausdorff-hesitance degree-based distance to calculate the satisfaction degrees of these alternatives, it is reasonable that the changes of the ranking orders of some alternatives are very apparent.

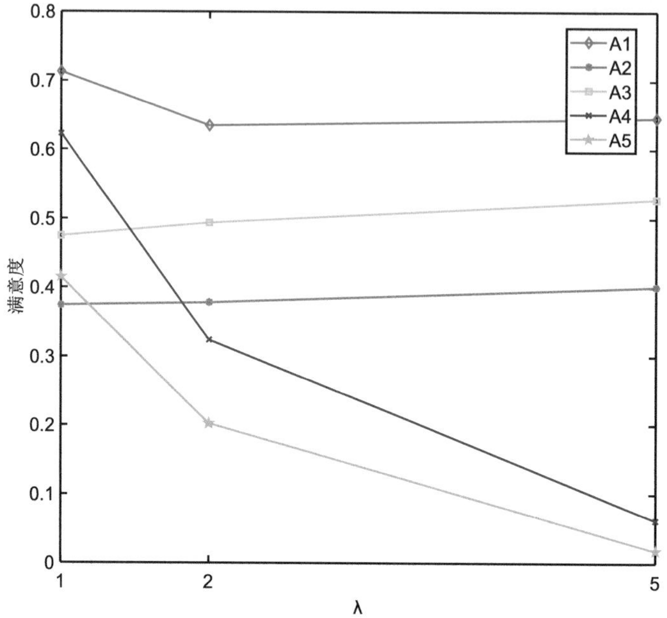

Fig. 2.4 The satisfaction degrees and ranking orders based on the generalized completely hybrid ordered weighted distance

References

Arevalillo-Herráez M, Ferri FJ, Domingo J (2013) A naive relevance feedback model for content-based image retrieval using multiple similarity measures. Pattern Recogn 43(3):9–629

Biswas A, Sarkar B (2019) Pythagorean fuzzy TOPSIS for multicriteria group decision-making with unknown weight information through entropy measure. Int J Intell Syst 34(6):1108–1128

Chen YS, Wang WH, Juang JG (2010) Application of intelligent computing to autonomous vehicle control. IEEE Cong Evolut Comput 1–8

Danielsson PE (1980) Euclidean distance Mapping. Comput Graphics Image Proc 14:227–248

Farhadinia B (2014) Distance and similarity measures for higher order hesitant fuzzy sets. Knowl-Based Syst 55:43–48

Gao W, Farahani MR, Aslam A, Hosamani S (2017) Distance learning techniques for ontology similarity measuring and ontology mapping. Cluster Comput 20(2):959–968

Gao W, Sarlak V, Parsaei MR, Ferdosi M (2018) Combination of fuzzy based on a meta-heuristic algorithm to predict electricity price in an electricity markets. Chem Eng Res Des 131:333–345

Gao W, Zhu LL, Wang KY (2015) Ontology sparse vector learning algorithm for ontology similarity measuring and ontology mapping via ADAL technology. Int J Bifurcat Chaos 25 (14):1540034

Gou XJ, Liao HC, Xu ZS, Herrera F (2017) Double hierarchy hesitant fuzzy linguistic term set and MULTIMOORA method: a case of study to evaluate the implementation status of haze controlling measures. Inf Fusion 38:22–34

Gou XJ, Xu ZS, Liao HC, Herrera F (2018) Multiple criteria decision making based on distance and similarity measures with double hierarchy hesitant fuzzy linguistic environment. Comput Ind Eng 126:516–530

Grzegorzewski P (2004) Distances between intuitionistic fuzzy sets and/or interval-valued fuzzy sets based on the Hausdorff metric. Fuzzy Sets Syst 148:319–328

Hafezalkotob A, Hafezalkotob A, Liao HC, Herrera F (2019) An overview of MULTIMOORA for multi-criteria decision-making: theory, developments, applications, and challenges. Inf Fusion 51:145–177

Hamming RW (1950) Error detecting and error correcting codes. Bell Syst Tech J 29(2):147–160

Hausdorff F (1957) Set theory. Chelsea, New York

Hung WL, Yang MS (2004) Similarity measures of intuitionistic fuzzy sets based on Hausdorff distance. Pattern Recogn Lett 25:1603–1611

Hussian Z, Yang MS (2019) Distance and similarity measures of Pythagorean fuzzy sets based on the Hausdorff metric with application to fuzzy TOPSIS. Int J Intell Syst 34(10):2633–2654

Li X, Hall NS, Humphreys GW (1993) Discrete distance and similarity measures for pattern candidate selection. Pattern Recogn 26(6):843–851

Li DQ, Zeng WY, Li JH (2015) New distance and similarity measures on hesitant fuzzy sets and their applications in multiple criteria decision making. Eng Appl Artif Intell 40:11–16

Liao HC, Wu XL, Liang XD, Yang JB, Xu DL, Herrera F (2018) A continuous interval-valued linguistic ORESTE method for multi-criteria group decision making. Knowl-Based Syst 153:65–77

Liao HC, Xu ZS (2015) Approaches to manage hesitant fuzzy linguistic information based on the cosine distance and similarity measures for HFLTSs and their application in qualitative decision making. Expert Syst Appl 42(12):5328–5336

Liao HC, Xu ZS, Zeng XJ (2014) Distance and similarity measures for hesitant fuzzy linguistic term sets and their application in multi-criteria decision making. Inf Sci 271:125–142

Liao HC, Xu ZS, Zeng XJ (2015a) Hesitant fuzzy linguistic VIKOR method and its application in qualitative multiple criteria decision making. IEEE Trans Fuzzy Syst 23(5):1343–1355

Liao HC, Xu ZS, Zeng XJ (2015b) Novel correlation coefficients between hesitant fuzzy sets and their application in decision-making. Knowl-Based Syst 82:115–127

Liu DH, Liu YY, Wang LZ (2019) Distance measure for Fermatean fuzzy linguistic term sets based on linguistic scale function: an illustration of the TODIM and TOPSIS methods. Int J Intell Syst 34(11):2807–2834

Merigó JM, Gil-Lafuente AM (2010) New decision-making techniques and their application in the selection of financial products. Inf Sci 180:2085–2094

Ren PJ, Xu ZS, Gou XJ (2016) Pythagorean fuzzy TODIM approach to multi-criteria decision making. Appl Soft Comput 42:246–259

Ren ZL, Xu ZS, Wang H (2017) Dual hesitant fuzzy VIKOR method for multi-criteria group decision-making based on fuzzy measure and new comparison method. Inf Sci 396:1–16

Tan QY, Wei CP, Liu Q, Feng XQ (2016) The hesitant fuzzy linguistic TOPSIS method based on novel information measures. Asia Pac J Oper Res 33(5):1–22

Wang XD, Gou XJ, Xu ZS (2020) Assessment of Traffic congestion with ORESTE method under double hierarchy hesitant fuzzy linguistic term set. Appl Soft Comput 86:105864

Wang W, Pang YF (2019) Hesitant interval-valued Pythagorean fuzzy VIKOR method. Int J Intell Syst 34(5):754–789

Wei CP, Ren ZL, Rodríguez RM (2015) A hesitant fuzzy linguistic TODIM method based on a score function. Int J Comput Intell Syst 8(4):701–712

Wu XL, Liao HC (2018) An approach to quality function deployment based on probabilistic linguistic term sets and ORESTE method for multi-expert multi-criteria decision making. Inf Fusion 43:13–26

Wu ZB, Xu JP, Jiang XL, Zhong L (2019) Two MAGDM models based on hesitant fuzzy linguistic term sets with possibility distributions: VIKOR and TOPSIS. Inf Sci 473:101–120

Xian SD, Liu Z, Gou XL, Wan WH (2020) Interval 2-tuple Pythagorean fuzzy linguistic MULTIMOORA method with CIA and their application to MCGDM. Int J Intell Syst. https://doi.org/10.1002/int.22221

Xu ZS (2005) An approach based on similarity measure to multiple attribute decision making with trapezoid fuzzy linguistic variables. Fuzzy Syst Knowl Dis Lect Notes Comput Sci 3613: 110–117

Xu ZS (2012) Fuzzy ordered distance measures. Fuzzy Optim Decis Mak 11:73–97

Xu ZS, Chen J (2008a) Ordered weighted distance measure. J Syst Sci Syst Eng 16:529–555

Xu ZS, Chen J (2008b) An overview of distance and similarity measures of intuitionistic fuzzy sets. Int J Uncertain Fuzziness Knowl-Based Syst 16:529–555

Xu YJ, Wang HM (2011) Distance measure for linguistic decision making. Syst Eng Proc 1:450–456

Xu ZS, Xia MM (2011) Distance and similarity measures for hesitant fuzzy sets. Inf Sci 181:2128–2138

Xue WT, Xu ZS, Zhang XL, Tian XL (2018) Pythagorean fuzzy LINMAP method based on the entropy theory for railway project investment decision making. Int J Intell Syst 33(1):93–125

Yager RR (1988) On ordered weighted averaging aggregation operators in multi-criteria decision-making. IEEE Trans Syst Man Cybern 18:183–190

Yager RR (2004) Generalized OWA aggregation operators. Fuzzy Optim Decis Mak 3:93–107

Yager RR (2010) Norms induced from OWA operators. IEEE Trans Fuzzy Syst 18:57–66

Zavadskas EK, Mardani A, Turskis Z, Jusoh A, Nor KM (2016) Development of TOPSIS method to solve complicated decision-making problems—an overview on developments from 2000 to 2015. Int J Inf Tech Decis Mak 15(3):645–682

Zavadskas EK, Turskis Z, Volvaciovas R, Kildiene S (2013) Multi-criteria assessment model of technologies. Stud Inform Control 22(4):249–258

Zhang CZ, Chen C, Streimikiene D, Balezentis T (2019) Intuitionistic fuzzy MULTIMOORA approach for multi-criteria assessment of the energy storage technologies. Appl Soft Comput 79:410–423

Chapter 3
Consistency Theory Framework of DHHFLPRs

In decision-making processes, preference relations are popular and powerful techniques for expert preference modeling (Ureña et al. 2015). Consistency measures of preference relations are the vital basis of group decision-making (GDM) and have been studied extensively, which show that the supplied preferences satisfy some transitive properties (Wu and Xu 2016). Consistency measures mainly consist of two parts: (1) judging whether each preference relation is of acceptable consistency; (2) improving the preference relation with unacceptable consistency. Based on double hierarchy hesitant fuzzy linguistic preference information, Gou et al. (2019, 2020) discussed the additive consistency measures and multiplicative consistency measures for DHHFLPRs, respectively. In this chapter, we will deeply discuss these two consistency measures for DHHFLPRs.

3.1 Additive Consistency Measures for DHHFLPRs

Up to now, two critical defects of existing consistency measures are being more and more apparent:

(1) It is common that the normalization procedure is very necessary for making calculations expediently. But almost all methods complete it by adding or deleting some linguistic terms (Zhu and Xu 2014). Obviously, these methods may cause the original information loss and make calculations complex.
(2) Considering that there are some unreasonable places in the calculations of consistency thresholds under linguistic preference information environment, it is necessary to improve the existing consistency thresholds as the novel references for consistency improving processes.

X. Gou and Z. Xu, *Double Hierarchy Linguistic Term Set and Its Extensions*, Studies in Fuzziness and Soft Computing 396,
https://doi.org/10.1007/978-3-030-51320-7_3

Under double hierarchy hesitant fuzzy linguistic environment, to solve these two defects successfully, and to avoid the occurrence of some self-contradictory situations, it is very important to carry out the new consistency checking and improving process for a DHHFLPR in a decision-making process. In this subsection, we will discuss some additive consistency measures for DHHFLPRs and the main contributions are summarized as follows:

(1) We develop a new normalization method by the linguistic expected-value of each DHHFLE to transform the DHHFLPR into the normalized DHHFLPR equivalently. The linguistic expected-value of DHHFLE can be obtained by aggregating all elements of a DHHFLE into a DHLT. With this method, we will not lose any linguistic terms and can make the calculations simpler.
(2) For the purpose of judging whether a DHHFLPR is of acceptable consistency or not, we define a consistency index of the DHHFLPR and develop a novel method to calculate the consistency thresholds by improving the existing methods. Then we present two convergent consistency repairing algorithms based on automatic improving method and feedback improving method respectively to improve the consistency index of a given DHHFLPR with unacceptable consistency.
(3) We propose a weight-determining method for obtaining the weight information of each expert, and then develop an algorithm to deal with the GDM problem with double hierarchy hesitant fuzzy linguistic preference information.
(4) We apply the proposed method to deal with a practical GDM problem which is to evaluate the water resource situations of some important cities in Sichuan province.

3.1.1 Additive Consistency Measure Method of DHHFLPRs

For the purpose of judging whether a DHHFLPR is of acceptable consistency or not, Gou et al. (2019) defined an additive consistency measure method for DHHFLPR. To do so, the normalization of DHHFLPR is the first and very important step considering it is very common that some DHHFLEs have different numbers of DHLTs. To carry out the normalization process in a more reasonable manner and keep all linguistic information intact, we develop a concept of linguistic expected-value for DHHFLE.

Definition 3.1 (Gou et al. 2019). Let $\bar{S}_O = \{s_{t<o_k^t>} | t = [-\tau, \ldots, \tau]; k = [-\varsigma, \ldots, \varsigma]\}$ be the continuous DHLTS, $h_{S_O} = \{s_{\phi_l<o_{\varphi_l}>} | s_{\phi_l<o_{\varphi_l}>} \in \bar{S}_O; l = 1, 2, \ldots, \#h_{S_O}\}$ be a DHHFLE, $\Phi \times \Psi$ be the set of all DHHFLEs over $\bar{S}_O$. Then a linguistic expected-value of h_{S_O}, denoted as $le(h_{S_O})$, is obtained by

$$e : \Phi \times \Psi \to \bar{S}_O, \quad le(h_{S_O}) = \frac{1}{\#h_{S_O}} \mathop{\oplus}_{l=1}^{\#h_{S_O}} s_{\phi_l<o_{\varphi_l}>} = s_{\frac{1}{\#h_{S_O}} \sum_{l=1}^{\#h_{S_O}} \phi_l < o_{\frac{1}{\#h_{S_O}} \sum_{l=1}^{\#h_{S_O}} \varphi_l} >} \tag{3.1}$$

Additionally, the normalized DHHFLPR of a DHHFLPR $\tilde{H}_{S_O} = (h_{S_{O_{ij}}})_{m\times m}$ can be obtained, denoted by $\tilde{H}^N_{S_O} = (h^N_{S_{O_{ij}}})_{m\times m}$, satisfying

$$h^N_{S_{O_{ij}}} = le(h_{S_{O_{ij}}}), \quad i,j = 1,2,\ldots,m \tag{3.2}$$

From Eq. (3.1), it is obvious that $le(h_{S_O})$ is a DHLT. Thus, every basic element included in the normalized DHHFLPR of one DHHFLPR is also a DHLT. Then the DHHFLPR included in Remark 1.3 can be normalized by Eq. (3.2) and shown as:

$$\tilde{H}^N_{S_O} = \begin{pmatrix} \{s_{0<o_0>}\} & \{s_{0<o_1>}\} & \left\{s_{-1/2<o_{-1/2}>}\right\} \\ \{s_{0<o_{-1}>}\} & \{s_{0<o_0>}\} & \{s_{2<o_1>}\} \\ \left\{s_{1/2<o_{1/2}>}\right\} & \{s_{-2<o_{-1}>}\} & \{s_{0<o_0>}\} \end{pmatrix}$$

Next, the definition of additive consistency measure for DHHFLPR can be given.

Definition 3.2 (Gou et al. 2019). Let $\tilde{H}_{S_O} = (h_{S_{O_{ij}}})_{m\times m}$ be a DHHFLPR and $\tilde{H}^N_{S_O} = (h^N_{S_{O_{ij}}})_{m\times m}$ be its normalized DHHFLPR, then we call $\tilde{H}_{S_O}$ an additive consistent DHHFLPR if it satisfies

$$h^N_{S_{O_{ij}}} = h^N_{S_{O_{i\rho}}} \oplus h^N_{S_{O_{\rho j}}} (i,j,\rho = 1,2,\ldots,m; i \neq j) \tag{3.3}$$

Theorem 3.1 (Gou et al. 2019). Let $\tilde{H}_{S_O} = (h_{S_{O_{ij}}})_{m\times m}$ be a DHHFLPR and $\tilde{H}^N_{S_O} = (h^N_{S_{O_{ij}}})_{m\times m}$ be its normalized DHHFLPR. If $\bar{h}^N_{S_{O_{ij}}} = \frac{1}{m}(\overset{m}{\underset{\rho=1}{\oplus}}(h^N_{S_{O_{i\rho}}} \oplus h^N_{S_{O_{\rho j}}}))$ for $i,j,\rho = 1,2,\ldots,m; i \neq j$, then $\tilde{H}_{S_O}$ is an additive consistent DHHFLPR, and $\bar{H}^N_{S_O} = (\bar{h}^N_{S_{O_{ij}}})_{m\times m}$ is an additive consistent normalized DHHFLPR.

Proof Since

$$\begin{aligned} \bar{h}^N_{S_{O_{i\rho}}} \oplus \bar{h}^N_{S_{O_{\rho j}}} &= \left(\frac{1}{m}\left(\sum_{b=1}^{m}(h^N_{S_{O_{ib}}} \oplus h^N_{S_{O_{b\rho}}})\right) \oplus \frac{1}{m}\left(\overset{m}{\underset{b=1}{\oplus}}(h^N_{S_{O_{\rho b}}} \oplus h^N_{S_{O_{bj}}})\right)\right) \\ &= \left(\frac{1}{m}\left(\overset{m}{\underset{b=1}{\oplus}}(h^N_{S_{O_{ib}}} \oplus h^N_{S_{O_{bj}}}) \oplus h^N_{S_{O_{b\rho}}} \oplus h^N_{S_{O_{\rho b}}}\right)\right) \end{aligned}$$

and considering that $\tilde{H}^N_{S_O}$ is a normalized DHHFLPR, which satisfies $h^N_{S_{O_{b\rho}}} \oplus h^N_{S_{O_{\rho b}}} = \{s_{0<o_0>}\}$. Therefore,

$$\bar{h}^N_{S_{O_{ip}}} \oplus \bar{h}^N_{S_{O_{pj}}} = \left(\frac{1}{m}\left(\overset{m}{\underset{b=1}{\oplus}}(h^N_{S_{O_{ib}}} \oplus h^N_{S_{O_{bj}}}) \oplus s_{0<o_0>}\right)\right) = \left(\frac{1}{m}\overset{m}{\underset{b=1}{\oplus}}(h^N_{S_{O_{ib}}} \oplus h^N_{S_{O_{bj}}})\right)$$
$$= \bar{h}^N_{S_{O_{ij}}}$$

Based on Definition 3.2, $\bar{H}^N_{S_O} = (\bar{h}^N_{S_{O_{ij}}})_{m\times m}$ is an additive consistent normalized DHHFLPR, which completes the proof of Theorem 3.1. ■

Remark 3.1 (Gou et al. 2019). Theorem 3.1 mainly provides the method that obtains the additive consistent normalized DHHFLPR. Meanwhile, it also gives a necessary condition which can be used to judge whether a normalized DHHFLPR is the additive consistent normalized DHHFLPR. Considering that checking the consistency of a DHHFLPR is very important when dealing with double hierarchy hesitant fuzzy linguistic preference information, so Theorem 3.1 is the most critical foundation of this subsection.

Example 3.1 (Gou et al. 2019). Let $S_O = \{s_{t<o_k>} | t = -4, \ldots, 4;\ k = -4, \ldots, 4\}$ be a DHLTS. For two DHHFLPRs

$$\tilde{H}^1_{S_O} = \begin{pmatrix} \{s_{0<o_0>}\} & \{s_{-1<o_1>}, s_{0<o_2>}\} & \{s_{1<o_{-2}>}, s_{2<o_1>}\} \\ \{s_{1<o_{-1}>}, s_{0<o_{-2}>}\} & \{s_{0<o_0>}\} & \{s_{-1<o_{-2}>}\} \\ \{s_{-1<o_2>}, s_{-2<o_{-1}>}\} & \{s_{-1<o_2>}\} & \{s_{0<o_0>}\} \end{pmatrix}$$

$$\tilde{H}^2_{S_O} = \begin{pmatrix} \{s_{0<o_0>}\} & \{s_{-1<o_1>}, s_{0<o_2>}\} & \{s_{-1<o_{-2}>}, s_0, s_{1<o_1>}\} & \{s_{1<o_{-2}>}\} \\ \{s_{1<o_{-1}>}, s_{0<o_{-2}>}\} & \{s_{0<o_0>}\} & \{s_{0<o_{-1}>}, s_{1<o_1>}\} & \{s_{1<o_1>}, s_{2<o_2>}\} \\ \{s_{1<o_2>}, s_0, s_{-1<o_{-1}>}\} & \{s_{0<o_1>}, s_{-1<o_{-1}>}\} & \{s_{0<o_0>}\} & \{s_{0<o_{-3}>}, s_1, s_{2<o_1>}\} \\ \{s_{-1<o_2>}\} & \{s_{-1<o_{-1}>}, s_{-2<o_{-2}>}\} & \{s_{0<o_3>}, s_{-1}, s_{-2<o_{-1}>}\} & \{s_{0<o_0>}\} \end{pmatrix}$$

The normalized DHHFLPRs $\tilde{H}^{1N}_{S_O} = (h^{1N}_{S_{O_{ij}}})_{3\times 3}$ and $\tilde{H}^{2N}_{S_O} = (h^{2N}_{S_{O_{ij}}})_{4\times 4}$ can be obtained:

$$\tilde{H}^{1N}_{S_O} = \begin{pmatrix} \{s_{0<o_0>}\} & \{s_{-1/2<o_{3/2}>}\} & \{s_{3/2<o_{-1/2}>}\} \\ \{s_{1/2<o_{-3/2}>}\} & \{s_{0<o_0>}\} & \{s_{-1<o_{-2}>}\} \\ \{s_{-3/2<o_{1/2}>}\} & \{s_{1<o_2>}\} & \{s_{0<o_0>}\} \end{pmatrix}$$

$$\tilde{H}^{2N}_{S_O} = \begin{pmatrix} \{s_{0<o_0>}\} & \{s_{-1/2<o_{3/2}>}\} & \{s_{0<o_{-1/3}>}\} & \{s_{1<o_{-2}>}\} \\ \{s_{1/2<o_{-3/2}>}\} & \{s_{0<o_0>}\} & \{s_{1/2<o_0>}\} & \{s_{3/2<o_{3/2}>}\} \\ \{s_{0<o_{1/3}>}\} & \{s_{-1/2<o_0>}\} & \{s_{0<o_0>}\} & \{s_{1<o_{-2/3}>}\} \\ \{s_{-1<o_2>}\} & \{s_{-3/2<o_{-3/2}>}\} & \{s_{-1<o_{2/3}>}\} & \{s_{0<o_0>}\} \end{pmatrix}$$

We utilize Theorem 3.1 to obtain the additive consistent normalized DHHFLPRs $\bar{H}^{1N}_{S_O} = (\bar{h}^{1N}_{S_{O_{ij}}})_{3\times 3}$ and $\bar{H}^{2N}_{S_O} = (\bar{h}^{2N}_{S_{O_{ij}}})_{4\times 4}$, respectively:

$$\bar{H}_{S_O}^{1N} = \begin{pmatrix} \{s_{0<o_0>}\} & \{s_{1/2<o_{3/2}>}\} & \{s_{1/2<o_{-1/2}>}\} \\ \{s_{-1/2<o_{-3/2}>}\} & \{s_{0<o_0>}\} & \{s_{0<o_{-2}>}\} \\ \{s_{-1/2<o_{1/2}>}\} & \{s_{0<o_2>}\} & \{s_{0<o_0>}\} \end{pmatrix}$$

$$\bar{H}_{S_O}^{2N} = \begin{pmatrix} \{s_{0<o_0>}\} & \{s_{-1/2<o_{-5/24}>}\} & \{s_{0<o_{-1/8}>}\} & \{s_{1<o_{-1/2}>}\} \\ \{s_{1/2<o_{5/24}>}\} & \{s_{0<o_0>}\} & \{s_{1/2<o_{1/12}>}\} & \{s_{3/2<o_{-7/24}>}\} \\ \{s_{0<o_{1/8}>}\} & \{s_{-1/2<o_{-1/12}>}\} & \{s_{0<o_0>}\} & \{s_{1<o_{-3/8}>}\} \\ \{s_{-1<o_{1/2}>}\} & \{s_{-3/2<o_{7/24}>}\} & \{s_{-1<o_{3/8}>}\} & \{s_{0<o_0>}\} \end{pmatrix}$$

Remark 3.2 (Gou et al. 2019). To compare the inconsistent DHHFLPRs and the additive consistent DHHFLPRs more intuitively, we can further utilize the visual method "Figure of area", which is a function of MATLAB drawing toolbar. Then we obtain Fig. 3.1. Based on the areas of different DHHFLPRs, the area that is more regular is clearly distinguished. For example, in Fig. 3.1a, b because the changes in the areas of different colors in Fig. 3.1b are more regular than the corresponding changes in Fig. 3.1a, we consider that the additive consistent DHHFLPR $\bar{H}_{S_O}^{1N}$ is more regular with respect to the areas in different colors than the inconsistent DHHFLPR $\tilde{H}_{S_O}^{1N}$. Similarly, the additive consistent DHHFLPR $\bar{H}_{S_O}^{2N}$ is more regular with respect to the areas in different colors than the inconsistent DHHFLPR $\tilde{H}_{S_O}^{2N}$ based on Fig. 3.1c, d.

3.1.2 Additive Consistency Index of DHHFLPRs

When dealing with DHHFLPRs, judging whether a DHHFLPR is of acceptable consistency or not is of great importance. Therefore, how to calculate the consistency and judge whether it can be accepted is the focus of this subsection. Here, we introduce an additive consistency index for DHHFLPRs on the basis of distance measure (Gou et al. 2018). Meanwhile, we develop a novel method to improve the existing method for calculating the consistency thresholds.

Firstly, one kind of distance measure of DHHFLEs can be shown as follows:

Definition 3.3 (Gou et al. 2019). Let S_O be a DHLTS, $h_{S_{O_i}} = \{h_{S_{O_i}}^{(l)} | h_{S_{O_i}}^{(l)} \in S_O; l = 1, 2, \ldots, h_{S_{O_i}}\}$ $(i = 1, 2)$ be two DHHFLEs, $le(h_{S_{O_1}})$ and $le(h_{S_{O_2}})$ be the linguistic expected-value of $h_{S_{O_1}}$ and $h_{S_{O_2}}$, respectively. Then we call

$$d(h_{S_{O_1}}, h_{S_{O_2}}) = \left| f(le(h_{S_{O_1}})) - f(le(h_{S_{O_2}})) \right| \tag{3.4}$$

the distance measure between $h_{S_{O_1}}$ and $h_{S_{O_2}}$.

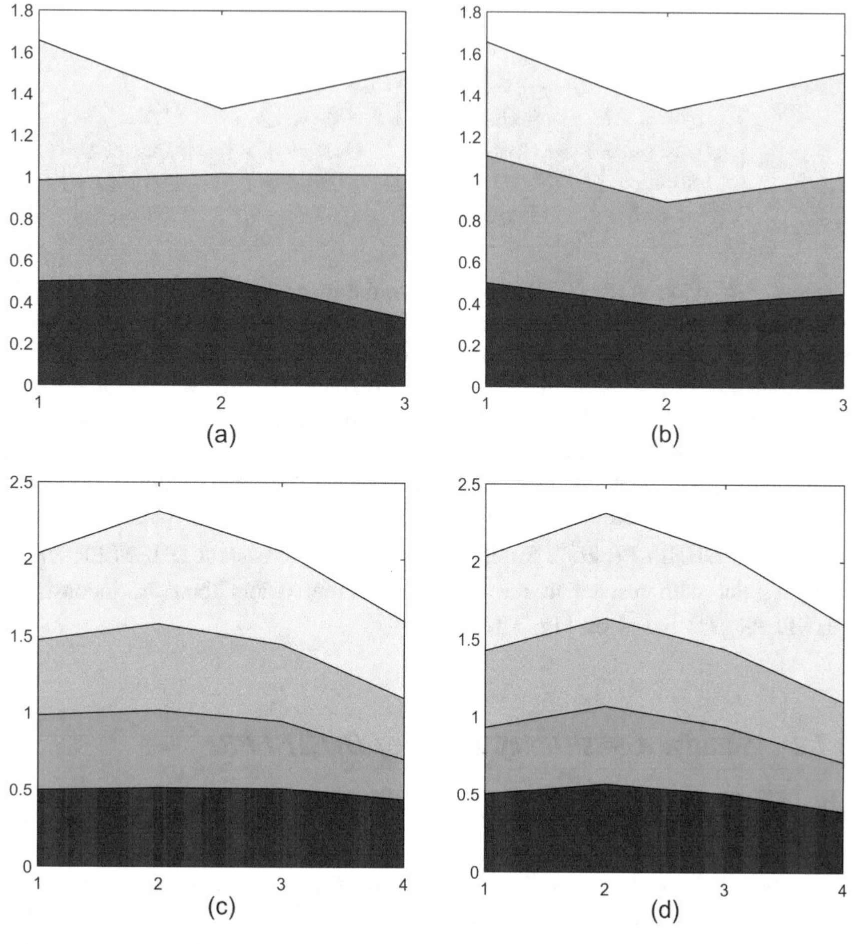

Fig. 3.1 The figures of area of $\tilde{H}_{S_O}^{1N}$, $\bar{H}_{S_O}^{1N}$, $\tilde{H}_{S_O}^{2N}$ and $\bar{H}_{S_O}^{2N}$

Given two DHHFLPRs $\tilde{H}_{S_O}^1 = (h_{S_{O_{ij}}}^1)_{m\times m}$ and $\tilde{H}_{S_O}^2 = (h_{S_{O_{ij}}}^2)_{m\times m}$, $\tilde{H}_{S_O}^{1N} = (h_{S_{O_{ij}}}^{1N})_{m\times m}$ and $\tilde{H}_{S_O}^{2N} = (h_{S_{O_{ij}}}^{2N})_{m\times m}$ are their corresponding normalized DHHFLPRs. Then the distance measure between $\tilde{H}_{S_O}^1$ and $\tilde{H}_{S_O}^2$ is:

$$\begin{aligned} d(\tilde{H}_{S_O}^1, \tilde{H}_{S_O}^2) &= \left(\frac{2}{m(m-1)}\sum_{i<j}^{m}(d(h_{S_{O_{ij}}}^{1N}, h_{S_{O_{ij}}}^{2N}))^2\right)^{1/2} \\ &= \left(\frac{2}{m(m-1)}\sum_{i<j}^{m}(f(le(h_{S_{O_1}})) - f(le(h_{S_{O_2}}))^2\right)^{1/2} \end{aligned} \tag{3.5}$$

Obviously, the distance measure $d(\tilde{H}^1_{S_O}, \tilde{H}^2_{S_O})$ satisfies the following properties:

(1) $0 \le d(\tilde{H}^1_{S_O}, \tilde{H}^2_{S_O}) \le 1$;
(2) $d(\tilde{H}^1_{S_O}, \tilde{H}^2_{S_O}) = 0$ if and only if $\tilde{H}^1_{S_O} = \tilde{H}^2_{S_O}$;
(3) $d(\tilde{H}^1_{S_O}, \tilde{H}^2_{S_O}) = d(\tilde{H}^2_{S_O}, \tilde{H}^1_{S_O})$.

Example 3.2 (Continued with Example 3.1) (Gou et al. 2019). Based on Eq. (3.5), we obtain

$$d(\tilde{H}^{1N}_{S_O}, \bar{H}^{1N}_{S_O}) = \left(\frac{2}{2\times 3}\sum_{i<j}^{3}(d(h^{1N}_{S_{O_{ij}}}, \bar{h}^{1N}_{S_{O_{ij}}}))^2\right)^{1/2} = 0.1250;$$

$$d(\tilde{H}^{2N}_{S_O}, \bar{H}^{2N}_{S_O}) = \left(\frac{2}{3\times 4}\sum_{i<j}^{4}(d(h^{2N}_{S_{O_{ij}}}, \bar{h}^{2N}_{S_{O_{ij}}}))^2\right)^{1/2} = 0.0371.$$

As we know, no matter what kind of preference relation, consistency index is a necessary tool to check whether a preference relation is of acceptable consistency or not. Similarly, it is necessary to develop a consistency index for DHHFLPR.

Definition 3.4 (Gou et al. 2019). Let $\tilde{H}_{S_O} = (h_{S_{O_{ij}}})_{m\times m}$ be a DHHFLPR. $\tilde{H}^N_{S_O} = (h^N_{S_{O_{ij}}})_{m\times m}$ and $\bar{H}^N_{S_O} = (\bar{h}^N_{S_{O_{ij}}})_{m\times m}$ are its normalized DHHFLPR and additive consistent normalized DHHFLPR, respectively. A consistency index of $\tilde{H}_{S_O}$ can be denoted as:

$$CI(\tilde{H}_{S_O}) = d(\tilde{H}^N_{S_O}, \bar{H}^N_{S_O}) \tag{3.6}$$

The consistency index $CI(\tilde{H}_{S_O})$ satisfies $0 \le CI(\tilde{H}_{S_O}) \le 1$. Additionally, the smaller the consistency index $CI(\tilde{H}_{S_O})$ is, the more consistent the DHHFLPR $\tilde{H}_{S_O}$ should be.

Dong et al. (2008) proposed some consistency thresholds to check whether a preference relation with linguistic preference information is of acceptable consistency. Here we introduce a novel and reasonable method to improve these consistency thresholds. Firstly, in order to make the method more clear, it is necessary to transform the function f into a new form. Let T be the number of linguistic terms in the first hierarchy LTS S. Obviously, we get $2\tau = T - 1$. Then the function f is equal to

$$f : [-\tau, \tau] \times [-\varsigma, \varsigma] \to [0, 1], f\phi_l, \varphi_l) = \frac{\varphi_l + (\tau + \phi_l)\varsigma}{(T-1)\varsigma} \tag{3.7}$$

Let $\frac{\varphi_l + (\tau+\phi_l)\varsigma}{\varsigma} = \Delta^l$, then $f(\phi_l, \varphi_l) = \frac{\Delta^l}{T-1}$. Considering that $le(h_{S_O})$ is a DHLT, then Eq. (3.4) can be rewritten as

$$d(h_{S_{O_1}}, h_{S_{O_2}}) = \left| f(le(h_{S_{O_1}})) - f(le(h_{S_{O_2}})) \right| = \left| \frac{\Delta^1}{T-1} - \frac{\Delta^2}{T-1} \right| = \left| \frac{\Delta^1 - \Delta^2}{T-1} \right| \quad (3.8)$$

Therefore, Eq. (3.5) can be developed into

$$CI(\tilde{H}_{S_O}) = d(\tilde{H}_{S_O}^N, \bar{H}_{S_O}^N) = \frac{1}{(T-1)} \left(\frac{2}{m(m-1)} \sum_{i<j}^{m} \left| \Delta_{ij} - \bar{\Delta}_{ij} \right|^2 \right)^{1/2} \quad (3.9)$$

Let $\mathbb{R}_{ij} = \left| \Delta_{ij} - \bar{\Delta}_{ij} \right|$. Then $CI(\tilde{H}_{S_O}) = \frac{1}{(T-1)} \left(\frac{2}{m(m-1)} \sum_{i<j}^{m} \mathbb{R}_{ij}^2 \right)^{1/2}$. Considering that the value of $\mathbb{R}_{ij}(i<j)$ is independent normally distributed with a mean of 0 and standard deviation of ϑ, similar to the analyses of Dong et al. (2008), we can obtain that $\frac{m(m-1)}{2} \left((T-1) \times \frac{1}{\vartheta} \times CI(\tilde{H}_{S_O}) \right)^2$ is a chi-square distribution with freedom degree $\frac{m(m-1)}{2}$, i.e., $\frac{m(m-1)}{2} \left((T-1) \times \frac{1}{\vartheta} \times CI(\tilde{H}_{S_O}) \right)^2 \sim \chi^2 \left(\frac{m(m-1)}{2} \right)$, on the condition that $\mathbb{R}_{ij}(i<j)$ is independent normally distributed with a mean of 0 and standard deviation of ϑ, namely, $\mathbb{R}_{ij} \sim N(0, \vartheta^2)$. As we know, the freedom degree of $\chi^2 = \sum_{i<j}^{m} \left(\frac{\mathbb{R}_{ij}}{\vartheta} \right)^2$ is $\frac{m(m-1)}{2}$. This is a one-sided right-tailed test. At a significance level α, the critical value of χ^2 is λ_α. Let

$$\bar{CI}(\tilde{H}_{S_O}) = \frac{\vartheta}{(T-1)} \left(\frac{2}{m(m-1)} \lambda_\alpha \right)^{1/2} \quad (3.10)$$

be the consistency threshold. Therefore, if $CI(\tilde{H}_{S_O}) \leq \bar{CI}(\tilde{H}_{S_O})$, then $\tilde{H}_{S_O}$ is a DHHFLPR of acceptable consistency; Otherwise, if $CI(\tilde{H}_{S_O}) > \bar{CI}(\tilde{H}_{S_O})$, then $\tilde{H}_{S_O}$ is a DHHFLPR of unacceptable consistency.

As we discussed above, the parameters α and ϑ are decided by the experts or according to practical situations. We let $\alpha = 0.1$ and $\vartheta = 2$, and then calculate the values of consistency thresholds $\bar{CI}(\tilde{H}_{S_O})$ for different m and T, which can be shown in Table 3.1.

In Example 3.2, we obtain $CI(\tilde{H}_{S_O}^1) = 0.1250$ and $CI(\tilde{H}_{S_O}^2) = 0.0371$. In Table 3.1, $\bar{CI}(\tilde{H}_{S_O}^1) = 0.1103$ and $\bar{CI}(\tilde{H}_{S_O}^2) = 0.1515$. We obtain $CI(\tilde{H}_{S_O}^1) > \bar{CI}(\tilde{H}_{S_O}^1)$ and

Table 3.1 The values of consistency thresholds based on different m and T

	$m = 3$	$m = 4$	$m = 5$	$m = 6$	$m = 7$	$m = 8$
$T = 5$	0.2207	0.3030	0.3488	0.3774	0.3970	0.4112
$T = 9$	0.1103	0.1515	0.1744	0.1887	0.1985	0.2056
$T = 17$	0.0552	0.0758	0.0872	0.0944	0.0993	0.1028

$CI(\tilde{H}_{S_O}^2) < \bar{CI}(\tilde{H}_{S_O}^2)$. Therefore, $\tilde{H}_{S_O}^2$ is a DHHFLPR with acceptable consistency, and $\tilde{H}_{S_O}^1$ is a DHHFLPR with unacceptable consistency.

Remark 3.3 (Gou et al. 2019). In this subsection, the additive consistency index for DHHFLPR is proposed. And then a novel specific calculation process of consistency thresholds is given, which is a novel method and more reasonable than the existing method based on the more correct parameter found in Eq. (3.10). Therefore, the results of Table 3.1 can be used as important references when judging whether a preference relation with linguistic information is of acceptable consistency. Additionally, according to the additive consistency index and consistency thresholds, if a DHHFLPR is of acceptable consistency, then there is no need for it to be optimized. Otherwise, it is necessary to develop some methods to improve it, which will be discussed in the next subsection.

3.1.3 Consistency Repairing Algorithms

In practical decision-making processes with DHHFLPRs, it is common for there to be a DHHFLPR $\tilde{H}_{S_O} = (h_{S_{O_{ij}}})_{m\times m}$ of unacceptable consistency, namely, $CI(\tilde{H}_{S_O}) > \bar{CI}(\tilde{H}_{S_O})$. In this case, we need to repair the DHHFLPR $\tilde{H}_{S_O}$ until it reaches the consistency threshold. To improve the consistency, two existing methods have been developed: the automatic method (Zhu and Xu 2014) and the feedback-based method (Alonso et al. 2010; Fu and Yang 2010; Herrera-Viedma et al. 2007; Zhu and Xu 2014). Similarly, we establish two consistency repairing algorithms based on the automatic improving method and feedback improving method respectively to repair the DHHFLPR with unacceptable consistency.

(a) **Consistency repairing algorithm based on automatic optimization method**

Considering that the automatic improving method is time-saving, effective, and practical without the interaction of the experts, so we develop a consistency repairing algorithm based on the automatic optimization method that can repair the DHHFLPR with unacceptable consistency by automatic iterative operations. Additionally, we analyze the convergence of repair results. Finally, we establish an optimization model which can be used to obtain the DHHFLPR is of acceptable consistency directly.

Algorithm 3.1 (Gou et al. 2019). Consistency repairing algorithm based on the automatic optimization method

Step 1. Let $(\tilde{H}_{S_O})^{(\mathbb{Z})} = ((h_{S_{O_{ij}}})_{m\times m})^{(\mathbb{Z})}$ ($\mathbb{Z} = 0$, $(\tilde{H}_{S_O})^{(\mathbb{Z})}$ expresses the $\mathbb{Z}$-th power of $\tilde{H}_{S_O}$, indicating the number of iterations). Based on Eqs. (3.1)–(3.2) and Theorem 3.1, we can calculate the normalized DHHFLPR $(\tilde{H}_{S_O}^N)^{(\mathbb{Z})} = ((h_{S_{O_{ij}}}^N)_{m\times m})^{(\mathbb{Z})}$ and the consistent normalized DHHFLPR $(\bar{H}_{S_O}^N)^{(\mathbb{Z})} = ((\bar{h}_{S_{O_{ij}}}^N)_{m\times m})^{(\mathbb{Z})}$, respectively.

Step 2. Calculate $C\bar{I}(\tilde{H}_{S_O})$ based on Eq. (3.10) or Table 3.1.
Step 3. Calculate $CI((\tilde{H}_{S_O})^{(\mathbb{Z})}) = d((\tilde{H}^N_{S_O})^{(\mathbb{Z})}, (\bar{H}^N_{S_O})^{(\mathbb{Z})})$ based on Eq. (3.6). If $CI((\tilde{H}_{S_O})^{(\mathbb{Z})}) \le C\bar{I}(\tilde{H}_{S_O})$, then go to step 5; If $CI((\tilde{H}_{S_O})^{(\mathbb{Z})}) > C\bar{I}(\tilde{H}_{S_O})$, then go to Step 4.
Step 4. Let θ $(0 \le \theta \le 1)$ be an adjusted parameter. Utilize the formula

$$(h^N_{S_{O_{ij}}})^{(\mathbb{Z}+1)} = (1-\theta)(h^N_{S_{O_{ij}}})^{(\mathbb{Z})} \oplus \theta(\bar{h}^N_{S_{O_{ij}}})^{(\mathbb{Z})} \quad (i,j = 1,2,\ldots,m;\ i \ne j) \tag{3.11}$$

to obtain the modified normalized DHHFLPR $(\tilde{H}^N_{S_O})^{(\mathbb{Z}+1)} = ((h^N_{S_{O_{ij}}})_{m\times m})^{(\mathbb{Z})}$. Let $\mathbb{Z} = \mathbb{Z}+1$ and go back to Step 3.
Step 5. Let $*\tilde{H}_{S_O} = (\tilde{H}^N_{S_O})^{(\mathbb{Z})}$ and output the modified normalized DHHFLPR $*H_{S_O}$.
Step 6. End

Based on Eq. (3.11), it is obvious that all the consistent normalized DHHFLPRs are same no matter what the value of $\mathbb{Z}$ is. Namely, $(\bar{H}^N_{S_O})^{(0)} = (\bar{H}^N_{S_O})^{(1)} = (\bar{H}^N_{S_O})^{(2)} = \cdots$.

Considering that the presented algorithm is convergent, so we can get a more consistent DHHFLPR after the consistency repairing process. The following theorem shows the convergence.

Theorem 3.2 (Gou et al. 2019). Let $\tilde{H}_{S_O} = (h_{S_{O_{ij}}})_{m\times m}$ be a DHHFLPR, θ $(0 \le \theta \le 1)$ be the adjusted parameter, and $*\tilde{H}_{S_O}$ be the modified normalized DHHFLPR obtained by Algorithm 3.1. Then $CI(*\tilde{H}_{S_O}) < CI(\tilde{H}_{S_O})$.

Proof From Algorithm 3.1, $(\tilde{H}^N_{S_O})^{(\mathbb{Z})} = ((h^N_{S_{O_{ij}}})_{m\times m})^{(\mathbb{Z})}$ is the modified normalized DHHFLPR in the $\mathbb{Z}$-th power of $\tilde{H}_{S_O}$. Suppose that the modified normalized DHHFLPR in the $\mathbb{Z}+1$-th power of $\tilde{H}_{S_O}$ is $(\tilde{H}^N_{S_O})^{(\mathbb{Z}+1)} = ((h^N_{S_{O_{ij}}})_{m\times m})^{(\mathbb{Z}+1)}$. Based on Eq. (3.11), we obtain

$$(h^N_{S_{O_{ij}}})^{(\mathbb{Z}+1)} = (1-\theta)(h^N_{S_{O_{ij}}})^{(\mathbb{Z})} \oplus \theta(\bar{h}^N_{S_{O_{ij}}})^{(\mathbb{Z})} = (1-\theta)(h^N_{S_{O_{ij}}})^{(\mathbb{Z})} \oplus \theta\bar{h}^N_{S_{O_{ij}}} \tag{3.12}$$

Then,

$$\begin{aligned} d((h^N_{S_{O_{ij}}})^{(\mathbb{Z}+1)}, \bar{h}^N_{S_{O_{ij}}}) &= \left| \left(\frac{\Delta_{ij}}{T-1}\right)^{(\mathbb{Z}+1)} - \frac{\bar{\Delta}_{ij}}{T-1} \right| = \left| (1-\theta)\left(\frac{\Delta_{ij}}{T-1}\right)^{(\mathbb{Z})} + \theta\left(\frac{\bar{\Delta}_{ij}}{T-1}\right) - \frac{\bar{\Delta}_{ij}}{T-1} \right| \\ &= \left| (1-\theta)\left(\frac{\Delta_{ij}}{T-1}\right)^{(\mathbb{Z})} - (1-\theta)\left(\frac{\bar{\Delta}_{ij}}{T-1}\right) \right| = (1-\theta)\left| \left(\frac{\Delta_{ij}}{T-1}\right)^{(\mathbb{Z})} - \left(\frac{\bar{\Delta}_{ij}}{T-1}\right) \right| \end{aligned} \tag{3.13}$$

Considering $0 \leq (1-\theta) \leq 1$, we can obtain

$$(1-\theta)\left|\left(\frac{\Delta_{ij}}{T-1}\right)^{(\mathbb{Z})} - \left(\frac{\bar{\Delta}_{ij}}{T-1}\right)\right| \leq \left|\left(\frac{\Delta_{ij}}{T-1}\right)^{(\mathbb{Z})} - \left(\frac{\bar{\Delta}_{ij}}{T-1}\right)\right| \tag{3.14}$$

Combining Eq. (3.13) and Eq. (3.14), we have

$$\left|\left(\frac{\Delta_{ij}}{T-1}\right)^{(\mathbb{Z}+1)} - \left(\frac{\bar{\Delta}_{ij}}{T-1}\right)\right| \leq \left|\left(\frac{\Delta_{ij}}{T-1}\right)^{(\mathbb{Z})} - \left(\frac{\bar{\Delta}_{ij}}{T-1}\right)\right| \tag{3.15}$$

Then we have $\frac{1}{T-1}\left(\frac{2}{m(m-1)}\sum_{i<j}^{m}\left((\Delta_{ij})^{(\mathbb{Z}+1)} - (\bar{\Delta}_{ij})\right)^2\right)^{1/2} \leq \frac{1}{T-1}\left(\frac{2}{m(m-1)}\sum_{i<j}^{m}\left((\Delta_{ij})^{(\mathbb{Z})} - (\bar{\Delta}_{ij})\right)^2\right)^{1/2}$. Therefore, we obtain $CI((\tilde{H}_{S_O})^{(\mathbb{Z}+1)}) \leq CI((\tilde{H}_{S_O})^{(\mathbb{Z})})$.

In a similar way, we can also get $CI((\tilde{H}_{S_O})^{(\mathbb{Z})}) \leq CI(\tilde{H}_{S_O})$, i.e., $CI(*\tilde{H}_{S_O}) < CI(\tilde{H}_{S_O})$. Which completes the proof of Theorem 3.2. ■

Remark 3.4 (Gou et al. 2019). Firstly, Theorem 3.2 mainly shows that utilizing the consistency repairing algorithm based on the automatic optimization method, the consistency index of the repaired DHHFLPR is always smaller than the original DHHFLPR. Secondly, for Algorithm 3.1, the adjusted parameter θ $(0 \leq \theta \leq 1)$ is very important. It determines the number of iterations and the accuracy of modification to the original HFLPR. Therefore, it is very important to choose a proper value of θ to reduce the number of iterations and simultaneously let the modified normalized DHHFLPR be close to its original normalized DHHFLPR as much as possible. Zhu and Xu (2014) calculated the average iterations of θ when $T = 9$, $\alpha = 0.1$ and $\vartheta = 2$. The results and the corresponding values of $\bar{CI}(\tilde{H}_{S_O})$ are listed in Table 3.2.

Next, we can set up an example to show the working process of Algorithm 3.1.

Example 3.3 (Continued with Example 3.1) (Gou et al. 2019). The DHHFLPR $\tilde{H}^1_{S_O}$ needs to be improved.

Table 3.2 The averaged values of iterations in Algorithm 3.1 ($T = 9$, $\alpha = 0.1$, $\vartheta = 2$)

θ	$m = 3$	$m = 4$	$m = 5$	$m = 6$	$m = 7$	$m = 8$
0.20	0.88	0.98	1.00	1.05	1.04	1.06
0.10	1.65	1.63	1.69	1.74	1.75	1.76
0.08	1.96	1.88	2.12	2.14	2.09	2.13
0.05	2.95	3.22	3.03	3.08	3.05	3.16
0.01	13.78	13.81	13.02	14.14	13.66	14.129
$\bar{CI}$	0.1103	0.1515	0.1744	0.1887	0.1985	0.2056

Firstly, based on Table 3.1, there is $\bar{CI}(\tilde{H}^1_{S_O}) = 0.1103$. From Example 3.2 and Eq. (3.6), we know $CI(\tilde{H}^1_{S_O}) = d(\tilde{H}^{1N}_{S_O}, \bar{H}^{1N}_{S_O}) = 0.1250 > \bar{CI}(\tilde{H}^1_{S_O})$. Suppose $\theta = 0.2$, and based on Eq. (3.11), we can get the modified normalized DHHFLPR $(\tilde{H}^{1N}_{S_O})^{(1)}$.

$$(\tilde{H}^{1N}_{S_O})^{(1)} = \begin{pmatrix} \{s_{0<o_0>}\} & \{s_{-3/10<o_{3/2}>}\} & \{s_{13/10<o_{-1/2}>}\} \\ \{s_{3/10<o_{-3/2}>}\} & \{s_{0<o_0>}\} & \{s_{-4/5<o_{-2}>}\} \\ \{s_{-13/10<o_{1/2}>}\} & \{s_{4/5<o_2>}\} & \{s_{0<o_0>}\} \end{pmatrix}$$

Go back and calculate $CI((\tilde{H}^{1N}_{S_O})^{(1)}) = d((\tilde{H}^{1N}_{S_O})^{(1)}, (\bar{H}^{1N}_{S_O})^{(1)}) = 0.1000 < \bar{CI}(\tilde{H}^1_{S_O})$, so the normalized DHHFLPR $(\tilde{H}^{1N}_{S_O})^{(1)}$ is of acceptable consistency. Let $*\tilde{H}^1_{S_O} = (\tilde{H}^{1N}_{S_O})^{(1)}$ and output $*\tilde{H}^1_{S_O}$.

Additionally, based on the modeling method proposed by Dong et al. (2008) and an optimization model of HFLPR introduced by Zhu and Xu (2014), we can develop an optimization model of the DHHFLPR to improve its consistency. Suppose that $\tilde{H}_{S_O} = (h_{S_{O_{ij}}})_{m\times m}$ is a DHHFLPR with unacceptable consistency, $\tilde{H}^N_{S_O} = (h^N_{S_{O_{ij}}})_{m\times m}$ being its normalized DHHFLPR. To obtain the modified normalized DHHFLPR $*\tilde{H}_{S_O} = (*h_{S_{O_{ij}}})_{m\times m}$ with acceptable consistency and reduce the loss of original information, we set up $*h_{S_{O_{ij}}} = h^N_{S_{O_{ij}}} \oplus y_{ij}$, in which y_{ij} $(i,j = 1,2,\ldots,m;\ i<j)$ are the adjusted DHHFLEs. An optimization model can be established as follows:

$$\begin{aligned} &\min_y \left(\frac{2}{m(m-1)} \sum_{i<j}^{m} |f(y_{ij})| \right) \\ &s.t. \begin{cases} f(y_{ij}) + f(y_{ji}) = 0 \\ CI(*\tilde{H}_{S_O}) \le \bar{CI}(\tilde{H}_{S_O}) \end{cases} \end{aligned} \tag{3.16}$$

where $CI(*\tilde{H}_{S_O}) = \left(\frac{2}{m(m-1)} \sum_{i<j}^m (d(*h^N_{S_{O_{ij}}}, *\bar{h}^N_{S_{O_{ij}}}))^2 \right)^{1/2}$.

Based on this model, we can optimize $*\tilde{H}_{S_O}$ as discussed in Example 3.3. All adjusted DHHFLEs y_{ij} $(i,j = 1,2,3)$ are obtained and the modified normalized DHHFLPR $*\tilde{H}^{1'}_{S_O} = (*h^{1'}_{S_{O_{ij}}})_{3\times 3}$ is established:

$$*\tilde{H}^{1'}_{S_O} = \begin{pmatrix} \{s_{0<o_0>}\} & \{s_{-173/518<o_{1424/783}>}\} & \{s_{3/2<o_{-1/2}>}\} \\ \{s_{173/518<o_{-1424/783}>}\} & \{s_{0<o_0>}\} & \{s_{-216/259<o_{-1254/265}>}\} \\ \{s_{-3/2<o_{1/2}>}\} & \{s_{216/259<o_{1254/265}>}\} & \{s_{0<o_0>}\} \end{pmatrix}$$

Then, the additive consistency index $CI(*\tilde{H}^{1'}_{S_O}) = 0.1103 = \bar{CI}(\tilde{H}^1_{S_O})$.

Remark 3.5 (Gou et al. 2019). From Algorithm 3.1, we can determine that the modified normalized DHHFLPR $*H^1_{S_O}$ is of acceptable consistency from several iterations. Simultaneously, the number of iterations can be controlled by different values of the adjusted parameter θ. Additionally, using the above optimization model, we only need to calculate it one time to obtain the modified normalized DHHFLPR $*\tilde{H}^{1'}_{S_O}$ with acceptable consistency, but the calculation is complex. Thus, if the DHHFLPR is simple, the optimization model is suitable; otherwise, we can use Algorithm 3.1.

(b) **Consistency repairing algorithm based on the feedback method**

Considering that sometimes the experts are more likely to modify their preference relations by themselves, then Algorithm 3.1 is not suitable any more. In the existing research, lots of scholars have developed some feedback methods under other preference circumstances (Alonso et al. 2010; Fu and Yang 2010; Herrera-Viedma et al. 2007; Zhu and Xu 2014), and the feedback method can feed suggestions back to the experts and help them to improve their preferences. Therefore, this subsection establishes a consistency repairing algorithm based on the feedback method under DHHFLPR. Firstly, a novel concept of interval-valued DHHFLPR is defined.

Definition 3.5 (Gou et al. 2019). An interval-valued DHHFLPR on X, $H_{\bar{S}_{\bar{O}}}$, is given in the mathematical form of

$$H_{\bar{S}_{\bar{O}}} = \{<x_i, h_{\bar{S}_{\bar{O}}}(x_i)> |x_i \in X\} \tag{3.17}$$

where $h_{\bar{S}_{\bar{O}}}$ is a set of some values in S_O, denoted by $h_{\bar{S}_{\bar{O}}} = \{h^{(l)}_{\bar{S}_{\bar{O}}}|l = 1, 2, \ldots, h_{\bar{S}_{\bar{O}}}\}$. We call $h_{\bar{S}_{\bar{O}}}$ interval-valued DHHFLE, and call $h^{(l)}_{\bar{S}_{\bar{O}}} = [(h^{(l)}_{\bar{S}_{\bar{O}}})^{\mu}, (h^{(l)}_{\bar{S}_{\bar{O}}})^{\nu}]$ interval-valued DHLT, which satisfies $(h^{(l)}_{\bar{S}_{\bar{O}}})^{\mu}$, $(h^{(l)}_{\bar{S}_{\bar{O}}})^{\nu} \in S_O$ and $(h^{(l)}_{\bar{S}_{\bar{O}}})^{\mu} \leq (h^{(l)}_{\bar{S}_{\bar{O}}})^{\nu}$.

Then an interval-valued DHHFLPR can be defined as follows:

Definition 3.6 (Gou et al. 2019). An interval-valued DHHFLPR $\tilde{H}_{S_O}$ is presented by a matrix

$$\tilde{H}_{\bar{S}_{\bar{O}}} = (h_{\bar{S}_{\bar{O}_{ij}}})_{m \times m} \subset A \times A \tag{3.18}$$

where $h_{\bar{S}_{\bar{O}_{ij}}} = \{h^{(l)}_{\bar{S}_{\bar{O}_{ij}}} |l = 1, 2, \ldots, \# h_{\bar{S}_{\bar{O}_{ij}}}\}$ is an interval-valued DHHFLE indicating the preferences in an interval to which A_i over A_j. For all $i, j = 1, 2, \ldots, m$, $h_{\bar{S}_{\bar{O}_{ij}}}(i<j)$ should satisfy that:

$$(h^{(l)}_{\bar{S}_{\bar{O}_{ij}}})^{\mu} \oplus (h^{(l)}_{\bar{S}_{\bar{O}_{ji}}})^{\nu} = (h^{(l)}_{\bar{S}_{\bar{O}_{ij}}})^{\nu} \oplus (h^{(l)}_{\bar{S}_{\bar{O}_{ji}}})^{\mu} = s_{0<o_0>},\ h_{\bar{S}_{\bar{O}_{ii}}} = \{s_{0<o_0>}\},\ \# h_{\bar{S}_{\bar{O}_{ij}}} = \# h_{\bar{S}_{\bar{O}_{ji}}} \tag{3.19}$$

and

$$h_{\bar{S}_{\bar{O}_{ij}}}^{(l)} \leq h_{\bar{S}_{\bar{O}_{ij}}}^{(l)}, \; h_{\bar{S}_{\bar{O}_{ji}}}^{(l)} \leq h_{\bar{S}_{\bar{O}_{ji}}}^{(l)} \tag{3.20}$$

where $h_{\bar{S}_{\bar{O}_{ij}}}^{(l)}$ is the l-th interval-valued DHLT in $h_{\bar{S}_{\bar{O}_{ij}}}$.

Then, an algorithm is established to show the feedback-based improving method:

Algorithm 3.2 (Gou et al. 2019). Consistency repairing algorithm based on the feedback method

Step 1. It is same as the Step 1 in Algorithm 3.1.
Step 2. It is same as the Step 2 in Algorithm 3.1.
Step 3. Calculate $CI((\tilde{H}_{S_O})^{(\mathbb{Z})}) = d((\tilde{H}_{S_O}^N)^{(\mathbb{Z})}, (\bar{H}_{S_O}^N)^{(\mathbb{Z})})$ based on Eq. (3.6). If $CI((\tilde{H}_{S_O})^{(\mathbb{Z})}) \leq \bar{CI}(\tilde{H}_{S_O})$, then go to Step 6; If $CI((\tilde{H}_{S_O})^{(\mathbb{Z})}) > \bar{CI}(\tilde{H}_{S_O})$, then go to Step 4.
Step 4. Construct the interval-valued DHHFLPR $\tilde{H}_{\bar{S}_{\bar{O}}} = ((h_{\bar{S}_{\bar{O}_{ij}}})_{m\times m})^{(\mathbb{Z})} = ((\{h_{\bar{S}_{\bar{O}_{ij}}}^{(l)} | l = 1, 2, \ldots, \# h_{\bar{S}_{\bar{O}_{ij}}}\})_{m\times m})^{(\mathbb{Z})}$, where $(h_{\bar{S}_{\bar{O}_{ij}}}^{(l)})^{(\mathbb{Z})} = [\min\{(h_{S_{O_{ij}}}^{(l)})^{(\mathbb{Z})}, (\bar{h}_{S_{O_{ij}}}^{(l)})^{(\mathbb{Z})}\}, \max\{(h_{S_{O_{ij}}}^{(l)})^{(\mathbb{Z})}, (\bar{h}_{S_{O_{ij}}}^{(l)})^{(\mathbb{Z})}\}]$. Then we return $\tilde{H}_{\bar{S}_{\bar{O}}}$ to the expert and ask him to provide new preference information.
Step 5. Receive all the preference information of the experts and establish the modified normalized DHHFLPR $(\tilde{H}_{S_O}^N)^{(\mathbb{Z}+1)} = ((h_{S_{O_{ij}}}^N)_{m\times m})^{(\mathbb{Z}+1)}$. Let $\mathbb{Z} = \mathbb{Z} + 1$. Go back to Step 3.
Step 6. Let $*\tilde{H}_{S_O} = (\tilde{H}_{S_O}^N)^{(\mathbb{Z})}$, and output the modified normalized DHHFLPR $*\tilde{H}_{S_O}$.
Step 7. End.

Figure 3.2 shows the consistency improving process of Algorithm 3.2.

Similar to Theorem 3.2, we can give the following theorem:

Theorem 3.3 (Gou et al. 2019). Let $\tilde{H}_{S_O} = (h_{S_{O_{ij}}})_{m\times m}$ be a DHHFLPR, θ $(0 \leq \theta \leq 1)$ be the adjusted parameter, and $*H_{S_O}^N$ be the modified normalized DHHFLPR obtained by Algorithm 3.2. Then $CI(*H_{S_O}^N) < CI(\tilde{H}_{S_O})$.

As we discussed above, Theorem 3.3 mainly shows that utilizing the consistency repairing algorithm based on the feedback method, the consistency index of the repaired DHHFLPR is also smaller than the original DHHFLPR. Considering that Theorem 3.3 is similar to Theorem 3.2, so its proof can be omitted.

Example 3.4 (Continued with Example 3.2) (Gou et al. 2019). $\tilde{H}_{S_O}^1$ is a DHHFLPR of unacceptable consistency and thus it needs to be improved. Let $\tilde{H}_{S_O}^1 = (\tilde{H}_{S_O}^1)^{(0)}$, and we can get the normalized DHHFLPR $(\tilde{H}_{S_O}^{1N})^{(0)} = ((h_{S_{O_{ij}}}^{1N})_{3\times 3})^{(0)}$ and the

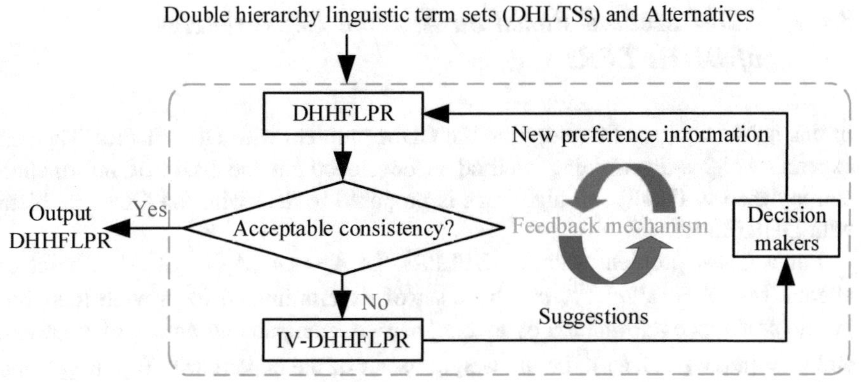

Fig. 3.2 The feedback-based improving method

consistent normalized DHHFLPR $(\bar{H}^{1N}_{S_O})^{(0)} = ((\bar{h}^{1N}_{S_{O_{ij}}})_{3\times3})^{(0)}$, respectively. Then we get $CI((\tilde{H}^{1}_{S_O})^{(0)}) = 0.1250 > \bar{CI}(\tilde{H}^{1}_{S_O})$. So we construct an interval-valued DHHFLPR $\tilde{H}_{\bar{S}_{\bar{O}}} = ((h_{\bar{S}_{\bar{O}_{ij}}})_{3\times3})^{(0)}$ based on $(\tilde{H}^{1N}_{S_O})^{(0)}$ and $(\bar{H}^{1N}_{S_O})^{(0)}$:

$$\tilde{H}_{\bar{S}_{\bar{O}}} = \begin{pmatrix} \{s_{0<o_0>}\} & \{[s_{-1/2<o_{3/2}>}, s_{1/2<o_{3/2}>}]\} & \{[s_{1/2<o_{-1/2}>}, s_{3/2<o_{-1/2}>}]\} \\ \{[s_{-1/2<o_{-3/2}>}, s_{1/2<o_{-3/2}>}]\} & \{s_{0<o_0>}\} & \{[s_{-1<o_{-2}>}, s_{0<o_{-2}>}]\} \\ \{[s_{-3/2<o_{1/2}>}, s_{-1/2<o_{1/2}>}]\} & \{[s_{0<o_2>}, s_{1<o_2>}]\} & \{s_{0<o_0>}\} \end{pmatrix}$$

Then, we return $\tilde{H}_{\bar{S}_{\bar{O}}}$ to the expert and ask him to propose new preference information, and collect all preferences to establish the modified normalized DHHFLPR $(\tilde{H}^{1N}_{S_O})^{(1)} = ((h^{1N}_{S_{O_{ij}}})_{3\times3})^{(1)}$. Suppose that the experts give a modified normalized DHHFLPR $(\tilde{H}^{1N}_{S_O})^{(1)}$ as:

$$(\tilde{H}^{1N}_{S_O})^{(1)} = \begin{pmatrix} \{s_{0<o_0>}\} & \{s_{0<o_1>}\} & \{s_{1<o_{-1}>}\} \\ \{s_{0<o_{-1}>}\} & \{s_{0<o_0>}\} & \{s_{1<o_1>}\} \\ \{s_{-1<o_1>}\} & \{s_{-1<o_{-1}>}\} & \{s_{0<o_0>}\} \end{pmatrix}$$

Then we obtain $CI((\tilde{H}^{1N}_{S_O})^{(1)}) = d((\tilde{H}^{1N}_{S_O})^{(1)}, (\bar{H}^{1N}_{S_O})^{(1)}) = 0.0313 \leq \bar{CI}(\tilde{H}^{1}_{S_O})$ $= 0.1103$. Let $*\tilde{H}^{1''}_{S_O} = (\tilde{H}^{1}_{S_O})^{(1)}$, and output the modified normalized DHHFLPR $*\tilde{H}^{1''}_{S_O}$.

3.1.4 GDM Method Based on Additive Consistencies of DHHFLPRs

In this subsection, we first describe the GDM problem with DHHFLPRs. Then an experts' weights-determining method is developed on the basis of information entropy theory. Finally, an algorithm is proposed to deal with the GDM problem with DHHFLPRs.

For a GDM problem with DHHFLPRs, let $A = \{A_1, A_2, \ldots, A_m\}$ be a set of alternatives, $E = \{e^1, e^2, \ldots, e^R\}$ be a set of experts invited to provide their linguistic preference information by making pairwise comparisons among alternatives, and $w = (w_1, w_2, \ldots, w_R)^T$ be the weight vector of the experts with $0 \leq w_r \leq 1$ and $\sum_{r=1}^{R} w_r = 1$. Experts' linguistic preference information can be established by DHHFLPRs and denoted as $\tilde{H}^r_{S_O} = (h^r_{S_{O_{ij}}})_{m \times m} (r = 1, 2, \ldots, R)$.

When developing the GDM method with DHHFLPRs, determining the experts' weights becomes an important step. Thus, a weights-determining method is developed to obtain the weights of experts at first (Gou et al. 2019).

We mainly utilize the information entropy theory to determine the weights of the experts. The first step is to obtain each expert's ordering vector $U^r = (u^r_1, u^r_2, \ldots, u^r_m)^T$ $(r = 1, 2, \ldots, R)$ for all alternatives, which can be calculated by

$$u^r_i = \left(\sum_{j=1}^{m} f(le(h^r_{S_{O_{ij}}})) \right) \Bigg/ \sum_{i=1}^{m} \sum_{j=1}^{m} f(le(h^r_{S_{O_{ij}}})), \quad i = 1, 2, \ldots, m \tag{3.21}$$

And then the information entropy of each expert e^r can be obtained by

$$IE(U^r) = -\frac{1}{\log_2 m} \cdot \sum_{i=1}^{m} u^r_i \log_2 u^r_i \tag{3.22}$$

Information entropy indicates the uncertainty degree and randomness of evaluation information. Therefore, the smaller the information entropy, the bigger the certainty degree of the evaluation information, which means that this expert plays an significant role and then we need to give him a bigger weight. So let w_r be weight of the r-th expert, then

$$w_r = (IE(U^r))^{-1} \Bigg/ \sum_{r=1}^{R} (IE(U^r))^{-1} \tag{3.23}$$

Furthermore, a double hierarchy hesitant fuzzy linguistic weighted averaging operator (DHHFLWA) can be defined.

Definition 3.7 (Gou et al. 2019). Suppose that $h_{S_{O_i}} = \{s_{\phi^i_l<o_{\varphi^i_l}>} | s_{\phi^i_l<o_{\varphi^i_l}>} \in S_O;$ $l = 1,2,\ldots,\#h_{S_{O_i}}\}(i = 1,2,\ldots,n)$ is a collection of DHHFLEs. A DHHFLWA operator is a mapping $M^R \to M$, such that

$$\text{DHHFLWA}(h_{S_{O_1}}, h_{S_{O_2}}, \ldots, h_{S_{O_n}}) = \sum_{i=1}^{n} (w_i \cdot le(h_{S_{O_i}})) \tag{3.24}$$

where $w = (w_1, w_2, \ldots, w_n)^T$ is the weight vector of $h_{S_{O_i}}$ (i = 1, 2, ..., n) with $0 \le w_i \le 1$ and $\sum_{i=1}^{n} w_i = 1$.

Theorem 3.4 (Gou et al. 2019). Let $h_{S_{O_i}} = \{s_{\phi^i_l<o_{\varphi^i_l}>} | s_{\phi^i_l<o_{\varphi^i_l}>} \in S_O; l = 1,2,\ldots,\#h_{S_{O_i}}\}$ $(i = 1,2,\ldots,n)$ be a collection of DHHFLEs, and $le(h_{S_{O_i}})$ $= \frac{1}{\#h_{S_{O_i}}} \sum_{l=1}^{\#h_{S_{O_i}}} s_{\phi^i_l<o_{\varphi^i_l}>} = s_{\phi^i<o_{\varphi^i}>}$, $w = (w_1, w_2, \ldots, w_n)^T$ be the weight vector of $h_{S_{O_i}}(i = 1,2,\ldots,n)$ with $0 \le w_i \le 1$ and $\sum_{i=1}^{n} w_i = 1$. Then

$$\text{DHHFLWA}(h^1_{S_O}, h^2_{S_O}, \ldots, h^n_{S_O}) = s_{\sum_{i=1}^{n} w_i\phi^i<o_{\sum_{i=1}^{n} w_i\varphi^i}>} \tag{3.25}$$

Remark 3.6 (Gou et al. 2019). Based on the operational laws of DHHFLEs and considering that every DHHFLE only contains a DHLT, we can sum all linguistic terms included in the first hierarchy and the second hierarchy respectively and obtain the aggregation result. In GDM problem with DHHFLPRs, this aggregation method is very suitable in the decision-making processes. Meanwhile, this aggregation method can be used as the most basic tool for the following GDM model based on additive consistencies of DHHFLPRs.

Then, a GDM model based on additive consistencies of DHHFLPRs is established.

Algorithm 3.3 (Gou et al. 2019). GDM model based on additive consistencies of DHHFLPRs

Step 1. Let $A = \{A_1, A_2, \ldots, A_m\}$ be a set of alternatives, $E = \{e^1, e^2, \ldots, e^R\}$ be a set of experts and their preference information can establish some DHHFLPRs $\tilde{H}^r_{S_O} = (h^r_{S_{O_{ij}}})_{m\times m}$ $(r = 1,2,\ldots,R)$.

Step 2. Calculate the normalized DHHFLPRs $\tilde{H}^{rN}_{S_O} = (h^{rN}_{S_{O_{ij}}})_{m\times m}$ $(r = 1,2,\ldots,R)$ and the consistent normalized DHHFLPR $\bar{H}^{rN}_{S_O} = (\bar{h}^{rN}_{S_{O_{ij}}})_{m\times m}$ $(r = 1,2,\ldots,R)$, respectively.

Step 3. Utilize Algorithm 3.1 or Algorithm 3.2 to ensure that each normalized DHHFLPR is of acceptable consistency.
Step 4. Calculate the experts' weight vector $w = (w_1, w_2, \ldots, w_R)^T$ based on Eqs. (3.21)–(3.23).
Step 5. Aggregate all of the normalized DHHFLPRs into a synthetical normalized DHHFLPR using the DHHFLWA operator, denoted as $\overset{\frown}{H}{}^N_{S_O} = (\overset{\frown}{h}{}^N_{S_{O_{ij}}})_{m\times m}$.
Step 6. Calculate the synthetical value of each alternative $SV(A_i) = \sum_{j=1}^{m} E(\overset{\frown}{h}{}^N_{S_{O_{ij}}})$ $(i = 1, 2, \ldots, m)$ based on Eq. (1.15).
Step 7. Rank all the alternatives based on the values of $SV(A_i)$ $(i = 1, 2, \ldots, m)$.
Step 8. End.

Remark 3.7 (Gou et al. 2019). At the end of Algorithm 3.3, it is necessary to develop a rank-reversal experiment to check the effectiveness of this algorithm by adding some other alternatives based on (Sałabun 2014, 2015). In this experiment, if the ranking order of the original alternative is not changed, then this algorithm is effective. Otherwise, the algorithm should be improved.

3.1.5 Application of the GDM Method Based on Additive Consistencies of DHHFLPRs

In this subsection, the proposed GDM method is validated by a practical application which is to evaluate the water resource situations of some cities in Sichuan Province.

Example 3.5 (Gou et al. 2019). Water resources require indispensable solutions to sustain human life and that of all living things. In China, the average volume of renewable water is estimated to be about 2.812 trillion cubic meters per year, ranked the fifth in the word. Meanwhile, Sichuan water resources are very abundant and prominent in China. As one of the upper reaches of the Yangtze River system, Sichuan water resources are important water systems in China. The protection of water quality of Sichuan water resources has become a crucial issue of economic and social stability and the rapid development of China. However, in recent years, with the development of the society's productivity and industrialization, urbanization in China has accelerated. The problems of Sichuan water resource development, protection and management are facing an increasingly severe test, and the grim reality of global climate change has made these problems more urgent. At present, the main problems that are being faced include sustainable utilization of regional water resources, rational development of water resources, water condition detection, rational exploration and utilization of water resources, integrated management of water resources, the harmonious development between economy and environment, etc. To solve the problems of water resource development, protection and management, a lot of experts and scholars carried out research and some achievements have been made (Shang et al. 2017; Wang et al. 2017; Winz et al. 2009). In some ways, these studies have solved some problems, but the reality has been unsatisfactory. For example, in 2016, the amount of water was once again insufficient in the irrigation period of Dujiangyan, Sichuan province; the water quality in Liangshan state still cannot reach the national average. Therefore, these realities have prompted the authorities to think about other ways to solve the problems of water resource development, protection and management in Sichuan province.

Because of this, the water resources of each city in the Sichuan province will be assessed annually. Additionally, amounts of studies have utilized the definite data to make analyses and calculations when dealing with water resource development, protection and management problems. However, in reality, it is very difficult to measure the key indicators of water resource management such as maintaining the quantity of water resources, the water quality and so on, and complex uncertainties often arise. Therefore, double hierarchy hesitant fuzzy linguistic information can be used to express some immeasurable phenomenon. Based on these water resource comprehensive assessment indices (criteria) and the DHLTS $S_O = \{s_{t<o_k>} | t = -4, \ldots, -1, 0, 1, \ldots, 4; k = -4, \ldots, -1, 0, 1, \ldots, 4\}$ with $S = \{s_{-4} = extremely\ bad,\ s_{-3} = very\ bad,\ s_{-2} = bad,\ s_{-1} = slightly\ bad,$ $s_0 = medium,\ s_1 = slightly\ good,\ s_2 = good,\ s_3 = very\ good,\ s_4 = extremely\ good\}$ and $O = \{o_{-4} = far\,from, o_{-3} = scarcely, o_{-2} = only\,a\,little, o_{-1} = a\,little,$ $o_0 = just\,right, o_1 = much, o_2 = very\,much, o_3 = extremely\ much, o_4 = entirely\}$, five experts $E = \{e^1, e^2, \ldots, e^5\}$ are invited to evaluate the water resource situations of four typical and important cities in the Sichuan province, including Chengdu (A_1), Nanchong (A_2), Panzhihua (A_3) and Dazhou (A_4). Collecting the linguistic preference information of each expert, five DHHFLPRs can be established and shown in Tables 3.3, 3.4, 3.5, 3.6 and 3.7.

We can utilize Algorithm 3.3 to solve this GDM problem. Considering that the first step has been discussed above, we start the decision-making process from Step 2 directly.

Step 2. Calculate the normalized DHHFLPR $\tilde{H}^{rN}_{S_O} = (h^{rN}_{S_{O_{ij}}})_{4\times4}$ $(r = 1, 2, \ldots, 5)$ and the consistent normalized DHHFLPR $\bar{H}^{rN}_{S_O} = (\bar{h}^{rN}_{S_{O_{ij}}})_{4\times4}$ $(r = 1, 2, \ldots, 5)$, respectively.

Step 3. Based on Eq. (3.6), the consistency indices of all experts' DHHFLPRs $CI(\tilde{H}^r_{S_O})$ $(r = 1, 2, \ldots, 5)$ can be obtained and shown in Table 3.8.

Clearly, $CI(\tilde{H}^1_{S_O})$, $CI(\tilde{H}^3_{S_O})$, $CI(\tilde{H}^5_{S_O}) > 0.1515$, which means the DHHFLPRs $\tilde{H}^1_{S_O}$, $\tilde{H}^3_{S_O}$, $\tilde{H}^5_{S_O}$ are of unacceptable consistency and the other DHHFLPRs are of acceptable consistency.

Table 3.3 The evaluation preference information of the expert e^1

	A_1	A_2	A_3	A_4
A_1	$\{s_{0<o_0>}\}$	$\{s_{-1<o_1>}, s_{0<o_2>}\}$	$\{s_{1<o_{-2}>}, s_{2<o_3>}\}$	$\{s_{-1<o_{-2}>}\}$
A_2	$\{s_{1<o_{-1}>}, s_{0<o_{-2}>}\}$	$\{s_{0<o_0>}\}$	$\{s_{0<o_{-1}>}, s_{1<o_1>}\}$	$\{s_{-2<o_1>}, s_{-1<o_2>}\}$
A_3	$\{s_{-1<o_2>}, s_{-2<o_{-3}>}\}$	$\{s_{0<o_1>}, s_{-1<o_{-1}>}\}$	$\{s_{0<o_0>}\}$	$\{s_{2<o_{-3}>}, s_{3<o_3>}\}$
A_4	$\{s_{1<o_2>}\}$	$\{s_{2<o_{-1}>}, s_{1<o_{-2}>}\}$	$\{s_{-2<o_3>}, s_{-3<o_{-3}>}\}$	$\{s_{0<o_0>}\}$

Table 3.4 The evaluation preference information of the expert e^2

	A_1	A_2	A_3	A_4
A_1	$\{s_{0<o_0>}\}$	$\{s_{1<o_1>}\}$	$\{s_{2<o_2>}, s_{3<o_{-1}>}\}$	$\{s_{0<o_3>}\}$
A_2	$\{s_{-1<o_{-1}>}\}$	$\{s_{0<o_0>}\}$	$\{s_{1<o_{-1}>}, s_{2<o_1>}\}$	$\{s_{-1<o_3>}, s_{0<o_2>}\}$
A_3	$\{s_{-2<o_{-2}>}, s_{-3<o_1>}\}$	$\{s_{-1<o_1>}, s_{-2<o_{-1}>}\}$	$\{s_{0<o_0>}\}$	$\{s_{-2<o_1>}, s_{-1<o_3>}\}$
A_4	$\{s_{0<o_{-3}>}\}$	$\{s_{1<o_{-3}>}, s_{0<o_{-2}>}\}$	$\{s_{2<o_{-1}>}, s_{1<o_{-3}>}\}$	$\{s_{0<o_0>}\}$

Table 3.5 The evaluation preference information of the expert e^3

	A_1	A_2	A_3	A_4
A_1	$\{s_{0<o_0>}\}$	$\{s_{2<o_2>}\}$	$\{s_{-1<o_2>}, s_{-1<o_3>}\}$	$\{s_{-2<o_{-2}>}, s_{-1<o_{-2}>}\}$
A_2	$\{s_{-2<o_{-2}>}\}$	$\{s_{0<o_0>}\}$	$\{s_{0<o_{-1}>}, s_{0<o_1>}\}$	$\{s_{-1<o_1>}, s_{-1<o_2>}\}$
A_3	$\{s_{1<o_{-2}>}, s_{1<o_{-3}>}\}$	$\{s_{0<o_1>}, s_{0<o_{-1}>}\}$	$\{s_{0<o_0>}\}$	$\{s_{1<o_{-3}>}, s_{2<o_3>}\}$
A_4	$\{s_{2<o_2>}, s_{1<o_2>}\}$	$\{s_{1<o_{-1}>}, s_{1<o_{-2}>}\}$	$\{s_{-1<o_3>}, s_{-2<o_{-3}>}\}$	$\{s_{0<o_0>}\}$

Table 3.6 The evaluation preference information of the expert e^4

	A_1	A_2	A_3	A_4
A_1	$\{s_{0<o_0>}\}$	$\{s_{-1<o_{-1}>}\}$	$\{s_{1<o_{-2}>}, s_{2<o_1>}\}$	$\{s_{2<o_{-3}>}\}$
A_2	$\{s_{1<o_1>}\}$	$\{s_{0<o_0>}\}$	$\{s_{2<o_1>}, s_{3<o_{-1}>}\}$	$\{s_{3<o_3>}\}$
A_3	$\{s_{-1<o_2>}, s_{-2<o_{-1}>}\}$	$\{s_{-2<o_{-1}>}, s_{-3<o_1>}\}$	$\{s_{0<o_0>}\}$	$\{s_{-1<o_3>}\}$
A_4	$\{s_{-2<o_3>}\}$	$\{s_{-3<o_{-3}>}\}$	$\{s_{1<o_{-3}>}\}$	$\{s_{0<o_0>}\}$

Table 3.7 The evaluation preference information of the expert e^5

	A_1	A_2	A_3	A_4
A_1	$\{s_{0<o_0>}\}$	$\{s_{2<o_1>}, s_{3<o_1>}\}$	$\{s_{1<o_{-2}>}, s_{2<o_1>}\}$	$\{s_{-1<o_2>}\}$
A_2	$\{s_{-2<o_1>}, s_{-3<o_1>}\}$	$\{s_{0<o_0>}\}$	$\{s_{1<o_{-2}>}\}$	$\{s_{1<o_1>}, s_{2<o_{-1}>}\}$
A_3	$\{s_{-1<o_2>}, s_{-2<o_{-1}>}\}$	$\{s_{-1<o_2>}\}$	$\{s_{0<o_0>}\}$	$\{s_{2<o_3>}, s_{3<o_1>}\}$
A_4	$\{s_{1<o_{-2}>}\}$	$\{s_{-1<o_{-1}>}, s_{-2<o_1>}\}$	$\{s_{-2<o_{-3}>}, s_{-3<o_{-1}>}\}$	$\{s_{0<o_0>}\}$

Table 3.8 The consistency index of each expert's DHHFLPR

	e^1	e^2	e^3	e^4	e^5
$CI(\tilde{H}^r_{S_O})$	0.1809	0.0292	0.1564	0.0500	0.1872

(a) **Utilizing the consistency repairing algorithm based on the automatic optimization method**

Utilizing Algorithm 3.1, we can improve these three DHHFLPRs. The improved normalized DHHFLPRs can be obtained with the adjusted parameter $\theta = 0.2$ and the corresponding consistency indices are $CI(*\tilde{H}^1_{S_O}) = 0.1447$, $CI(*\tilde{H}^3_{S_O}) = 0.1251$, and $CI(*\tilde{H}^5_{S_O}) = 0.1498$.

(b) **Utilizing the consistency repairing algorithm based on the feedback improving method**

Utilizing Algorithm 3.2, we can establish three interval-valued DHHFLPRs $\tilde{H}^r_{S_{\bar{O}}} = (h^r_{S_{\bar{O}_{ij}}})_{4\times4}$ $(r = 1, 3, 5)$, and return them to the experts and ask them to provide their new preferences. Collecting the feedback information and obtaining three improved normalized DHHFLPRs shown in Tables 3.9, 3.10 and 3.11, and the corresponding consistency indices are $CI(*\tilde{H}^{1'}_{S_O}) = 0.0959$, $CI(*\tilde{H}^{3'}_{S_O}) = 0.0749$, and $CI(*\tilde{H}^{5'}_{S_O}) = 0.0832$.

Table 3.9 The improved DHHFLPR $\tilde{H}^{1'N}_{S_O}$

	A_1	A_2	A_3	A_4
A_1	$\{s_{0<o_0>}\}$	$\{s_{0<o_0>}\}$	$\{s_{1<o_{-1}>}\}$	$\{s_{0<o_{-2}>}\}$
A_2	$\{s_{0<o_0>}\}$	$\{s_{0<o_0>}\}$	$\{s_{0<o_1>}\}$	$\{s_{-1<o_1>}\}$
A_3	$\{s_{-1<o_1>}\}$	$\{s_{0<o_{-1}>}\}$	$\{s_{0<o_0>}\}$	$\{s_{1<o_2>}\}$
A_4	$\{s_{0<o_2>}\}$	$\{s_{1<o_{-1}>}\}$	$\{s_{-1<o_{-2}>}\}$	$\{s_{0<o_0>}\}$

Table 3.10 The improved DHHFLPR $\tilde{H}^{3'N}_{S_O}$

	A_1	A_2	A_3	A_4
A_1	$\{s_{0<o_0>}\}$	$\{s_{1<o_2>}\}$	$\{s_{-1<o_3>}\}$	$\{s_{-1<o_{-1}>}\}$
A_2	$\{s_{-1<o_{-2}>}\}$	$\{s_{0<o_0>}\}$	$\{s_{-1<o_1>}\}$	$\{s_{-1<o_1>}\}$
A_3	$\{s_{1<o_{-3}>}\}$	$\{s_{1<o_{-1}>}\}$	$\{s_{0<o_0>}\}$	$\{s_{1<o_{-1}>}\}$
A_4	$\{s_{1<o_1>}\}$	$\{s_{1<o_{-1}>}\}$	$\{s_{-1<o_1>}\}$	$\{s_{0<o_0>}\}$

Table 3.11 The improved DHHFLPR $\tilde{H}_{S_O}^{5'N}$

	A_1	A_2	A_3	A_4
A_1	$\{s_{0<o_0>}\}$	$\{s_{1<o_2>}\}$	$\{s_{1<o_{-1}>}\}$	$\{s_{1<o_{-2}>}\}$
A_2	$\{s_{-1<o_{-2}>}\}$	$\{s_{0<o_0>}\}$	$\{s_{0<o_1>}\}$	$\{s_{1<o_1>}\}$
A_3	$\{s_{-1<o_1>}\}$	$\{s_{0<o_{-1}>}\}$	$\{s_{0<o_0>}\}$	$\{s_{2<o_{-1}>}\}$
A_4	$\{s_{-1<o_2>}\}$	$\{s_{-1<o_{-1}>}\}$	$\{s_{-2<o_1>}\}$	$\{s_{0<o_0>}\}$

Step 4. Based on Eqs. (3.21)–(3.23), the weight vector of the experts can be calculated as:

(1) Under Algorithm 3.1: $w = (0.1973, 0.2009, 0.1983, 0.2041, 0.1994)^T$.
(2) Under Algorithm 3.2: $w = (0.1974, 0.2011, 0.1984, 0.2042, 0.1989)^T$.

Step 5. Aggregate all the normalized DHHFLPRs into the synthetical normalized DHHFLPRs $\overset{\frown}{H}{}_{S_O}^{N} = (\overset{\frown}{h}{}_{S_{O_{ij}}}^{N})_{4\times4}$ and $\overset{\frown}{H}{}_{S_O}^{N'} = (\overset{\frown}{h}{}_{S_{O_{ij}}}^{N'})_{4\times4}$ by the DHHFLWA operator. The aggregated results based on these two methods can be shown in Table 3.12 and Table 3.13, respectively.

Step 6. Calculate the synthetical value of each alternative

(1) Using Algorithm 3.1: $SV = \{2.2478, 2.0922, 1.8467, 1.8132\}$;
(2) Using Algorithm 3.2: $SV = \{2.2291, 2.1140, 1.8373, 1.8195\}$.

Step 7. Rank all the alternatives based on $SV(A_i)\,(i = 1,2,3,4)$: $A_1 \succ A_2 \succ A_3 \succ A_4$. Therefore, the ranking order of the water resource situations of these four cities is Chengdu $\succ$ Nanchong $\succ$ Panzhihua $\succ$ Dazhou.

Some discussions can be summarized as follows:

Table 3.12 The synthetical normalized DHHFLPR $\overset{\frown}{H}{}_{S_O}^{N} = (\overset{\frown}{h}{}_{S_{O_{ij}}}^{N})_{4\times4}$ based on Algorithm 3.1

	A_1	A_2	A_3	A_4
A_1	$\{s_{0<o_0>}\}$	$\{s_{0.69<o_{0.8}>}\}$	$\{s_{1.12<o_{0.53}>}\}$	$\{s_{-0.1<o_{-0.24}>}\}$
A_2	$\{s_{-0.69<o_{-0.8}>}\}$	$\{s_{0<o_0>}\}$	$\{s_{0.99<o_{-0.37}>}\}$	$\{s_{0.34<o_{1.58}>}\}$
A_3	$\{s_{-1.12<o_{-0.53}>}\}$	$\{s_{-0.99<o_{0.37}>}\}$	$\{s_{0<o_0>}\}$	$\{s_{0.58<o_{1.38}>}\}$
A_4	$\{s_{0.1<o_{0.24}>}\}$	$\{s_{-0.34<o_{-1.58}>}\}$	$\{s_{-0.58<o_{-1.38}>}\}$	$\{s_{0<o_0>}\}$

Table 3.13 The synthetical normalized DHHFLPR $\overset{\frown}{H}{}_{S_O}^{N'} = (\overset{\frown}{h}{}_{S_{O_{ij}}}^{N'})_{4\times4}$ based on Algorithm 3.2

	A_1	A_2	A_3	A_4
A_1	$\{s_{0<o_0>}\}$	$\{s_{0.39<o_{0.79}>}\}$	$\{s_{1.01<o_{0.3}>}\}$	$\{s_{0.41<o_{-1}>}\}$
A_2	$\{s_{-0.39<o_{-0.79}>}\}$	$\{s_{0<o_0>}\}$	$\{s_{0.61<o_{0.6}>}\}$	$\{s_{0.32<o_{1.71}>}\}$
A_3	$\{s_{-1.01<o_{-0.3}>}\}$	$\{s_{-0.61<o_{-0.6}>}\}$	$\{s_{0<o_0>}\}$	$\{s_{0.29<o_{1.01}>}\}$
A_4	$\{s_{-0.41<o_1>}\}$	$\{s_{-0.32<o_{-1.71}>}\}$	$\{s_{-0.29<o_{-1.01}>}\}$	$\{s_{0<o_0>}\}$

Firstly, the proposed normalization method mainly has two advantages: (1) The calculation becomes simple by transforming all DHHFLEs into the corresponding DHLTs. (2) The obtained DHLTs consist of all original linguistic information, so the proposed normalization method can avoid the original information loss.

Secondly, considering that we can utilize the adjusted parameter θ $(0 \leq \theta \leq 1)$ to adjust the number of iterations and to improve the accuracy of modification to the original DHHFLPR, the reasonable value of θ can be chosen based on Table 3.2. Therefore, if we choose the reasonable value of θ, based on MATLAB software, we can quickly obtain the modified normalized DHHFLPR of acceptable consistency on the basis of the automatic improving method. Additionally, the feedback-based improving method can feed the suggestions back to the experts and help them improve their preferences, so the consistency repairing algorithm based on the feedback method to satisfies the expert's willingness.

Thirdly, after the consistency checking and repairing processes, we can calculate the weight vector of all experts and aggregate all DHHFLPRs, then we obtain the final decision result. Obviously, the consistency indices of all repaired DHHFLPRs are different when using different methods, and the repaired places in each DHHFLPR are also different. But both of two methods obtain the same ranking of alternatives: $A_1 \succ A_2 \succ A_3 \succ A_4$. Therefore, there is no significant impact on the outcome based on these two different consistency checking and repairing methods.

3.2 Multiplicative Consistency Measures for DHHFLPRs

In the existing research, scholars are more inclined to utilize multiplicative transitivity considering that it is a special case of the cycle transitivity property (Baets et al. 2006). In addition, amounts of scholars have proven that the multiplicative transitivity is the most appropriate property for modeling cardinal consistency of preference relations because it can avoid some gaps such as the conflict with the given range used for providing the preference values (Chiclana et al. 2009). Therefore, the multiplicative consistency has been studied in different preferences such as the multiplicative consistent intuitionistic preference relation (Xu et al. 2011), the multiplicative consistent fuzzy preference relations (Xia et al. 2013), the perfect multiplicative consistent interval reciprocal relation (Xia and Xu 2011), the multiplicative consistent linguistic preference relation (Xia et al. 2014), the multiplicative consistent hesitant fuzzy preference relation (Zhu et al. 2014; Li et al. 2019), and multiplicative consistent probabilistic hesitant fuzzy preference relation (Li and Wang 2019), etc.

As the discussion in (Gou et al. 2017), the DHLTS and DHHFLTS can be regarded as the extensions of fuzzy set in linguistic environment and the transformations between each other are equivalent (Yu et al. 2018). Therefore, the advantages of the multiplicative consistency of fuzzy preference relations even other preference relations also work for DHHFLPRs. Based on this, the first contribution of this subsection is to focus on investigating the multiplicative consistency of DHHFLPRs. As we mentioned above, it is unrealistic that the DHHFLPRs provided by experts are perfectly multiplicative consistent. To solve this defect, Gou et al. (2020) developed a concept of acceptable multiplicative consistent DHHFLPR, proposed a consistency checking method to judge whether a DHHFLPR is of acceptable consistency, and developed a feedback mechanism-based repairing method to improve the consistency of a DHHFLPR to fully respect and consider the opinions of experts.

3.2.1 Multiplicative Consistency Measure Method of DHHFLPRs

This subsection mainly discusses the multiplicative consistency of DHHFLPR. Firstly, as the basic tool, a normalization method of DHHFLTS is developed, the multiplicative consistency property of DHHFLPRs is defined, and a consistency index of DHHFLPRs is proposed. In addition, a consistency checking method is given to judge whether a DHHFLPR is of acceptable multiplicative consistency. Finally, a feedback mechanism-based method is developed to repair the DHHFLPR with unacceptable multiplicative consistency.

To introduce the consistency measure of DHHFLPR, the first step is to normalize DHHFLPR and make sure that all DHHFLEs have the same length. Zhu and Xu (2014) developed two normalization methods, i.e., α-normalization and β-normalization. The former is to remove some elements and the latter is to add some elements. Considering the α-normalization may result in more loss of original information, this subsection mainly develops a normalization method for DHHFLPR on the basis of the β-normalization method.

Let $\tilde{H}_{S_O} = (h_{S_{O_{ij}}})_{m \times m}$ be a DHHFLPR, $h^-_{S_{O_{ij}}}$ and $h^+_{S_{O_{ij}}}$ be the minimum and maximum DHLT obtained by the above comparative method, respectively, $\varepsilon (0 \le \varepsilon \le 1)$ be an optimized parameter. Adding $h'_{S_{O_{ij}}} = \varepsilon h^+_{S_{O_{ij}}} + (1-\varepsilon) h^-_{S_{O_{ij}}}$ and $h'_{S_{O_{ji}}} = (1-\varepsilon) h^+_{S_{O_{ji}}} + \varepsilon h^-_{S_{O_{ji}}}$ to $h_{S_{O_{ij}}}$ and $h_{S_{O_{ji}}}$, respectively, then the normalized DHHFLPR can be defined as follows,

Definition 3.8 (Gou et al. 2020). Let $\tilde{H}_{S_O} = (h_{S_{O_{ij}}})_{m \times m} = (\{h^{(l)}_{S_{O_{ij}}} | l = 1, 2, \ldots, \#h_{S_{O_{ij}}}\})_{m \times m}$ be a DHHFLPR, $\varepsilon (0 \le \varepsilon \le 1)$ be an optimized parameter, and $\breve{h}_{S_{O_{ij}}}$ be the DHHFLE with the largest number of DHLTs. The normalized DHHFLPR, denoted as $\tilde{H}^N_{S_O} = (h^N_{S_{O_{ij}}})_{m \times m}$, can be established by ε and $(1-\varepsilon)$ to add the DHLTs to $h_{S_{O_{ij}}} (i<j)$ and $h_{S_{O_{ji}}} (i<j)$, respectively, and we can obtain that the number of DHLTs in all DHHFLEs is equal to $\breve{h}_{S_{O_{ij}}}$, i.e., $\#h^N_{S_{O_{12}}} = \cdots = \#h^N_{S_{O_{1m}}} = \cdots = \#h^N_{S_{O_{ij}}} = \cdots = \#h^N_{S_{O_{(m-2)m}}} = \#h^N_{S_{O_{(m-1)m}}} = \#\breve{h}_{S_{O_{ij}}}$.

Considering that the multiplicative transitivity property is characterized as the most appropriate property to model and measure consistency (Chiclana et al. 2009), Gou et al. (2020) defined the multiplicative consistency of DHHFLPR:

Definition 3.9 (Gou et al. 2020). Given a DHHFLPR $\tilde{H}_{S_O} = (h_{S_{O_{ij}}})_{m\times m}$ and its NDHHFLPR $\tilde{H}^N_{S_O} = (h^N_{S_{O_{ij}}})_{m\times m}$ with the parameter $\varepsilon(0 \le \varepsilon \le 1)$. If, for any $i, j, k = 1, 2, \ldots, m$,

$$f(h^{N(l)}_{S_{O_{ik}}})f(h^{N(l)}_{S_{O_{kj}}})f(h^{N(l)}_{S_{O_{ji}}}) = f(h^{N(l)}_{S_{O_{ki}}})f(h^{N(l)}_{S_{O_{jk}}})f(h^{N(l)}_{S_{O_{ij}}}) \tag{3.26}$$

where $h^{N(l)}_{S_{O_{ij}}}$ is the l-th DHLT in the DHHFLE $h^N_{S_{O_{ij}}}$, then $\tilde{H}_{S_O}$ is called a multiplicative consistent DHHFLPR with $\varepsilon(0 \le \varepsilon \le 1)$.

Next, two theorems are developed:

Theorem 3.5 (Gou et al. 2020). Given a DHHFLPR $\tilde{H}_{S_O} = (h_{S_{O_{ij}}})_{m\times m}$ and its NDHHFLPR $\tilde{H}^N_{S_O} = (h^N_{S_{O_{ij}}})_{m\times m}$ with the parameter $\varepsilon(0 \le \varepsilon \le 1)$, the following statements are equivalent:

(1) $\tilde{H}_{S_O}$ is multiplicative consistent.

(2)
$$f(h^{N(l)}_{S_{O_{ij}}}) = \frac{f(h^{N(l)}_{S_{O_{ik}}})f(h^{N(l)}_{S_{O_{kj}}})}{f(h^{N(l)}_{S_{O_{ik}}})f(h^{N(l)}_{S_{O_{kj}}}) + (1 - f(h^{N(l)}_{S_{O_{ik}}}))(1 - f(h^{N(l)}_{S_{O_{kj}}}))}. \tag{3.27}$$

(3)
$$f(h^{N(l)}_{S_{O_{ij}}}) = \frac{\sqrt[m]{\prod_{k=1}^{m} f(h^{N(l)}_{S_{O_{ik}}})f(h^{N(l)}_{S_{O_{kj}}})}}{\sqrt[m]{\prod_{k=1}^{m} f(h^{N(l)}_{S_{O_{ik}}})f(h^{N(l)}_{S_{O_{kj}}})} + \sqrt[m]{\prod_{k=1}^{m} (1 - f(h^{N(l)}_{S_{O_{ik}}}))(1 - f(h^{N(l)}_{S_{O_{kj}}}))}}. \tag{3.28}$$

Proof (1)⇔(2). Based on Eq. (3.26), we can obtain

$$\begin{aligned}
& f(h^{N(l)}_{S_{O_{ik}}})f(h^{N(l)}_{S_{O_{kj}}})f(h^{N(l)}_{S_{O_{ji}}}) = f(h^{N(l)}_{S_{O_{ki}}})f(h^{N(l)}_{S_{O_{jk}}})f(h^{N(l)}_{S_{O_{ij}}}) \\
& \Leftrightarrow f(h^{N(l)}_{S_{O_{ik}}})f(h^{N(l)}_{S_{O_{kj}}})(1 - f(h^{N(l)}_{S_{O_{ij}}})) = (1 - f(h^{N(l)}_{S_{O_{ik}}}))(1 - f(h^{N(l)}_{S_{O_{kj}}}))f\left(h^{N(l)}_{S_{O_{ij}}}\right) \\
& \Leftrightarrow f(h^{N(l)}_{S_{O_{ij}}}) = \frac{f(h^{N(l)}_{S_{O_{ik}}})f(h^{N(l)}_{S_{O_{kj}}})}{f(h^{N(l)}_{S_{O_{ik}}})f(h^{N(l)}_{S_{O_{kj}}}) + (1-f(h^{N(l)}_{S_{O_{ik}}}))(1-f(h^{N(l)}_{S_{O_{kj}}}))}
\end{aligned}$$

(1)⇔(3). Based on Eq. (3.26), for any $i,j,k = 1,2,\ldots,m$, we obtain

$$f(h_{S_{O_{ik}}}^{N(l)})f(h_{S_{O_{kj}}}^{N(l)})f(h_{S_{O_{ji}}}^{N(l)}) = f(h_{S_{O_{ki}}}^{N(l)})f(h_{S_{O_{jk}}}^{N(l)})f(h_{S_{O_{ij}}}^{N(l)})$$

For different value of k, for any $i,j = 1,2,\ldots,m$, we obtain

$$f(h_{S_{O_{i1}}}^{N(l)})f(h_{S_{O_{1j}}}^{N(l)})f(h_{S_{O_{ji}}}^{N(l)}) = f(h_{S_{O_{1i}}}^{N(l)})f(h_{S_{O_{j1}}}^{N(l)})f(h_{S_{O_{ij}}}^{N(l)}) \tag{3.29}$$

$$f(h_{S_{O_{i2}}}^{N(l)})f(h_{S_{O_{2j}}}^{N(l)})f(h_{S_{O_{ji}}}^{N(l)}) = f(h_{S_{O_{2i}}}^{N(l)})f(h_{S_{O_{j2}}}^{N(l)})f(h_{S_{O_{ij}}}^{N(l)}) \tag{3.30}$$

$$\vdots$$

$$f(h_{S_{O_{im}}}^{N(l)})f(h_{S_{O_{mj}}}^{N(l)})f(h_{S_{O_{ji}}}^{N(l)}) = f(h_{S_{O_{mi}}}^{N(l)})f(h_{S_{O_{jm}}}^{N(l)})f(h_{S_{O_{ij}}}^{N(l)}) \tag{3.31}$$

Multiply Eqs. (3.29)–(3.31), we obtain

$$\begin{aligned}
&(f(h_{S_{O_{ji}}}^{N(l)}))^m \prod_{k=1}^{m} f(h_{S_{O_{ik}}}^{N(l)})f(h_{S_{O_{kj}}}^{N(l)}) = (f(h_{S_{O_{ij}}}^{N(l)}))^m \prod_{k=1}^{m} f(h_{S_{O_{ki}}}^{N(l)})f(h_{S_{O_{jk}}}^{N(l)}) \\
&\Rightarrow (1 - f(h_{S_{O_{ij}}}^{N(l)})) \sqrt[m]{\prod_{k=1}^{m} f(h_{S_{O_{ik}}}^{N(l)})f(h_{S_{O_{kj}}}^{N(l)})} = (f(h_{S_{O_{ij}}}^{N(l)})) \sqrt[m]{\prod_{k=1}^{m} f(h_{S_{O_{ki}}}^{N(l)})f(h_{S_{O_{jk}}}^{N(l)})} \\
&\Rightarrow f(h_{S_{O_{ij}}}^{N(l)}) = \frac{\sqrt[m]{\prod_{k=1}^{m} f(h_{S_{O_{ik}}}^{N(l)})f(h_{S_{O_{kj}}}^{N(l)})}}{\sqrt[m]{\prod_{k=1}^{m} f(h_{S_{O_{ik}}}^{N(l)})f(h_{S_{O_{kj}}}^{N(l)})} + \sqrt[m]{\prod_{k=1}^{m} (1 - f(h_{S_{O_{ik}}}^{N(l)}))(1 - f(h_{S_{O_{kj}}}^{N(l)}))}}
\end{aligned} \tag{3.32}$$

(3)⇒(1). For any $i,j = 1,2,\ldots,m$, Eq. (3.32) is equal to

$$(f(h_{S_{O_{ji}}}^{N(l)}))^m \prod_{z=1}^{m} f(h_{S_{O_{iz}}}^{N(l)})f(h_{S_{O_{zj}}}^{N(l)}) = (f(h_{S_{O_{ij}}}^{N(l)}))^m \prod_{z=1}^{m} f(h_{S_{O_{zi}}}^{N(l)})f(h_{S_{O_{jz}}}^{N(l)}) \tag{3.33}$$

Similarly, for any $i,k = 1,2,\ldots,m$, we obtain

$$(f(h_{S_{O_{ki}}}^{N(l)}))^m \prod_{z=1}^{m} f(h_{S_{O_{iz}}}^{N(l)})f(h_{S_{O_{zk}}}^{N(l)}) = (f(h_{S_{O_{ik}}}^{N(l)}))^m \prod_{z=1}^{m} f(h_{S_{O_{zi}}}^{N(l)})f(h_{S_{O_{kz}}}^{N(l)}) \tag{3.34}$$

For any $k, j = 1, 2, \ldots, m$, we obtain

$$(f(h_{S_{O_{jk}}}^{N(l)}))^m \prod_{z=1}^{m} f(h_{S_{O_{kz}}}^{N(l)}) f(h_{S_{O_{zj}}}^{N(l)}) = (f(h_{S_{O_{kj}}}^{N(l)}))^m \prod_{z=1}^{m} f(h_{S_{O_{zk}}}^{N(l)}) f(h_{S_{O_{jz}}}^{N(l)}) \tag{3.35}$$

Multiply Eqs. (3.34) and (3.35), there is

$$(f(h_{S_{O_{ki}}}^{N(l)}))^m (f(h_{S_{O_{jk}}}^{N(l)}))^m \prod_{z=1}^{m} f(h_{S_{O_{iz}}}^{N(l)}) f(h_{S_{O_{zj}}}^{N(l)}) = (f(h_{S_{O_{ik}}}^{N(l)}))^m (f(h_{S_{O_{kj}}}^{N(l)}))^m \prod_{z=1}^{m} f(h_{S_{O_{zi}}}^{N(l)}) f(h_{S_{O_{jz}}}^{N(l)}) \tag{3.36}$$

Using Eq. (3.36) to divide Eq. (3.33), we have

$$\frac{(f(h_{S_{O_{ki}}}^{N(l)}))^m (f(h_{S_{O_{jk}}}^{N(l)}))^m}{(f(h_{S_{O_{ji}}}^{N(l)}))^m} = \frac{(f(h_{S_{O_{ik}}}^{N(l)}))^m (f(h_{S_{O_{kj}}}^{N(l)}))^m}{(f(h_{S_{O_{ij}}}^{N(l)}))^m}, \quad i, j, k = 1, 2, \ldots, m \tag{3.37}$$

Equation (3.37) is equal to $f(h_{S_{O_{ik}}}^{N(l)}) f(h_{S_{O_{kj}}}^{N(l)}) f(h_{S_{O_{ji}}}^{N(l)}) = f(h_{S_{O_{ki}}}^{N(l)}) f(h_{S_{O_{jk}}}^{N(l)}) f(h_{S_{O_{ij}}}^{N(l)})$. Therefore, based on Definition 3.9, $\tilde{H}_{S_O}$ is a multiplicative consistent DHHFLPR. Which completes the proof of Theorem 3.5. ■

Based on Eq. (3.27), it can be easily proven that

$$\begin{aligned} f(h_{S_{O_{ij}}}^{N(l)}) &= \frac{f(h_{S_{O_{ik}}}^{N(l)}) f(h_{S_{O_{kj}}}^{N(l)})}{f(h_{S_{O_{ik}}}^{N(l)}) f(h_{S_{O_{kj}}}^{N(l)}) + (1 - f(h_{S_{O_{ik}}}^{N(l)}))(1 - f(h_{S_{O_{kj}}}^{N(l)}))} \\ &= \frac{1}{(1 + (\frac{1}{f(h_{S_{O_{ik}}}^{N(l)})} - 1)(\frac{1}{f(h_{S_{O_{kj}}}^{N(l)})} - 1))} \end{aligned}$$

Therefore, $f(h_{S_{O_{ij}}}^{N(l)})$ is an increasing function about $f(h_{S_{O_{ik}}}^{N(l)})$ and $f(h_{S_{O_{kj}}}^{N(l)})$, and we can obtain

$$0 \leq 1 \Bigg/ 1 + \left(\left(\frac{1}{f(h_{S_{O_{ik}}}^{N(l)})} - 1 \right) \left(\frac{1}{f(h_{S_{O_{kj}}}^{N(l)})} - 1 \right) \right) \leq 1 \tag{3.38}$$

Which means that Eq. (3.27) produces a reasonable result because the result of $f(h_{S_{O_{ij}}}^{N(l)})$ is included in $[0, 1]$.

Theorem 3.6 (Gou et al. 2020). Given a DHHFLPR $\tilde{H}_{S_O} = (h_{S_{O_{ij}}})_{m \times m}$ and its NDHHFLPR $\tilde{H}_{S_O}^N = (h_{S_{O_{ij}}}^N)_{m \times m}$ with the parameter $\varepsilon (0 \leq \varepsilon \leq 1)$, for $i, j, k = 1, 2, \ldots, m$, let

$$\hat{h}_{S_{O_{ij}}}^{N(l)} = f^{-1}\left(\frac{\sqrt[m]{\prod_{k=1}^{m} f(h_{S_{O_{ik}}}^{N(l)}) f(h_{S_{O_{kj}}}^{N(l)})}}{\sqrt[m]{\prod_{k=1}^{m} f(h_{S_{O_{ik}}}^{N(l)}) f(h_{S_{O_{kj}}}^{N(l)})} + \sqrt[m]{\prod_{k=1}^{m} (1 - f(h_{S_{O_{ik}}}^{N(l)}))(1 - f(h_{S_{O_{kj}}}^{N(l)}))}}\right) \tag{3.39}$$

Then, $\hat{H}_{S_O} = (\hat{h}_{S_{O_{ij}}}^{N})_{m\times m} = (\{\hat{h}_{S_{O_{ij}}}^{N(l)} | l = 1, 2, \ldots, \#\hat{h}_{S_{O_{ij}}}^{N}\})_{m\times m}$ is a multiplicative consistent DHHFLPR with $\varepsilon(0 \le \varepsilon \le 1)$.

Proof Based on Eq. (3.39), for any $i, j, k = 1, 2, \ldots, m$, we have

$$f(\hat{h}_{S_{O_{ik}}}^{N(l)}) = \frac{\sqrt[m]{\prod_{z=1}^{m} f(h_{S_{O_{iz}}}^{N(l)}) f(h_{S_{O_{zk}}}^{N(l)})}}{\sqrt[m]{\prod_{z=1}^{m} f(h_{S_{O_{iz}}}^{N(l)}) f(h_{S_{O_{zk}}}^{N(l)})} + \sqrt[m]{\prod_{z=1}^{m} (1 - f(h_{S_{O_{iz}}}^{N(l)}))(1 - f(h_{S_{O_{zk}}}^{N(l)}))}}$$

$$f(\hat{h}_{S_{O_{kj}}}^{N(l)}) = \frac{\sqrt[m]{\prod_{z=1}^{m} f(h_{S_{O_{kz}}}^{N(l)}) f(h_{S_{O_{zj}}}^{N(l)})}}{\sqrt[m]{\prod_{z=1}^{m} f(h_{S_{O_{kz}}}^{N(l)}) f(h_{S_{O_{zj}}}^{N(l)})} + \sqrt[m]{\prod_{z=1}^{m} (1 - f(h_{S_{O_{kz}}}^{N(l)}))(1 - f(h_{S_{O_{zj}}}^{N(l)}))}}$$

Let

$$a = \sqrt[m]{\prod_{z=1}^{m} f(h_{S_{O_{iz}}}^{N(l)}) f(h_{S_{O_{zk}}}^{N(l)})} + \sqrt[m]{\prod_{z=1}^{m} (1 - f(h_{S_{O_{iz}}}^{N(l)}))(1 - f(h_{S_{O_{zk}}}^{N(l)}))}$$

$$b = \sqrt[m]{\prod_{z=1}^{m} f(h_{S_{O_{kz}}}^{N(l)}) f(h_{S_{O_{zj}}}^{N(l)})} + \sqrt[m]{\prod_{z=1}^{m} (1 - f(h_{S_{O_{kz}}}^{N(l)}))(1 - f(h_{S_{O_{zj}}}^{N(l)}))}$$

We have $f(\hat{h}_{S_{O_{ik}}}^{N(l)}) f(\hat{h}_{S_{O_{kj}}}^{N(l)}) = \frac{\sqrt[m]{\prod_{z=1}^{m} f(h_{S_{O_{iz}}}^{N(l)}) f(h_{S_{O_{zk}}}^{N(l)}) f(h_{S_{O_{kz}}}^{N(l)}) f(h_{S_{O_{zj}}}^{N(l)})}}{ab}$, and

$$\begin{aligned}
&f(\hat{h}_{S_{O_{ik}}}^{N(l)}) f(\hat{h}_{S_{O_{kj}}}^{N(l)}) + (1 - f(\hat{h}_{S_{O_{ik}}}^{N(l)}))(1 - f(\hat{h}_{S_{O_{kj}}}^{N(l)}))\\
&= \frac{\sqrt[m]{\prod_{z=1}^{m} f(h_{S_{O_{iz}}}^{N(l)}) f(h_{S_{O_{zk}}}^{N(l)}) f(h_{S_{O_{kz}}}^{N(l)}) f(h_{S_{O_{zj}}}^{N(l)})}}{ab} + \left(1 - \frac{\sqrt[m]{\prod_{z=1}^{m} f(h_{S_{O_{iz}}}^{N(l)}) f(h_{S_{O_{zk}}}^{N(l)})}}{a}\right)\left(1 - \frac{\sqrt[m]{\prod_{z=1}^{m} f(h_{S_{O_{kz}}}^{N(l)}) f(h_{S_{O_{zj}}}^{N(l)})}}{b}\right)\\
&= \frac{\sqrt[m]{\prod_{z=1}^{m} f(h_{S_{O_{iz}}}^{N(l)}) f(h_{S_{O_{zk}}}^{N(l)}) f(h_{S_{O_{kz}}}^{N(l)}) f(h_{S_{O_{zj}}}^{N(l)})} + \sqrt[m]{\prod_{z=1}^{m} (1 - f(h_{S_{O_{iz}}}^{N(l)}))(1 - f(h_{S_{O_{zk}}}^{N(l)}))(1 - f(h_{S_{O_{kz}}}^{N(l)}))(1 - f(h_{S_{O_{zj}}}^{N(l)}))}}{ab}
\end{aligned}$$

Because $\prod_{z=1}^{m} f(h_{S_{O_{zk}}}^{N(l)}) f(h_{S_{O_{kz}}}^{N(l)}) = \prod_{z=1}^{m} (1 - f(h_{S_{O_{zk}}}^{N(l)}))(1 - f(h_{S_{O_{kz}}}^{N(l)}))$, we have

$$\frac{f(\hat{h}^{N(l)}_{S_{O_{ik}}})f(\hat{h}^{N(l)}_{S_{O_{kj}}})}{f(\hat{h}^{N(l)}_{S_{O_{ik}}})f(\hat{h}^{N(l)}_{S_{O_{kj}}})+(1-f(\hat{h}^{N(l)}_{S_{O_{ik}}}))(1-f(\hat{h}^{N(l)}_{S_{O_{kj}}}))}$$

$$=\frac{\sqrt[m]{\prod_{z=1}^{m}f(h^{N(l)}_{S_{O_{iz}}})f(h^{N(l)}_{S_{O_{zk}}})f(h^{N(l)}_{S_{O_{kz}}})f(h^{N(l)}_{S_{O_{zj}}})}}{\sqrt[m]{\prod_{z=1}^{m}f(h^{N(l)}_{S_{O_{iz}}})f(h^{N(l)}_{S_{O_{zk}}})f(h^{N(l)}_{S_{O_{kz}}})f(h^{N(l)}_{S_{O_{zj}}})}+\sqrt[m]{\prod_{z=1}^{m}(1-f(h^{N(l)}_{S_{O_{iz}}}))(1-f(h^{N(l)}_{S_{O_{zk}}}))(1-f(h^{N(l)}_{S_{O_{kz}}}))(1-f(h^{N(l)}_{S_{O_{zj}}}))}}$$

$$=\frac{\sqrt[m]{\prod_{z=1}^{m}f(h^{N(l)}_{S_{O_{iz}}})f(h^{N(l)}_{S_{O_{zj}}})}}{\sqrt[m]{\prod_{z=1}^{m}f(h^{N(l)}_{S_{O_{iz}}})f(h^{N(l)}_{S_{O_{zj}}})}+\sqrt[m]{\prod_{z=1}^{m}(1-f(h^{N(l)}_{S_{O_{iz}}}))(1-f(h^{N(l)}_{S_{O_{zj}}}))}}=f(\hat{h}^{N(l)}_{S_{O_{ij}}})$$

Additionally, we have

$$f(\hat{h}^{N(l)}_{S_{O_{ij}}})+f(\hat{h}^{N(l)}_{S_{O_{ji}}})=\frac{\sqrt[m]{\prod_{k=1}^{m}f(h^{N(l)}_{S_{O_{ik}}})f(h^{N(l)}_{S_{O_{kj}}})}}{\sqrt[m]{\prod_{k=1}^{m}f(h^{N(l)}_{S_{O_{ik}}})f(h^{N(l)}_{S_{O_{kj}}})}+\sqrt[m]{\prod_{k=1}^{m}(1-f(h^{N(l)}_{S_{O_{ik}}}))(1-f(h^{N(l)}_{S_{O_{kj}}}))}}$$
$$+\frac{\sqrt[m]{\prod_{k=1}^{m}f(h^{N(l)}_{S_{O_{jk}}})f(h^{N(l)}_{S_{O_{ki}}})}}{\sqrt[m]{\prod_{k=1}^{m}f(h^{N(l)}_{S_{O_{jk}}})f(h^{N(l)}_{S_{O_{ki}}})}+\sqrt[m]{\prod_{k=1}^{m}(1-f(h^{N(l)}_{S_{O_{jk}}}))(1-f(h^{N(l)}_{S_{O_{ki}}}))}}$$
$$=\frac{\sqrt[m]{\prod_{k=1}^{m}f(h^{N(l)}_{S_{O_{ik}}})f(h^{N(l)}_{S_{O_{kj}}})}}{\sqrt[m]{\prod_{k=1}^{m}f(h^{N(l)}_{S_{O_{ik}}})f(h^{N(l)}_{S_{O_{kj}}})}+\sqrt[m]{\prod_{k=1}^{m}(1-f(h^{N(l)}_{S_{O_{ik}}}))(1-f(h^{N(l)}_{S_{O_{kj}}}))}}$$
$$+\frac{\sqrt[m]{\prod_{k=1}^{m}(1-f(h^{N(l)}_{S_{O_{ik}}}))(1-f(h^{N(l)}_{S_{O_{kj}}}))}}{\sqrt[m]{\prod_{k=1}^{m}(1-f(h^{N(l)}_{S_{O_{ik}}}))(1-f(h^{N(l)}_{S_{O_{kj}}}))}+\sqrt[m]{\prod_{k=1}^{m}f(h^{N(l)}_{S_{O_{ik}}})f(h^{N(l)}_{S_{O_{kj}}})}}=1$$

Based on Definition 1.10, $\hat{H}_{S_O}=(\hat{h}^{N}_{S_{O_{ij}}})_{m\times m}$ is a multiplicative consistent DHHFLPR with $\varepsilon(0\leq\varepsilon\leq 1)$. Which completes the proof of Theorem 3.6. ■

Remark 3.8 (Gou et al. 2020). Theorem 3.5 mainly discusses some properties of the multiplicative consistency of DHHFLPR and Theorem 3.6 gives a method to calculate the multiplicative consistent DHHFLPR. Therefore, based on the normalization method and Theorem 3.6, it is easy to obtain the multiplicative consistent DHHFLPR from an original DHHFLPR. And then we can check and repair the DHHFLPR with unacceptable consistency, which will be discussed in Sect. 3.2.2 and 3.2.3.

3.2.2 Multiplicative Consistency Index of DHHFLPRs

Under most circumstances, the consistency of the DHHFLPR provided by expert may be unacceptable. Thus, it is important to check and repair the DHHFLPR with unacceptable consistency. To do so, Gou et al. (2020) defined the multiplicative consistency index for DHHFLPR.

Firstly, Eq. (3.5) has proposed the distance measure between two DHHFLPRs. However, considering that the normalization method given in this subsection is different from that shown in Subsection 3.1.1, so the distance between two DHHFLPRs can be redefined as follows:

Definition 3.10 (Gou et al. 2020). Let $\tilde{H}^k_{S_O} = (h^k_{S_{O_{ij}}})_{m\times m} = (\{h^{k(l)}_{S_{O_{ij}}} | l = 1,2,\ldots, \#h^k_{S_{O_{ij}}}\})_{m\times m} (k = 1,2)$ be two DHHFLPRs, and $\tilde{H}^{kN}_{S_O} = (h^{kN}_{S_{O_{ij}}})_{m\times m} = (\{h^{kN(l)}_{S_{O_{ij}}} | l = 1,2,\ldots, \#h^{kN}_{S_{O_{ij}}}\})_{m\times m} (k = 1,2)$ be their corresponding NDHHFLPRs with $\varepsilon (0 \le \varepsilon \le 1)$, where $\#h^N_{S_{O_{ij}}} = \#h^{1N}_{S_{O_{ij}}} = \#h^{2N}_{S_{O_{ij}}}$. Then the distance between $\tilde{H}^1_{S_O}$ and $\tilde{H}^2_{S_O}$ is defined as:

$$d(\tilde{H}^1_{S_O}, \tilde{H}^2_{S_O}) = \left(\frac{2}{m(m-1)} \sum_{i<j}^{m} \left(\frac{1}{\#h^N_{S_{O_{ij}}}} \sum_{l=1}^{\#h^N_{S_{O_{ij}}}} \left(f(h^{1N(l)}_{S_{O_{ij}}}) - f(h^{2N(l)}_{S_{O_{ij}}}) \right)^2 \right) \right)^{1/2} \tag{3.40}$$

Obviously, $d(\tilde{H}^1_{S_O}, \tilde{H}^2_{S_O})$ satisfies:

(1) $0 \le d(\tilde{H}^1_{S_O}, \tilde{H}^2_{S_O}) \le 1$;
(2) $d(\tilde{H}^1_{S_O}, \tilde{H}^2_{S_O}) = 0$ if and only if $\tilde{H}^1_{S_O} = \tilde{H}^2_{S_O}$;
(3) $d(\tilde{H}^1_{S_O}, \tilde{H}^2_{S_O}) = d(\tilde{H}^2_{S_O}, \tilde{H}^1_{S_O})$.

Based on the distance measure of DHHFLPRs, the multiplicative consistency index of DHHFLPR is defined as follows.

Definition 3.11 (Gou et al. 2020). Given a DHHFLPR $\tilde{H}_{S_O} = (h_{S_{O_{ij}}})_{m\times m} = (\{h^{(l)}_{S_{O_{ij}}} | l = 1,2,\ldots, \#h_{S_{O_{ij}}}\})_{m\times m}$, its NDHHFLPR $\tilde{H}^N_{S_O} = (h^N_{S_{O_{ij}}})_{m\times m} = (\{h^{N(l)}_{S_{O_{ij}}} | l = 1,2,\ldots, L\})_{m\times m}$ with $\varepsilon (0 \le \varepsilon \le 1)$ $(L = max\{\#h_{S_{O_{ij}}} | i,j = 1,2,\ldots,m; i<j\}$, and its multiplicative consistent DHHFLPR $\hat{H}^N_{S_O} = (\hat{h}^N_{S_{O_{ij}}})_{m\times m} = (\{\hat{h}^{N(l)}_{S_{O_{ij}}} | l = 1,2,\ldots, L\})_{m\times m}$ with ε. Then, the consistency index (CI) of $\tilde{H}_{S_O}$ is obtained by

$$\begin{aligned} CI(\tilde{H}_{S_O}) &= 1 - d(\tilde{H}^N_{S_O}, \hat{H}^N_{S_O}) \\ &= 1 - \left(\frac{2}{m(m-1)} \sum_{i<j}^{m} \left(\frac{1}{L} \sum_{l=1}^{L} \left(f(h^{N(l)}_{S_{O_{ij}}}) - f(\hat{h}^{N(l)}_{S_{O_{ij}}}) \right)^2 \right) \right)^{1/2} \end{aligned} \tag{3.41}$$

$CI(\tilde{H}_{S_O})$ can be taken as a similarity measure between $\tilde{H}_{S_O}$ and its multiplicative consistent DHHFLPR $\hat{H}_{S_O}$. Therefore, the larger the value of $CI(\tilde{H}_{S_O})$ is, the more consistent $\tilde{H}_{S_O}$ will be. Especially, $CI(\tilde{H}_{S_O}) = 1$ if and only if $\tilde{H}_{S_O}$ is a complete multiplicative consistent DHHFLPR.

From Definition 3.11, it is obvious that the value of $CI(\tilde{H}_{S_O})$ is decided by $\varepsilon(0 \le \varepsilon \le 1)$, which reflects the expert's risk preferences. Considering that the consistent DHHFLPR will be helpful to obtain meaningful results, a nonlinear optimization model is developed to obtain the optimal $CI(\tilde{H}_{S_O})$ and the corresponding optimal parameter $\varepsilon(0 \le \varepsilon \le 1)$:

$$\max CI(\tilde{H}_{S_O}) = 1 - d(\tilde{H}_{S_O}^N, \hat{H}_{S_O}^N)$$
$$s.t. \begin{cases} d(\tilde{H}_{S_O}^N, \hat{H}_{S_O}^N) = \left(\frac{2}{m(m-1)} \sum_{i<j}^{m} \left(\frac{1}{L} \sum_{l=1}^{L} \left(f(h_{S_{O_{ij}}}^{N(l)}) - f(\hat{h}_{S_{O_{ij}}}^{N(l)}) \right)^2 \right) \right)^{1/2} \\ 0 \le \varepsilon \le 1 \end{cases}$$

Solving this model, we can obtain the optimal parameter $\varepsilon(0 \le \varepsilon \le 1)$, the unique multiplicative consistent DHHFLPR $\hat{H}_{S_O}$ and the unique consistency index $CI(\tilde{H}_{S_O})$ with the highest consistency level.

Example 3.6 (Gou et al. 2020). Let $S_O = \{s_{t<o_k>} | t = -4, \ldots, 4;\ k = -4, \ldots, 4\}$ be a DHLTS, $\tilde{H}_{S_O} = (h_{S_{O_{ij}}})_{4 \times 4}$ be a DHHFLPR as:

$$\tilde{H}_{S_O} = \begin{pmatrix} \{s_{0<o_0>}\} & \{s_{-1<o_1>}, s_0, s_{1<o_2>}\} & \{s_{0<o_{-2}>}, s_{1<o_2>}\} & \{s_{3<o_2>}\} \\ \{s_{1<o_{-1}>}, s_0, s_{-1<o_{-2}>}\} & \{s_{0<o_0>}\} & \{s_{2<o_0>}\} & \{s_{-1<o_{-2}>}\} \\ \{s_{0<o_2>}, s_{-1<o_{-2}>}\} & \{s_{-2<o_0>}\} & \{s_{0<o_0>}\} & \{s_{-2<o_0>}, s_{-1<o_1>}, s_{0<o_1>}\} \\ \{s_{-3<o_{-2}>}\} & \{s_{1<o_2>}\} & \{s_{2<o_0>}, s_{1<o_{-1}>}, s_{0<o_{-1}>}\} & \{s_{0<o_0>}\} \end{pmatrix}$$

Based on the normalization method with $\varepsilon = 1$, we obtain $CI(\tilde{H}_{S_O}) = 0.8127$. However, the consistent index of $\tilde{H}_{S_O}$ may be different based on different optimized parameter. Therefore, based on the above model, the variation of the distances between $\tilde{H}_{S_O}$ and $\hat{H}_{S_O}$ associated to the parameter $\varepsilon(0 \le \varepsilon \le 1)$ can be shown in Fig. 3.3. Then we obtain the maximum consistent index of $\tilde{H}_{S_O}$ as $CI(\tilde{H}_{S_O}) = 0.8127$ with $\varepsilon = 1$.

As the calculation results obtained in Example 3.6, it is common that some DHHFLPRs may be of unacceptable multiplicative consistencies. Then, an acceptably consistent DHHFLPR is developed:

Definition 3.12 (Gou et al. 2020). Let $\tilde{H}_{S_O} = (h_{S_{O_{ij}}})_{m \times m}$ be a DHHFLPR. Given a threshold value $\overline{CI}$, if

$$CI(\tilde{H}_{S_O}) \ge \overline{CI} \tag{3.42}$$

then we call $\tilde{H}_{S_O}$ the DHHFLPR with acceptable multiplicative consistency.

The value of $\overline{CI}$ can be determined based on the DHLTS and the number of the alternatives in GDM. Let S_O be a DHLTS, and $T = 2\tau + 1$. Gou et al. (2019) proposed the admissible bounds based on the values of T and m for checking the consistency of $\overline{CI}$. However, the threshold values developed by Gou et al. (2019)

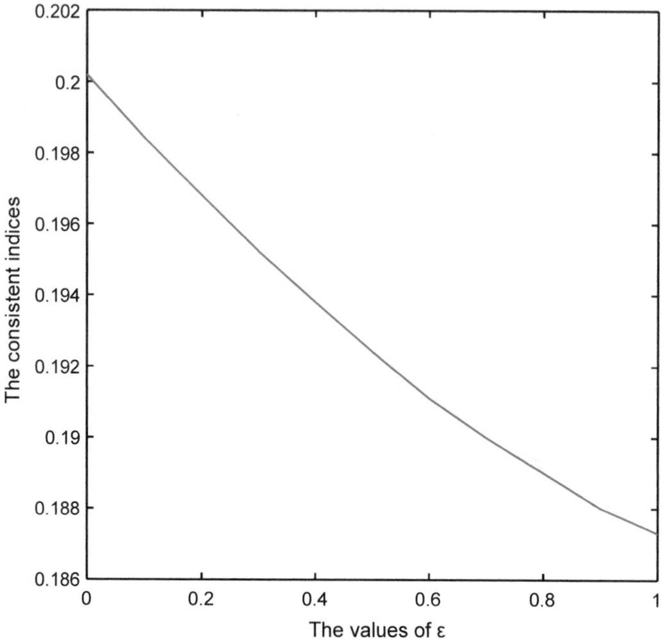

Fig. 3.3 The distances between $\tilde{H}_{S_O}$ and $\hat{H}_{S_O}$ with respect to ε

are based on the distance measure. So, we need to transform them into reasonable values with similarity measure and they are listed in Table 3.14.

In Example 3.6, we obtain that $CI(\tilde{H}_{S_O}) = 0.8127$. Considering that $T = 9$ and $m = 4$, we can get $CI(\tilde{H}_{S_O}) = 0.8127 < \overline{CI}(\tilde{H}_{S_O}) = 0.8485$. Therefore, the DHHFLPR discussed in Example 3.6 is a DHHFLPR with unacceptable multiplicative consistency.

3.2.3 *Feedback Mechanism-Based Consistency Repairing Method*

As we mentioned above, it is common that a DHHFLPR $\tilde{H}_{S_O} = (h_{S_{O_{ij}}})_{m \times m}$ is of unacceptably consistency, i.e., $CI(\tilde{H}_{S_O}) < \overline{CI}(\tilde{H}_{S_O})$. In this case, we need to repair $\tilde{H}_{S_O}$ until it reaches the given threshold. In Sect. 3.1.3, Gou et al. (2020) proposed two consistency repairing methods to improve the DHHFLPR with unacceptable additive consistency. Considering that the automatic optimization method may lead to unreasonable result because it does not adequately consider the opinions of experts, and the feedback mechanism can avoid this gap very well. Therefore, this subsection develops a feedback mechanism-based consistency repairing method,

Table 3.14 The values of consistency thresholds $\overline{CI}(\tilde{H}_{S_O})$ for different values of m and T

	$m = 3$	$m = 4$	$m = 5$	$m = 6$	$m = 7$	$m = 8$
$T = 5$	0.8793	0.6970	0.6512	0.6226	0.6030	0.5888
$T = 9$	0.8897	0.8485	0.8256	0.8113	0.8015	0.7944
$T = 17$	0.9448	0.9242	0.9128	0.9056	0.9007	0.8972

which can feedback suggestions to experts and help them to improve their preferences.

Then, the feedback mechanism-based consistency repairing method is developed as follows:

Algorithm 3.4 (Gou et al. 2020). Feedback mechanism-based consistency repairing method

Step 1. Let $(\tilde{H}_{S_O})^{(\mathbb{Z})} = ((h_{S_{O_{ij}}})_{m\times m})^{(\mathbb{Z})}$ ($\mathbb{Z} = 0$, $(\tilde{H}_{S_O})^{(\mathbb{Z})}$ express the $\mathbb{Z}$-th power of $\tilde{H}_{S_O}$, indicating the number of iterations) be a DHHFLPR. Based on Definition 3.8 and Theorem 3.6, calculate the NDHHFLPR $(\tilde{H}^N_{S_O})^{(\mathbb{Z})} = ((h^N_{S_{O_{ij}}})_{m\times m})^{(\mathbb{Z})}$ and the multiplicative consistent DHHFLPR $(\hat{H}^N_{S_O})^{(\mathbb{Z})} = ((\hat{h}^N_{S_{O_{ij}}})_{m\times m})^{(\mathbb{Z})}$, respectively.
Step 2. Obtain consistency threshold $\overline{CI}(\tilde{H}_{S_O})$ based on Table 3.14.
Step 3. Calculate consistency index $CI((\tilde{H}_{S_O})^{(\mathbb{Z})}) = 1 - d((\tilde{H}^N_{S_O})^{(\mathbb{Z})}, (\hat{H}^N_{S_O})^{(\mathbb{Z})})$ based on Eq. (3.41). If $CI((\tilde{H}_{S_O})^{(\mathbb{Z})}) \geq \overline{CI}(\tilde{H}_{S_O})$, then go to Step 6; otherwise, go to Step 4.
Step 4. Construct the interval-valued DHHFLPR $\tilde{H}_{\bar{S}_{\bar{O}}} = ((h_{\bar{S}_{\bar{O}_{ij}}})_{m\times m})^{(\mathbb{Z})}$ $= ((\{h^{(l)}_{\bar{S}_{\bar{O}_{ij}}} | l = 1, 2, \ldots, \#h_{\bar{S}_{\bar{O}_{ij}}}\})_{m\times m})^{(\mathbb{Z})}$, where $(h^{(l)}_{\bar{S}_{\bar{O}_{ij}}})^{(\mathbb{Z})} = [\min\{(h^{(l)}_{S_{O_{ij}}})^{(\mathbb{Z})}, (\hat{h}^{(l)}_{S_{O_{ij}}})^{(\mathbb{Z})}\}, \max\{(h^{(l)}_{S_{O_{ij}}})^{(\mathbb{Z})}, (\hat{h}^{(l)}_{S_{O_{ij}}})^{(\mathbb{Z})}\}]$. Take $\tilde{H}_{\bar{S}_{\bar{O}}}$ as a reference and feedback it to the expert, then the expert provides preference information again, and then go to Step 5.
Step 5. Collect all new preference information and then form the repaired DHHFLPR $(\tilde{H}^N_{S_O})^{(\mathbb{Z}+1)} = ((h^N_{S_{O_{ij}}})_{m\times m})^{(\mathbb{Z}+1)}$. Let $\mathbb{Z} = \mathbb{Z}+1$. Go back to Step 3.
Step 6. Let $*\tilde{H}_{S_O} = (\tilde{H}^N_{S_O})^{(\mathbb{Z})}$. Output the repaired normalized DHHFLPR $*\tilde{H}_{S_O}$.
Step 7. End.

Considering that the consistency index of the repaired DHHFLPR should be superior to the original DHHFLPR, then an important property about the repaired DHHFLPR is obtained.

Theorem 3.7 (Gou et al. 2020). Given a DHHFLPR $\tilde{H}_{S_O}$, $*\tilde{H}_{S_O}$ is the repaired DHHFLPR obtained by Algorithm 3.4. Then, $CI(*\tilde{H}_{S_O}) \geq CI(\tilde{H}_{S_O})$.

Proof Since the repaired DHHFLPR $(\tilde{H}_{S_O})^{(\mathbb{Z}+1)} = ((h_{S_{O_{ij}}})_{m\times m})^{(\mathbb{Z}+1)}$ is constructed by the experts' preferences with an interval-valued DHHFLPR, $\tilde{H}_{\bar{S}_{\bar{O}}} = ((h_{\bar{S}_{\bar{O}_{ij}}})_{m\times m})^{(\mathbb{Z})}$. Then we have

$$(h_{S_{O_{ij}}}^{(l)})^{(\mathbb{Z}+1)} \in [\min\{(h_{S_{O_{ij}}}^{(l)})^{(\mathbb{Z})}, (\hat{h}_{S_{O_{ij}}}^{N(l)})^{(\mathbb{Z})}\}, \max\{(h_{S_{O_{ij}}}^{(l)})^{(\mathbb{Z})}, (\hat{h}_{S_{O_{ij}}}^{N(l)})^{(\mathbb{Z})}\}] \tag{3.43}$$

Thus,

$$|f((h_{S_{O_{ij}}}^{(l)})^{(\mathbb{Z}+1)}) - f((\hat{h}_{S_{O_{ij}}}^{(l)})^{(\mathbb{Z})})| \le |f((h_{S_{O_{ij}}}^{(l)})^{(\mathbb{Z})}) - f((\hat{h}_{S_{O_{ij}}}^{(l)})^{(\mathbb{Z})})| \tag{3.44}$$

Based on Eqs. (3.43) and (3.44), and Eq. (3.44), we have

$$\begin{aligned} & 1 - \left(\frac{2}{m(m-1)}\sum_{i<j}^{m}\left(\frac{1}{L}\sum_{l=1}^{L}\left(f((h_{S_{O_{ij}}}^{(l)})^{(\mathbb{Z}+1)}) - f((\hat{h}_{S_{O_{ij}}}^{(l)})^{(\mathbb{Z})})\right)^2\right)\right)^{1/2} \\ & \ge 1 - \left(\frac{2}{m(m-1)}\sum_{i<j}^{m}\left(\frac{1}{L}\sum_{l=1}^{L}\left(f((h_{S_{O_{ij}}}^{(l)})^{(\mathbb{Z})}) - f((\hat{h}_{S_{O_{ij}}}^{(l)})^{(\mathbb{Z})})\right)^2\right)\right)^{1/2} \end{aligned} \tag{3.45}$$

Then, we get $CI((\tilde{H}_{S_O})^{(\mathbb{Z}+1)}) \ge CI((\tilde{H}_{S_O})^{(\mathbb{Z})})$, i.e., $CI(*\tilde{H}_{S_O}) \ge CI(\tilde{H}_{S_O})$. This completes the proof of Theorem 3.7. ■

3.2.4 GDM Method Based on Multiplicative Consistencies of DHHFLPRs

To clarify the proposed multiplicative consistency repairing method, Gou et al. (2020) developed a GDM method based on multiplicative consistency of DHHFLPRs.

Algorithm 3.5 (Gou et al. 2020). GDM method based on multiplicative consistencies of DHHFLPRs

Step 1. Collect the linguistic preferences of experts and establish their individual DHHFLPRs $\tilde{H}_{S_O}^r = (h_{S_{O_{ij}}}^r)_{m\times m}(r = 1, 2, \ldots, R)$. Go to Step 2.

Step 2. Calculate the normalized DHHFLPRs $\tilde{H}_{S_O}^{rN} = (h_{S_{O_{ij}}}^{rN})_{m\times m}$ $(r = 1, 2, \ldots, R)$ and the consistent normalized DHHFLPR $\hat{H}_{S_O}^{rN} = (\hat{h}_{S_{O_{ij}}}^{rN})_{m\times m}$ $(r = 1, 2, \ldots, R)$, respectively.

Step 3. Utilize Algorithm 3.4 to ensure that each normalized DHHFLPR is of acceptable consistency.

Step 4. It is same as the Step 4 in Algorithm 3.3.

Step 5. It is same as the Step 5 in Algorithm 3.3.

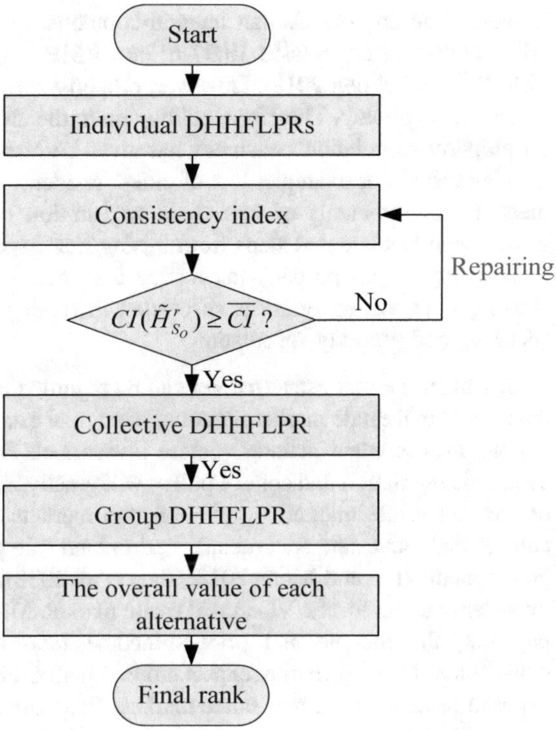

Fig. 3.4 The GDM method based on multiplicative consistencies of DHHFLPRs

Step 6. It is same as the Step 6 in Algorithm 3.3.
Step 7. It is same as the Step 7 in Algorithm 3.3.
Step 8. End (Fig. 3.4).

3.2.5 Application of the GDM Method Based on Multiplicative Consistency of DHHFLPRs

In this subsection, we apply the GDM method based on multiplicative consistencies of DHHFLPRs to a practical GDM problem which is to find a VC project about real estate market.

Example 3.7 (Gou et al. 2020). In recent years, especially since 2016, with the implementation of talent introduction policy and real estate destocking, housing prices have exploded in many Chinese cities such as Nanjing, Chengdu, Xi'an, etc., as well as more and more people believe that property speculation is a huge business opportunity. Meanwhile, lots of real estate developers have sprung up such as Evergrande Real Estate Group, Country Garden Holdings, China Vanke, Greenland, etc. As we know, the real estate market is not only a key part of the

Chinese economy but also an integral component of China's financial system. In 2017, housing sales totaled 13.37 trillion RMB, equivalent to 16.4% of China's GDP (Liu and Xiong 2018). However, property speculation has had a huge impact on the real economy. Therefore, holding on to the idea that "house is for living, not for property speculation", each city has given corresponding policies to regulate the housing price. For example, lots of cities' residents cannot buy a house until they meet the requirements of household registration or social security, the central government has adopted steps from raising mortgage rates and down-payments to making buyers pay property tax for the first time. Now, China's economy is also slowing after the government succeeded in reining in credit growth, investment spending and property speculation.

In China, the real estate market can be regarded as a typical VC. In the face of the current real estate market situation, many real estate developers need to consider various factors when making venture investment. For instance, the government's economic regulation and control policy will greatly affect the development direction of the real estate market. In addition, the markets for residential properties and commercial real estate are crucially tied to land sale policies and strategies of local governments (Liu and Xiong 2018; Chen et al. 2018), so the land prices are also the important factors for the VC in real estate market. Moreover, the households' down payment, the income and price-to-income ratio of mortgage borrowers, and households' housing market expectations. Finally, Firms in China are also heavily exposed to risks of the real estate market. There are at least two key channels that expose firms to the real estate market. One is that the real estate assets are the most widely used collateral for firms to borrow from banks, and the other one is that there is also another speculative channel through which firms actively seek real estate exposure during China's ongoing real estate boom (Liu and Xiong 2018; Chen et al. 2018).

Based on these factors, suppose that a firm wants to find a VC project about real estate market in a city of China, and four famous cities are the alternatives as Chengdu (A_1), Nanjing (A_2), Xi'an (A_3) and Zhengzhou (A_4). To do so, four experts $\{e^1, e^2, e^3, e^4\}$ with the same importance are invited to form a group and to provide their evaluations. Then, the DHHFLPR $\tilde{H}^r_{S_O} = (h^r_{S_{O_{ij}}})_{4\times 4}$ of each expert can be established based on a DHLTS S_O, where the first and second LTSs in S_O are respectively denoted as:

$$
\begin{aligned}
S = \{ & s_{-4} = \textit{extremely bad}, s_{-3} = \textit{very bad}, s_{-2} = \textit{bad}, s_{-1} = \textit{slightly bad}, \\
& s_0 = \textit{equal}, s_1 = \textit{slightly good}, s_2 = \textit{good}, s_3 = \textit{very good}, s_4 = \textit{extremely good}\}
\end{aligned}
$$

$$O = \{o_{-4} = \textit{far from}, o_{-3} = \textit{scarcely}, o_{-2} = \textit{only a littlle}, o_{-1} = \textit{a little}, o_0 = \textit{just right}, o_1 = \textit{much}, o_2 = \textit{very much}, o_3 = \textit{extremely much}, o_4 = \textit{entirely}\}$$

$$\tilde{H}^1_{S_O} = \begin{pmatrix} \{s_{0<o_0>}\} & \{s_{-1<o_1>}, s_{0<o_1>}\} & \{s_{1<o_{-2}>}, s_{1<o_2>}\} & \{s_{0<o_{-1}>}, s_{0<o_1>}\} \\ \{s_{1<o_{-1}>}, s_{0<o_{-1}>}\} & \{s_{0<o_0>}\} & \{s_{0<o_0>}\} & \{s_{0<o_1>}\} \\ \{s_{-1<o_2>}, s_{-1<o_{-2}>}\} & \{s_{0<o_0>}\} & \{s_{0<o_0>}\} & \{s_{3<o_{-3}>}\} \\ \{s_{0<o_1>}, s_{0<o_{-1}>}\} & \{s_{0<o_{-1}>}\} & \{s_{-3<o_3>}\} & \{s_{0<o_0>}\} \end{pmatrix}$$

$$\tilde{H}^2_{S_O} = \begin{pmatrix} \{s_{0<o_0>}\} & \{s_{0<o_{-2}>}, s_{0<o_1>}\} & \{s_{1<o_{-2}>}, s_{1<o_1>}\} & \{s_{0<o_{-2}>}, s_{1<o_1>}\} \\ \{s_{0<o_2>}, s_{0<o_{-1}>}\} & \{s_{0<o_0>}\} & \{s_{-1<o_3>}, s_{0<o_0>}\} & \{s_{0<o_1>}, s_{1<o_{-2}>}\} \\ \{s_{-1<o_2>}, s_{-1<o_{-1}>}\} & \{s_{1<o_{-3}>}, s_{0<o_0>}\} & \{s_{0<o_0>}\} & \{s_{3<o_{-3}>}\} \\ \{s_{0<o_2>}, s_{-1<o_{-1}>}\} & \{s_{0<o_{-1}>}, s_{-1<o_2>}\} & \{s_{-3<o_3>}\} & \{s_{0<o_0>}\} \end{pmatrix}$$

$$\tilde{H}^3_{S_O} = \begin{pmatrix} \{s_{0<o_0>}\} & \{s_{-1<o_{-2}>}, s_{0<o_3>}\} & \{s_{1<o_{-2}>}, s_{1<o_1>}\} & \{s_{0<o_{-1}>}, s_{0<o_0>}\} \\ \{s_{1<o_2>}, s_{0<o_{-3}>}\} & \{s_{0<o_0>}\} & \{s_{0<o_{-2}>}, s_{1<o_{-1}>}\} & \{s_{0<o_1>}\} \\ \{s_{-1<o_2>}, s_{-1<o_{-1}>}\} & \{s_{0<o_2>}, s_{-1<o_1>}\} & \{s_{0<o_0>}\} & \{s_{3<o_1>}, s_{3<o_2>}\} \\ \{s_{0<o_1>}, s_{0<o_0>}\} & \{s_{0<o_{-1}>}\} & \{s_{3<o_{-1}>}, s_{3<o_{-2}>}\} & \{s_{0<o_0>}\} \end{pmatrix}$$

$$\tilde{H}^4_{S_O} = \begin{pmatrix} \{s_{0<o_0>}\} & \{s_{-1<o_1>}, s_{0<o_1>}\} & \{s_{1<o_{-2}>}, s_{2<o_{-2}>}\} & \{s_{0<o_{-1}>}, s_{0<o_3>}\} \\ \{s_{1<o_{-1}>}, s_{0<o_{-1}>}\} & \{s_{0<o_0>}\} & \{s_{0<o_0>}\} & \{s_{0<o_0>}, s_{0<o_1>}\} \\ \{s_{-1<o_2>}, s_{-2<o_2>}\} & \{s_{0<o_0>}\} & \{s_{0<o_0>}\} & \{s_{3<o_1>}, s_{3<o_2>}\} \\ \{s_{0<o_1>}, s_{0<o_{-3}>}\} & \{s_{0<o_0>}, s_{0<o_{-1}>}\} & \{s_{-3<o_{-1}>}, s_{-3<o_{-2}>}\} & \{s_{0<o_0>}\} \end{pmatrix}$$

Clearly, this is a GDM problem under double hierarchy hesitant fuzzy linguistic environment. Based on Algorithm 3.5, this GDM problem can be solved.

Since Step 1 has been given above, we start to handle this case from Step 2.

Step 2. Calculate the consistency indices $\tilde{H}^r_{S_O} = (h^r_{S_{O_{ij}}})_{4\times 4} (r = 1, 2, 3, 4)$ of all DHHFLPRs and show them in Table 3.15.

Based on Table 3.15, the threshold value is $\overline{CI} = 0.8485$. Because $CI(\tilde{H}^{(3)}_{S_O}) < \overline{CI}$, then it is necessary to repair the unacceptably consistent DHHFLPR $\tilde{H}^3_{S_O}$.

Step 3. Utilize Algorithm 3.4 to repair $\tilde{H}^3_{S_O}$. Firstly, we obtain the interval-valued DHHFLPR $\tilde{H}^3_{\bar{S}_{\bar{O}}}$ as:

Table 3.15 The consistency index of each DHHFLPR $CI(\tilde{H}^r_{S_O})$

	$\tilde{H}^1_{S_O}$	$\tilde{H}^2_{S_O}$	$\tilde{H}^3_{S_O}$	$\tilde{H}^4_{S_O}$
CI	0.8967	0.9003	0.8393	0.8502

$$\begin{pmatrix} \{s_{0<o_0>}\} & \{[s_{-1<o_{-2}>}, s_{-1<o_{1.37}>}], [s_{0<o_{1.79}>}, s_{0<o_3>}]\} \\ \{[s_{1<o_{-1.37}>}, s_{1<o_2>}], [s_{0<o_{-3}>}, s_{0<o_{-1.79}>}]\} & \{s_{0<o_0>}\} \\ \{[s_{-1<o_2>}, s_{1<o_{1.61}>}], [s_{-1<o_{-1}>}, s_{0<o_{1.31}>}]\} & \{[s_{0<o_2>}, s_{0<o_{3.17}>}], [s_{-1<o_1>}, s_{0<o_{3.07}>}]\} \\ \{[s_{0<o_{-3.17}>}, s_{0<o_1>}], [s_{-1<o_{-3.17}>}, s_{0<o_0>}]\} & \{[s_{-1<o_{-1.61}>}, s_{0<o_{-1}>}], [s_{-1<o_{-1.66}>}, s_{0<o_{-1}>}]\} \\ \{[s_{-1<o_{-1.61}>}, s_{1<o_{-2}>}], [s_{0<o_{-1.31}>}, s_{1<o_1>}]\} & \{[s_{0<o_{-1}>}, s_{0<o_{3.17}>}], [s_{0<o_0>}, s_{1<o_{3.17}>}]\} \\ \{[s_{0<o_{-3.17}>}, s_{0<o_{-2}>}], [s_{0<o_{-3.07}>}, s_{1<o_{-1}>}]\} & \{[s_{0<o_1>}, s_{1<o_{1.61}>}], [s_{0<o_1>}, s_{1<o_{1.66}>}]\} \\ \{s_{0<o_0>}\} & \{[s_{2<o_{0.21}>}, s_{3<o_1>}], [s_{2<o_{0.18}>}, s_{3<o_2>}]\} \\ \{[s_{-3<o_{-1}>}, s_{-2<o_{-0.21}>}], [s_{-3<o_{-2}>}, s_{-2<o_{-0.18}>}]\} & \{s_{0<o_0>}\} \end{pmatrix}$$

Feedback $\tilde{H}^3_{\bar{S}_{\bar{O}}}$ to expert e^3 and ask him to provide preference again based on the suggestions. Then we obtain the new DHHFLPR $\tilde{H}'^3_{S_O}$ as:

$$\tilde{H}'^3_{S_O} = \begin{pmatrix} \{s_{0<o_0>}\} & \{s_{-1<o_0>}, s_{0<o_2>}\} & \{s_{-1<o_2>}, s_{0<o_2>}\} & \{s_{0<o_1>}, s_{0<o_2>}\} \\ \{s_{1<o_0>}, s_{0<o_{-2}>}\} & \{s_{0<o_0>}\} & \{s_{0<o_{-3}>}, s_{0<o_1>}\} & \{s_{0<o_3>}, s_{1<o_1>}\} \\ \{s_{1<o_{-2}>}, s_{0<o_{-2}>}\} & \{s_{0<o_3>}, s_{0<o_{-1}>}\} & \{s_{0<o_0>}\} & \{s_{3<o_{-1}>}, s_{3<o_1>}\} \\ \{s_{0<o_{-1}>}, s_{0<o_{-2}>}\} & \{s_{0<o_{-3}>}, s_{-1<o_{-1}>}\} & \{s_{-3<o_1>}, s_{-3<o_{-1}>}\} & \{s_{0<o_0>}\} \end{pmatrix}$$

Go back to Step 2, and we obtain $CI(\tilde{H}'^3_{S_O}) = 0.9029 > 0.8485$.

Step 4. Suppose that the weight vector of the experts is $w = (0.25, 0.25, 0.25, 0.25)^T$.

Step 5. Aggregate all of the normalized DHHFLPRs into a synthetical normalized DHHFLPR Based on Eq. (3.24), denoted as $\overset{\frown}{H}{}^N_{S_O} = (\overset{\frown}{h}{}^N_{S_{O_{ij}}})_{m \times m}$.

$$\overset{\frown}{H}{}^N_{S_O} = \begin{pmatrix} \{s_{0<o_0>}\} & \{s_{-0.75<o_{-0.11}>}, s_{0<o_{1.66}>}\} & \{s_{0.75<o_{-1.34}>}, s_{1.25<o_{-0.43}>}\} & \{s_{0<o_{-0.59}>}, s_{0<o_{2.12}>}\} \\ \{s_{0.75<o_{0.11}>}, s_{0<o_{-1.66}>}\} & \{s_{0<o_0>}\} & \{s_{-0.25<o_{-0.64}>}, s_{0<o_{0.48}>}\} & \{s_{0<o_{1.28}>}, s_{0.5<o_{-0.32}>}\} \\ \{s_{-0.75<o_{1.34}>}, s_{-1.25<o_{0.43}>}\} & \{s_{0.25<o_{0.64}>}, s_{0<o_{-0.48}>}\} & \{s_{0<o_0>}\} & \{s_{3<o_{-0.62}>}, s_{3<o_{0.06}>}\} \\ \{s_{0<o_{0.59}>}, s_{0<o_{-2.12}>}\} & \{s_{0<o_{-1.28}>}, s_{-0.5<o_{0.32}>}\} & \{s_{-3<o_{0.62}>}, s_{-3<o_{-0.06}>}\} & \{s_{0<o_0>}\} \end{pmatrix}$$

Step 6. Calculate the synthetical value of each alternative based on Eq. (1.15): $SV = \{2.1047, 2.0625, 2.2859, 1.5469\}$.

Step 7. Rank all the alternatives based on the values of SV: $A_3 \succ A_1 \succ A_2 \succ A_4$.

Next, Based on the method of Gou et al. (2019), the consistency index of each expert by the additive consistency-based method can be obtained and shown in Table 3.16.

Table 3.16 The additive consistency index of each DHHFLPR $CI(\tilde{H}^{(q)}_{S_O})$

	$\tilde{H}^{(1)}_{S_O}$	$\tilde{H}^{(2)}_{S_O}$	$\tilde{H}^{(3)}_{S_O}$	$\tilde{H}^{(4)}_{S_O}$
CI	0.8975	0.9011	0.8618	0.8619

In Table 3.16, the preference of each expert is the DHHFLPR with acceptable consistency. Therefore, it is not necessary to make any adjustment.

Step 3. Calculate the group DHHFLPR and the synthetical value of each alternative is obtained as $\{2.0860, 2.0469, 2.25, 1.6172\}$. Then, the ranking of all alternatives is $A_3 \succ A_1 \succ A_2 \succ A_4$. Similarly, Xi'an is also the optimal alternative to invest a VC project about real estate market.

All the above two methods show that the ranking of all alternatives is $A_3 \succ A_1 \succ A_2 \succ A_4$, which verifies the effectiveness of the proposed method. However, the synthetical value of each alternative, and the process of consistency checking and repairing are different.

(1) The comparative analyses of different consistency properties

In this subsection, we check and repair the consistencies of all DHHFLPRs on the basis of multiplicative consistency and additive consistency, respectively. However, the consistency index of each DHHFLPR obtained by different consistency checking method is different. Especially, by the multiplicative consistency, we can find that $\tilde{H}_{S_O}^{(3)}$ is a DHHFLPR with unacceptable multiplicative consistency, but all the DHHFLPRs are of acceptable additive consistency. Therefore, the multiplicative consistency property is much more precise to measure the consistency of DHHFLPR in some degree.

Additionally, we can further use the MATLAB Drawing toolbar to produce the "Figure of area" to give a visible description of the unacceptable consistent DHHFLPRs, the acceptably consistent DHHFLPRs, the multiplicative consistent DHHFLPRs, and the additive consistent DHHFLPRs, which are shown in Figs. 3.5, 3.6, 3.7 and 3.8. Clearly, both the multiplicative consistent and additive consistent DHHFLPRs are more regular than the original DHHFLPRs. We can also find that the every pair acceptably multiplicative consistent DHHFLPR and the

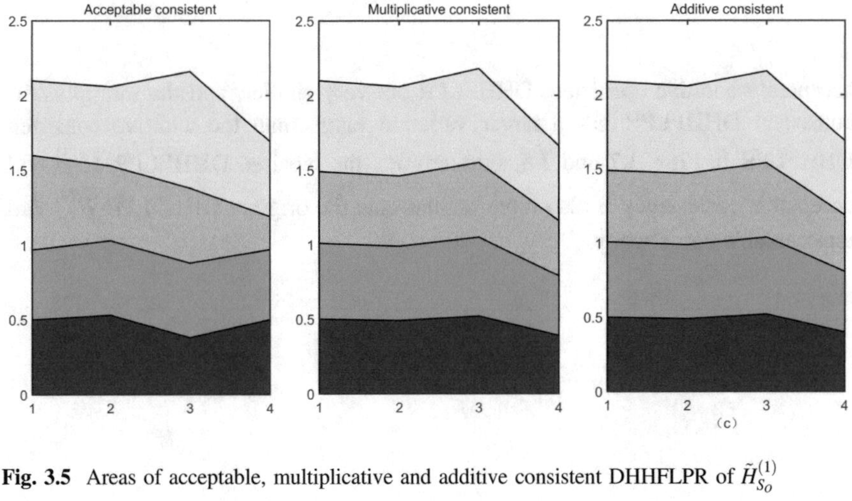

Fig. 3.5 Areas of acceptable, multiplicative and additive consistent DHHFLPR of $\tilde{H}_{S_O}^{(1)}$

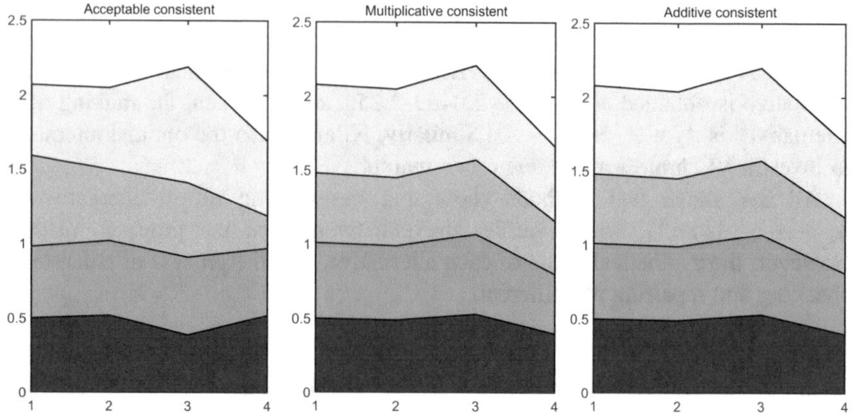

Fig. 3.6 Areas of acceptable, multiplicative and additive consistent DHHFLPR of $\tilde{H}_{S_O}^{(2)}$

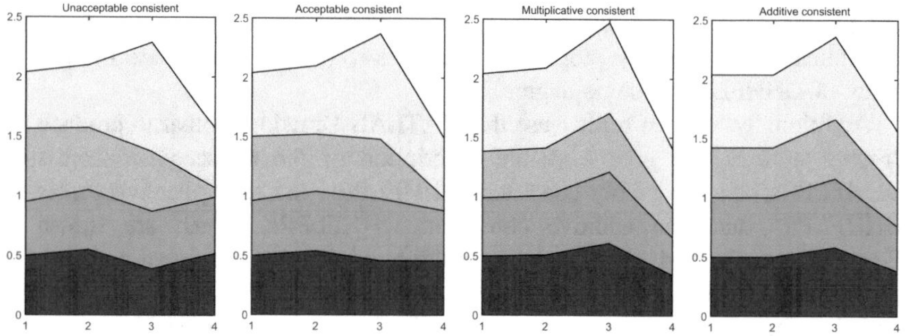

Fig. 3.7 Areas of unacceptable, acceptable, multiplicative and additive consistent DHHFLPR of $\tilde{H}_{S_O}^{(3)}$

acceptably additive consistent DHHFLPR are very similar, and the multiplicative consistent DHHFLPR has a larger variation range than the additive consistent DHHFLPR in Figs. 3.7 and 3.8. Furthermore, the repaired DHHFLPR $\tilde{H}_{S_O}^{\prime(3)}$ with acceptable consistency is also more regular than the original DHHFLPR $\tilde{H}_{S_O}^{(3)}$ with unacceptable consistency.

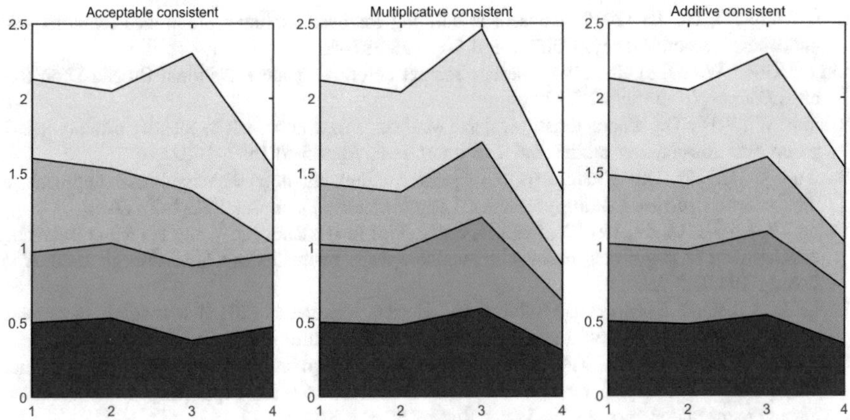

Fig. 3.8 Areas of acceptable, multiplicative and additive consistent DHHFLPR of $\tilde{H}_{S_O}^{(4)}$

References

Alonso S, Herrera-Viedma E, Chiclana F, Herrera F (2010) A web based consensus support system for group decision making problems and incomplete preferences. Inf Sci 180:4477–4495

Baets BD, Meyer HD, Schuymer BD, Jenei S (2006) Cyclic evaluation of transitivity of reciprocal relations. Soc Choice Welf 26:217–238

Chen T, Liu L, Xiong W, Zhou LA (2018) Real estate boom and misallocation of capital in China (Working paper). Princeton University

Chiclana F, Herrera-Viedma E, Alonso S, Herrera F (2009) Cardinal consistency of reciprocal preference relations: a characterization of multiplicative transitivity. IEEE Trans Fuzzy Syst 17 (1):14–23

Dong YC, Xu YF, Li HY (2008) On consistency measures of linguistic preference relations. Eur J Oper Res 189(2):430–444

Fu C, Yang SL (2010) The group consensus based evidential reasoning approach for multiple attributive group decision analysis. Eur J Oper Res 206(3):601–608

Gou XJ, Liao HC, Wang XX, Xu ZS, Herrera F (2020) Consensus based on multiplicative consistent double hierarchy linguistic preferences: Venture capital in real estate market. Int J Strateg Prop Manag 42(1):1–23

Gou XJ, Liao HC, Xu ZS, Herrera F (2017) Double hierarchy hesitant fuzzy linguistic term set and multimoora method: A case of study to evaluate the implementation status of haze controlling measures. Inf Fusion 38:22–34

Gou XJ, Liao HC, Xu ZS, Min R, Herrera F (2019) Group decision making with double hierarchy hesitant fuzzy linguistic preference relations: consistency based measures, index and repairing algorithms and decision model. Inf Sci 489:93–112

Gou XJ, Xu ZS, Liao HC, Herrera F (2018) Multiple criteria decision making based on distance and similarity measures with double hierarchy hesitant fuzzy linguistic environment. Comput Ind Eng 126:516–530

Herrera-Viedma E, Alonso S, Chiclana F, Herrera F (2007) A consensus model for group decision making with incomplete fuzzy preference relations. IEEE Trans Fuzzy Syst 15(5):863–877

Li J, Wang JQ (2019) Multi-criteria decision making with probabilistic hesitant fuzzy information based on expected multiplicative consistency. Neural Comput Appl 31:8897–8915

Li J, Wang JQ, Hu JH (2019) Consensus building for hesitant fuzzy preference relations with multiplicative consistency. Comput Ind Eng 128:387–400

Liu C, Xiong W (2018) China's real estate market (Working paper). National Bureau Econ Res. https://doi.org/10.3386/w25297

Sałabun W (2014) The Characteristic Objects Method: a new approach to identify a multi-criteria group decision-making model. Intl J Comput Tech Appl 5(5):1597–1602

Sałabun W (2015) The Characteristic Objects Method: A new distance-based approach to multicriteria decision-making problems. J Multi-Criteria Decis Anal 22(1–2):37–50

Shang TZ, Lu SB, Li XF, Hei PY, Lei XH, Gong JG, Liu JH, Zhai JQ, Wang H (2017) Balancing development of major coal bases with available water resources in China through 2020. Appl Energy 194:735–750

Ureña R, Chiclana F, Morente-Molinera JA, Herrera-Viedma E (2015) Managing incomplete preference relations in decision making: are view and future trends. Inf Sci 302:14–32

Wang BL, Cai YP, Yin XA, Tan Q, Hao Y (2017) An integrated approach of system dynamics, orthogonal experimental design and inexact optimization for supporting water resources management under uncertainty. Water Resour Manag 31(5):1665–1694

Winz I, Brierley G, Trowsdale S (2009) The use of system dynamics simulation in water resources management. Water Resour Manag 23(7):1301–1323

Wu ZB, Xu JP (2016) Managing consistency and consensus in group decision making with hesitant fuzzy linguistic preference relations. Omega 65(3):28–40

Xia MM, Xu ZS (2011) Some issues on multiplicative consistency of interval reciprocal relations. Int J Inf Tech Decis Mak 10(6):1043–1065

Xia MM, Xu ZS, Chen J (2013) Algorithms for improving consistency or consensus of reciprocal [0,1]-valued preference relations. Fuzzy Sets Syst 216:108–133

Xia MM, Xu ZS, Wang Z (2014) Multiplicative consistency-based decision support system for incomplete linguistic preference relations. Int J Syst Sci 45(3):625–636

Xu ZS, Cai XQ, Szmidt E (2011) Algorithms for estimating missing elements of incomplete intuitionistic preference relations. Int J Intell Syst 26(9):787–813

Yu DJ, Xu ZS, Wang W (2018) Bibliometric analysis of fuzzy theory research in China: a 30-year perspective. Knowl-Based Syst 141:188–199

Zhu B, Xu ZS (2014) Consistency measures for hesitant fuzzy linguistic preference relations. IEEE Trans Fuzzy Syst 22(1):35–45

Zhu B, Xu ZS, Xu JP (2014) Deriving a ranking from hesitant fuzzy preference relations under group decision making. IEEE Trans Cybern 44(8):1328–1337

Chapter 4
Group Consensus Decision-Making Methods with DHHFLPRs, LPOs and Self-confident DHLPRs

In GDM processes, after ensuring that all DHHFLPRs are of acceptable consistencies, we cannot overlook another important step: the consensus reaching process, which is an essential process in GDM for enabling sufficient communications among all experts and obtaining an accepted decision result. Therefore, in this chapter, we will first focus on developing methods to manage the consensus reaching process and ensure that all experts reach consensus.

Additionally, in Chap. 1, we have introduced the concepts of LPOs defined by Gou et al. (2020c), which is a novel preference ordering structure in which the ordering of alternatives and the relationships between two adjacent alternatives are fused well. Additionally, this chapter develops models to equivalently transform each LPO into the corresponding DHLPR with complete consistency, and proposes some consensus models with DHLPRs to obtain the final decision-making result which is equal to the decision-making result with LPO information.

Moreover, in recent years, some scholars have proposed a novel preference relation, which considers the self-confident degrees of the basic elements of the preference relation. The self-confident degrees can be used to depict the degrees of confidences that the experts have in their own evaluation information, as well as enrich the integrity of evaluation information. As we know, the DHLT is only a linguistic expression and cannot reflect the self-confident degree of expert. Considering that there is little research about the DHLPRs with self-confident degrees in the literature, and the experts' self-confident degrees in the DHLPR have to be perfect. Therefore, Gou et al. (2020b) defined a concept of self-confident DHLPR, in which the basic element consists of the DHLT and the self-confident degree simultaneously. Then, this chapter will develop a double hierarchy linguistic preference values and self-confident degrees modifying (DHSM)-based consensus model to manage the GDM problems with self-confident DHLPRs based on the priority ordering theory.

X. Gou and Z. Xu, *Double Hierarchy Linguistic Term Set and Its Extensions*, Studies in Fuzziness and Soft Computing 396,
https://doi.org/10.1007/978-3-030-51320-7_4

4.1 Group Consensus Decision-Making Method with DHHFLPRs

In most research, the similarity and distance measures between individual evaluations and group opinions have been used widely to measure the consensus degrees in GDM problems (Liao et al. 2016; Wu and Xu 2016). However, these two kinds of measures can only obtain the positive values which are regarded as the positive correlation coefficients. However, one most important defect of them is that they cannot be used to fully reflect the real relationships between individual evaluations and group opinions by considering only the positive correlation coefficients. Therefore, we dedicate to developing a novel correlation measure for DHHFLPRs by considering the positive correlation and negative correlation coefficients simultaneously from the statistical point of view, and then use it to measure the consensus degrees between individuals and collective opinions.

4.1.1 Correlation Coefficient Between DHHFLTSs

Gou et al. (2018) introduced some distance measures between DHHFLEs in Sect. 2.1. Based on the Euclidean distance between DHHFLEs, the correlation coefficient between two DHHFLTSs can be defined as follows:

Definition 4.1 (Gou et al. 2020a). Let $H_{S_{O_i}} = \{h_{S_{O_{i1}}}, h_{S_{O_{i2}}}, \ldots, h_{S_{O_{in}}}\}(i = 1, 2)$ be two DHHFLTSs, and $h_{S_{O_{ij}}} = \{h^{(l)}_{S_{O_{ij}}} | h^{(l)}_{S_{O_{ij}}} \in S_O; l = 1, 2, \ldots, \# h_{S_{O_{ij}}}\}$ $(i = 1, 2; j = 1, 2, \ldots, n)$. Then the correlation coefficient between $H_{S_{O_i}}(i = 1, 2)$ is defined as

$$C(H_{S_{O_1}}, H_{S_{O_2}}) = \frac{\sum_{j=1}^{n}\left(\left(\frac{d(h^+_{S_{O_1}}, h_{S_{O_{1j}}})}{d(h^+_{S_{O_1}}, h^-_{S_{O_1}})} - \frac{1}{n}\sum_{j=1}^{n}\frac{d(h^+_{S_{O_1}}, h_{S_{O_{1j}}})}{d(h^+_{S_{O_1}}, h^-_{S_{O_1}})}\right) \times \left(\frac{d(h^+_{S_{O_2}}, h_{S_{O_{2j}}})}{d(h^+_{S_{O_2}}, h^-_{S_{O_2}})} - \frac{1}{n}\sum_{j=1}^{n}\frac{d(h^+_{S_{O_2}}, h_{S_{O_{2j}}})}{d(h^+_{S_{O_2}}, h^-_{S_{O_2}})}\right)\right)}{\sqrt{\sum_{j=1}^{n}\left(\frac{d(h^+_{S_{O_1}}, h_{S_{O_{1j}}})}{d(h^+_{S_{O_1}}, h^-_{S_{O_1}})} - \frac{1}{n}\sum_{j=1}^{n}\frac{d(h^+_{S_{O_1}}, h_{S_{O_{1j}}})}{d(h^+_{S_{O_1}}, h^-_{S_{O_1}})}\right)^2} \times \sqrt{\sum_{j=1}^{n}\left(\frac{d(h^+_{S_{O_2}}, h_{S_{O_{2j}}})}{d(h^+_{S_{O_2}}, h^-_{S_{O_2}})} - \frac{1}{n}\sum_{j=1}^{n}\frac{d(h^+_{S_{O_2}}, h_{S_{O_{2j}}})}{d(h^+_{S_{O_2}}, h^-_{S_{O_2}})}\right)^2}} \tag{4.1}$$

where n is the number of DHHFLEs in each DHHFLTS; $h^+_{S_{O_i}} = \max_j h_{S_{O_{ij}}}$ and $h^-_{S_{O_i}} = \min_j h_{S_{O_{ij}}}$ are the biggest and smallest DHHFLEs in $H_{S_{O_i}}(i = 1, 2)$, respectively.

Theorem 4.1 (Gou et al. 2020a). *The double hierarchy hesitant fuzzy linguistic correlation coefficient satisfies the property of the Pearson correlation coefficient that* $C(H_{S_{O_1}}, H_{S_{O_2}}) \in [-1,1]$.

Proof Since $h^+_{S_{O_q}} = \max_i h^i_{S_{O_q}}$ and $h^-_{S_{O_q}} = \min_i h^i_{S_{O_q}}$, then we have $|E(h^+_{S_{O_i}}) - E(h^-_{S_{O_i}})| \geq |E(h^+_{S_{O_i}}) - E(h_{S_{O_{ij}}})|$. Thus, we can obtain $d(h^+_{S_{O_i}}, h^-_{S_{O_i}}) \geq d(h^+_{S_{O_i}}, h_{S_{O_{ij}}})$, $i = 1,2,\ldots,m$. Therefore, $d(h^+_{S_{O_i}}, h_{S_{O_{ij}}}) \big/ d(h^+_{S_{O_i}}, h^-_{S_{O_i}}) \leq 1$. Let

$$\Delta_{ij} = d(h^+_{S_{O_i}}, h_{S_{O_{ij}}}) \big/ d(h^+_{S_{O_i}}, h^-_{S_{O_i}}) - \frac{1}{n}\sum_{j=1}^{n}\left(d(h^+_{S_{O_i}}, h_{S_{O_{ij}}}) \big/ d(h^+_{S_{O_i}}, h^-_{S_{O_i}})\right), \quad i = 1,2.$$

Then $\Delta_{ij} \in [-1,1]$. Equation (4.1) can be transformed to

$$C(H_{S_{O_1}}, H_{S_{O_2}}) = \sum_{j=1}^{n}(\Delta_{1j} \times \Delta_{2j}) \Bigg/ \left(\sqrt{\sum_{j=1}^{n}(\Delta_{1j})^2} \times \sqrt{\sum_{j=1}^{n}(\Delta_{2j})^2}\right).$$

(1) When $j = 1$, $C(H_{S_{O_1}}, H_{S_{O_2}}) = 1 \in [-1,1]$.
(2) When $j = 2$, $C(H_{S_{O_1}}, H_{S_{O_2}}) = \frac{\Delta_{11}\Delta_{21} + \Delta_{12}\Delta_{22}}{\sqrt{(\Delta_{11})^2 + (\Delta_{12})^2} \times \sqrt{(\Delta_{21})^2 + (\Delta_{22})^2}}$. Then,

$$C^2(H_{S_{O_1}}, H_{S_{O_2}}) = \frac{(\Delta_{11}\Delta_{21})^2 + (\Delta_{12}\Delta_{22})^2 + 2\Delta_{11}\Delta_{21}\Delta_{12}\Delta_{22}}{(\Delta_{11}\Delta_{21})^2 + (\Delta_{11}\Delta_{22})^2 + (\Delta_{12}\Delta_{21})^2 + (\Delta_{12}\Delta_{22})^2}$$

According to average value inequality, we get

$$(\Delta_{11}\Delta_{21})^2 + (\Delta_{12}\Delta_{22})^2 + 2\Delta_{11}\Delta_{21}\Delta_{12}\Delta_{22} \leq (\Delta_{11}\Delta_{21})^2 + (\Delta_{11}\Delta_{22})^2 + (\Delta_{12}\Delta_{21})^2 + (\Delta_{12}\Delta_{22})^2$$

Then $C^2(H_{S_{O_1}}, H_{S_{O_2}}) \leq 1$, namely, $C(H_{S_{O_1}}, H_{S_{O_2}}) \in [-1,1]$.

(3) Suppose that $C(H_{S_{O_1}}, H_{S_{O_2}}) \in [-1,1]$ when $j = n$.
(4) When $j = n+1$, we have

$$C(H_{S_{O_1}}, H_{S_{O_2}}) = \frac{\sum_{j=1}^{n}(\Delta_{1j}\Delta_{2j}) + (\Delta_{1,n+1}\Delta_{2,n+1})}{\sqrt{\sum_{j=1}^{n}(\Delta_{1j})^2 + (\Delta_{1,n+1})^2} \times \sqrt{\sum_{j=1}^{n}(\Delta_{2j})^2 + (\Delta_{2,n+1})^2}}$$

Let $a = \sum_{j=1}^{n}(\Delta_{1j}\Delta_{2j})$, $b = \sum_{j=1}^{n}(\Delta_{1j})^2$, and $c = \sum_{j=1}^{n}(\Delta_{2j})^2$, where $a \in [-1,1]$, $b, c \in [0,1]$. Clearly, we have $b + c \geq 2a$ and $bc \geq a^2$.

Then, $C^2(H_{S_{O_1}}, H_{S_{O_2}}) = \frac{a^2 + 2a\Delta_{1,n+1}\Delta_{2,n+1} + (\Delta_{1,n+1}\Delta_{2,n+1})^2}{bc + b(\Delta_{2,n+1})^2 + c(\Delta_{1,n+1})^2 + (\Delta_{1,n+1})^2(\Delta_{2,n+1})^2}$, and

$$a^2 + 2a\Delta_{1,n+1}\Delta_{2,n+1} + (\Delta_{1,n+1}\Delta_{2,n+1})^2 \leq bc + b(\Delta_{2,n+1})^2 + c(\Delta_{1,n+1})^2 + (\Delta_{1,n+1})^2(\Delta_{2,n+1})^2$$

Therefore, $C^2(H_{S_{O_1}}, H_{S_{O_2}}) \leq 1$, namely, $C(H_{S_{O_1}}, H_{S_{O_2}}) \in [-1, 1]$.

According to the above proof, we get $C(H_{S_{O_1}}, H_{S_{O_2}}) \in [-1, 1]$ for all j. This completes the proof of Theorem 4.1. ■

Remark 4.1 (Gou et al. 2020a). Since $C(H_{S_{O_1}}, H_{S_{O_2}}) \in [-1, 1]$, there exists three situations:

(1) If $C(H_{S_{O_1}}, H_{S_{O_2}}) > 0$, then $H_{S_{O_1}}$ and $H_{S_{O_2}}$ have a positive correlation;
(2) If $C(H_{S_{O_1}}, H_{S_{O_2}}) < 0$, then $H_{S_{O_1}}$ and $H_{S_{O_2}}$ have a negative correlation;
(3) If $C(H_{S_{O_1}}, H_{S_{O_2}}) = 0$, then $H_{S_{O_1}}$ and $H_{S_{O_2}}$ have no any correlation.

In conclusion, the bigger the absolute value of $C(H_{S_{O_1}}, H_{S_{O_2}})$ is, the stronger the correlation between $H_{S_{O_1}}$ and $H_{S_{O_2}}$ should be. Considering that the correlation coefficient reflects the relation of two DHHFLPRs from both positive and negative angles. It can be regarded as a useful tool to represent the consensus degree of experts in GDM.

4.1.2 Consensus Reaching Process

The consensus reaching process mainly consists of two aspects: consensus checking process and consensus improving process.

(a) **Consensus checking process**

In the GDM process, the consensus measures can be used to depict the closeness degree among the experts' preferences. The most popular method for measuring the consensus degree is to calculate the deviation degree between each individual preference and aggregated opinion (Dong et al. 2010). Additionally, distance measure is very important to depict the deviations between the experts' preferences, and it has been utilized to measure the consensus degree in GDM (Gou et al. 2018). However, the consensus degree measured by distance cannot reflect the relationship between the individual preference and aggregated opinion from both of positive and negative angles, and may lead to some unreasonable consensus results. Motivated by Wu and Liao (2019), and considering that the correlation coefficient of DHHFLTSs discussed in Eq. (4.1) can overcome this gap. Therefore, this subsection develops a method to obtain the consensus degree of each expert based on the correlation coefficient of DHHFLTSs.

Firstly, Based on Eq. (3.24), the group DHHFLPR $\tilde{H}^c_{S_O} = (h^c_{S_{O_{ij}}})_{m\times m}$ can be established, and we call $h^c_{S_{O_{ij}}} = \{h^{c(l)}_{S_{O_{ij}}} | h^{c(l)}_{S_{O_{ij}}} = \sum_{r=1}^{R} w_r h^{r(l)}_{S_{O_{ij}}}; l = 1, 2, \ldots, \# h^c_{S_{O_{ij}}}\}$ the group preference element. Then, motivated by Eq. (4.1), the consensus degree between each DHHFLPR and the group DHHFLPR is developed as follows:

Definition 4.2 (Gou et al. 2020a). Let $\tilde{H}^r_{S_O} = (h^r_{S_{O_{ij}}})_{m\times m}$ be the individual DHHFLPR of the expert e^r, and $\tilde{H}^c_{S_O} = (h^c_{S_{O_{ij}}})_{m\times m}$ be the group DHHFLPR. Then the consensus degree of e^r is defined as:

$$CD(\tilde{H}^r_{S_O}) = \frac{1}{m}\sum_{j=1}^{m}\Theta^r_j,\ \Theta^r_j = \frac{\sum_{i=1}^{m}\left(\left(\frac{d^r_{ij}}{d^r_j} - \frac{1}{m}\sum_{i=1}^{m}\frac{d^r_{ij}}{d^r_j}\right)\times\left(\frac{d^c_{ij}}{d^G_j} - \frac{1}{m}\sum_{i=1}^{m}\frac{d^c_{ij}}{d^G_j}\right)\right)}{\sqrt{\sum_{i=1}^{m}\left(\frac{d^r_{ij}}{d^r_j} - \frac{1}{m}\sum_{i=1}^{m}\frac{d^r_{ij}}{d^r_j}\right)^2}\times\sqrt{\sum_{i=1}^{m}\left(\frac{d^c_{ij}}{d^G_j} - \frac{1}{m}\sum_{i=1}^{m}\frac{d^c_{ij}}{d^G_j}\right)^2}} \tag{4.2}$$

where $d^r_{ij} = d(h^{r+}_{S_{O_j}}, h^r_{S_{O_{ij}}})$, $d^r_j = d(h^{r+}_{S_{O_j}}, h^{r-}_{S_{O_j}})$, $d^c_{ij} = d(h^{c+}_{S_{O_j}}, h^c_{S_{O_{ij}}})$, $d^c_j = d(h^{c+}_{S_{O_j}}, h^{c-}_{S_{O_j}})$, $h^{r+}_{S_{O_j}} = \max_i h^r_{S_{O_{ij}}}$, $h^{r-}_{S_{O_j}} = \min_i h^r_{S_{O_{ij}}}$, $h^{c+}_{S_{O_j}} = \max_i h^c_{S_{O_{ij}}}$ and $h^{c-}_{S_{O_j}} = \min_i h^c_{S_{O_{ij}}}$.

According to Theorem 4.1, it is obvious that $\Theta^r_j \in [-1, 1]$. Then, $CD(\tilde{H}^r_{S_O}) \in [-1, 1]$. If $CD(\tilde{H}^r_{S_O}) > 0$, then, there exists consensus with different strengths. The larger the value of $CD(\tilde{H}^r_{S_O})$ is, the stronger the correlation consensus degree of the expert e^r to the group will be. However, if $CD(\tilde{H}^r_{S_O}) < 0$, which means negative correlation, so there exists no consensus of expert e^r with respect to the remaining experts. Based on Karplus and Diederichs (2012), the strength of consensus degree can be defined in Table 4.1.

(b) **Consensus modifying process**

After obtaining the consensus degrees of all experts, there may be some preferences with low consensus degrees. Therefore, it is necessary to improve the group

Table 4.1 The consensus degree of two DHHFLPRs based on the ranges of correlation strength

The range of $CD(\tilde{H}^r_{S_O})$	The correlation strength of two variables
(0.8, 1]	Extremely strong consensus
(0.6, 0.8]	Strong consensus
(0.4, 0.6]	Moderate degree consensus
(0.2, 0.4]	Weak consensus
(0, 0.2]	Extremely weak consensus or no consensus
[−1, 0]	No consensus

consensus degree by a consensus modifying process before making a decision. In this regard, two processes should be handled: one is to detect the individual DHHFLPR with the lowest consensus degree, and the other is to adjust the corresponding element of it. Considering that the automatic modification is a low cost but effective technique (Dong and Cooper 2016; Ma 2016), this subsection develops an automatic modification method to modify the group consensus degree:

Definition 4.3 (Gou et al. 2020a). Let $CD^* = CD(\tilde{H}_{S_O}^{\varphi}) = \min_r CD(\tilde{H}_{S_O}^{r})$ $(r = 1, 2, \ldots, R)$ be the minimum consensus degree of an expert in a group, ξ be the given consensus threshold, and Θ be the given correlation threshold. If $CD^* < \xi$, then the corresponding individual DHHFLPR $CD(\tilde{H}_{S_O}^{\varphi})$ with the minimum consensus degree $\tilde{H}_{S_O}^{\varphi}$ should be adjusted as $\tilde{H}_{S_O}^{\prime\varphi} = (h_{S_{O_{ij}}}^{\prime\varphi})_{m\times m}$, where

$$h_{S_{O_{ij}}}^{\prime\varphi} = \begin{cases} \frac{1}{2}\left(h_{S_{O_{ij}}}^{\varphi} \oplus h_{S_{O_{ij}}}^{c}\right), & if\ \Theta_j^{\varphi} < \Theta \\ h_{S_{O_{ij}}}^{\varphi}, & if\ \Theta_j^{\varphi} \geq \Theta \end{cases} \tag{4.3}$$

If the consensus degrees of all experts are larger than or equal to the given consensus threshold ξ, then the final group DHHFLPR $\tilde{H}_{S_O}^{*} = (h_{S_{O_{ij}}}^{*})_{m\times m}$ is obtained by aggregating all the individual DHHFLPRs.

Remark 4.2 (Gou et al. 2020a). Based on the strength of consensus degree given in Table 4.1, the values of the thresholds ξ and Θ can be determined as $\xi, \Theta \in [0.4, 0.8]$. Additionally, according to the practical situation of GDM, the values of the thresholds ξ and Θ can be adjusted. If the demand for consensus is strict, then the threshold should be given a higher value; Otherwise, a lower threshold value should be provided.

4.1.3 Group Consensus Decision-Making Model

In the process of GDM, when all DHHFLPRs are of acceptable consistencies and the consensus degrees of all DHHFLPRs reach the given consensus threshold, the next step is to make a decision.

Based on the additive consistency index and multiplicative consistency index proposed in Chapter 3, a group consensus decision-making method can be developed. We summarize the procedure as Algorithm 4.1 and illustrate it by Fig. 4.1.

Algorithm 4.1 (Gou et al. 2020a). Group consensus decision-making method

Step 1. Collect the linguistic preferences of experts and establish their individual DHHFLPRs $\tilde{H}_{S_O}^{r} = (h_{S_{O_{ij}}}^{r})_{m\times m}(r = 1, 2, \ldots, R)$. Go to Step 2.

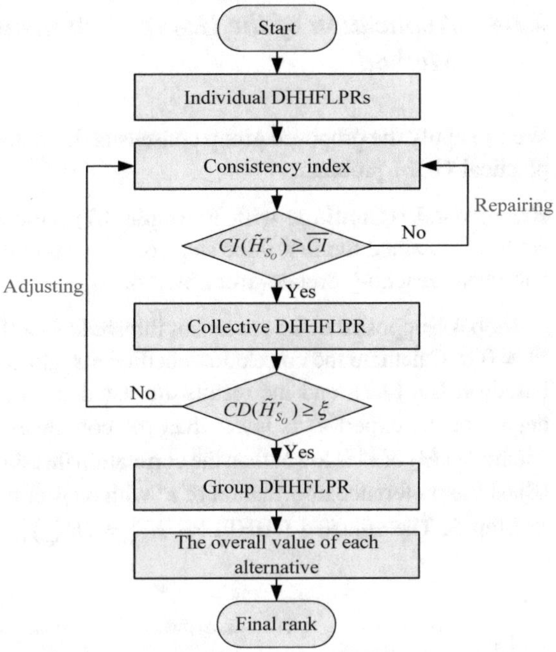

Fig. 4.1 The flowchart of the group consensus decision-making method with DHHFLPRs

Step 2. Based on Algorithm 3.3 and Algorithm 3.5, check and improve the DHHFLPRs with unacceptable consistencies, and ensure that all DHHFLPRs are of acceptable consistencies. Go to Step 3.

Step 3. Obtain the group DHHFLPR $\tilde{H}^c_{S_O} = (h^c_{S_{O_{ij}}})_{m\times m}$ by Eq. (3.24). Go to Step 4.

Step 4. Determine the consensus threshold ξ and the correlation threshold Θ, and calculate the consensus degree $CD(\tilde{H}^r_{S_O})$ by Eq. (4.2). If $CD(\tilde{H}^r_{S_O}) \geq \xi$, then go to Step 6; Otherwise, go to Step 5.

Step 5. Find out the DHHFLPR with $CD^* = CD(\tilde{H}^\varphi_{S_O}) = \min_r CD(\tilde{H}^r_{S_O})$ $(r = 1, 2, \ldots, R)$. Derive its adjusted DHHFLPR $\tilde{H}'^\varphi_{S_O} = (h'^\varphi_{S_{O_{ij}}})_{m\times m}$. Go back to Step 2.

Step 6. Obtain the final group DHHFLPR $\tilde{H}^*_{S_O} = (h^*_{S_{O_{ij}}})_{m\times m}$ by Eq. (3.24). Go to Step 7.

Step 7. Based on Eq. (1.15), calculate the synthetical value of each alternative $SV(A_i) = \sum_{j=1}^m E(h^*_{S_{O_{ij}}})(i = 1, 2, \ldots, m)$.

Step 8. Rank all the alternatives based on the values of $SV(A_i)$ $(i = 1, 2, \ldots, m)$.

Step 9. End.

4.1.4 Application of the Group Consensus Decision-Making Method

We can apply the proposed group consensus decision-making method to deal with a practical GDM problem.

Example 4.1 **(Continue with Example** 3.7) (Gou et al. 2020a). In Example 3.7, we have obtained the first three steps of Algorithm 4.1. Therefore, we can start the consensus reaching process from Step 4.

Step 4. Suppose that the consensus threshold $\xi = 0.9$ and the correlation threshold $\Theta = 0.6$. Calculate the correlation coefficients and consensus degrees of all experts based on Eq. (4.2), and the results are listed in Table 4.2. Clearly, the consensus degree of the expert e^4 is lower than the consensus threshold, and the correlation coefficient Θ_2 of e^4 is lower than the correlation threshold. Therefore, it is necessary to adjust the preference information of e^4 with respect to the second alternative A_2.

Step 5. The adjusted DHHFLPR $\tilde{H}'^{4}_{S_O} = (h'^{4}_{S_{O_{ij}}})_{4\times4}$ of e^4 is obtained as follows:

$$\tilde{H}'^{4}_{S_O} = \begin{pmatrix} \{s_{0<o_0>}\} & \{s_{-0.88<o_{0.65}>}, s_{0<o_{1.2}>}\} & \{s_{1<o_{-2}>}, s_{2<o_{-2}>}\} & \{s_{0<o_{-1}>}, s_{0<o_3>}\} \\ \{s_{0.88<o_{-0.65}>}, s_{0<o_{-1.2}>}\} & \{s_{0<o_0>}\} & \{s_{-0.13<o_{-0.45}>}, s_{0<o_{0.2}>}\} & \{s_{0<o_{0.75}>}, s_{0.25<o_{0.85}>}\} \\ \{s_{-1<o_2>}, s_{-2<o_2>}\} & \{s_{0.13<o_{0.45}>}, s_{0<o_{-0.2}>}\} & \{s_{0<o_0>}\} & \{s_{3<o_1>}, s_{3<o_2>}\} \\ \{s_{0<o_1>}, s_{0<o_{-3}>}\} & \{s_{0<o_{-0.75}>}, s_{-0.25<o_{-0.85}>}\} & \{s_{-3<o_{-1}>}, s_{-3<o_{-2}>}\} & \{s_{0<o_0>}\} \end{pmatrix}$$

Go back to Step 2. $\tilde{H}'^{4}_{S_O}$ is a DHHFLPR with the acceptable consistency. Then, we can obtain the group DHHFLPR $\tilde{H}'^{c}_{S_O} = (h'^{c}_{S_{O_{ij}}})_{4\times4}$:

$$\tilde{H}'^{c}_{S_O} = \begin{pmatrix} \{s_{0<o_0>}\} & \{s_{-0.72<o_{-0.2}>}, s_{0<o_{1.46}>}\} & \{s_{0.5<o_{-0.4}>}, s_{1<o_{0.7}>}\} & \{s_{0<o_{-0.3}>}, s_{0<o_2>}\} \\ \{s_{0.72<o_{0.2}>}, s_{0<o_{-1.46}>}\} & \{s_{0<o_0>}\} & \{s_{-0.28<o_{-1.04}>}, s_{0<o_{0.46}>}\} & \{s_{0<o_{1.73}>}, s_{0.56<o_{-0.66}>}\} \\ \{s_{-0.5<o_{0.4}>}, s_{-1<o_{-0.7}>}\} & \{s_{0.28<o_{1.04}>}, s_{0<o_{-0.46}>}\} & \{s_{0<o_0>}\} & \{s_{3<o_{-1}>}, s_{3<o_{0.1}>}\} \\ \{s_{0<o_{0.3}>}, s_{0<o_{-2}>}\} & \{s_{0<o_{-1.73}>}, s_{-0.56<o_{0.66}>}\} & \{s_{-3<o_1>}, s_{-3<o_{-0.1}>}\} & \{s_{0<o_0>}\} \end{pmatrix}$$

Additionally, Calculate the correlation coefficients and consensus degrees of all experts, and the results are listed in Table 4.3.

Table 4.2 The correlation coefficients and consensus degrees of each DHHFLPR $\tilde{H}^{r}_{S_O}$

Experts	Correlation coefficients				Consensus degree
	Θ_1	Θ_2	Θ_3	Θ_4	
e^1	0.9700	0.8803	0.9948	0.9970	0.9605
e^2	0.9224	0.9981	0.9969	0.9653	0.9707
e^3	0.9135	0.9917	0.9862	0.9543	0.9614
e^4	0.9998	−0.5766	0.9992	0.9863	0.6022

Table 4.3 The correlation coefficients and consensus degrees of each DHHFLPR $\tilde{H}^{r}_{S_o}$

Experts	Correlation coefficients				Consensus degree
	Θ_1	Θ_2	Θ_3	Θ_4	
e^1	0.9706	0.8511	0.9949	0.9954	0.9530
e^2	0.9239	0.9929	0.9971	0.9655	0.9699
e^3	0.9105	0.9841	0.9857	0.9487	0.9572
e^4	1	0.8414	0.9995	0.9873	0.9570

Table 4.4 The complementary preference relation and the rank of alternatives

The complementary preference relation				Overall values	Rank
0.5	0.4809	0.5984	0.5266	2.1059	$A_3 \succ A_1 \succ A_2 \succ A_4$
0.5191	0.5	0.4734	0.5723	2.0648	
0.4016	0.5266	0.5	0.8609	2.2891	
0.4734	0.4277	0.1391	0.5	1.5402	

Clearly, the group preference information reaches the consensus.

Step 6. $\tilde{H}^{\prime c}_{S_o}$ is the final group DHHFLPR denoted as $\tilde{H}^{*}_{S_o}$.

Step 7. Based on Eqs. (1.15), $\tilde{H}^{*}_{S_o}$ can be transformed into the complementary preference relation. and the ranking of all alternatives is obtained and shown in Table 4.4. Therefore, the optimal alternative is to invest a VC project about real estate market in Xi'an.

4.2 Group Consensus Decision-Making Method with LPOs

In this subsection, we will develop a group consensus decision-making method with LPOs by transforming all LPOs into the corresponding completely consistent DHLPRs. For achieving consensus among a group of experts, lots of consensus reaching approaches and models over preference relations have been developed in recent years (Del Moral et al. 2018; Gou et al. 2018; Kacprzyk and Fedrizzi 1988; Kamis et al. 2018; Liao et al. 2016; Morente-Molinera et al. 2018, 2019; Wu et al. 2018; Zhang and Chen 2019; Zhu and Xu 2018). However, most of the consensus models are generally iterative models that utilize heuristics as their calculation tools such as automatic improvement models (Gou et al. 2018; Liao et al. 2016; Zhang and Chen 2019) and feedback mechanism-based improvement models (Gou et al. 2018; Liao et al. 2016; Wu et al. 2018). Therefore, the most important defect of these consensus models mentioned above is that we have to go through trial and error before reaching a consensus or even failing to reach a consensus.

In contrast, some scholars developed another popular tool named as optimization model to help experts achieve consensus (Ben-Arieh and Easton 2007; Fan et al. 2006; Meng et al. 2019; Wan et al. 2018; Wu et al. 2019a, b; Xu et al. 2018; Yu and Xu 2019; Zhang et al. 2018, 2019; Zhang and Pedrycz 2018). As far as we know, the most important advantage of optimization models is that they can provide the adjusted preference solutions directly by setting goals and solving the established models. Therefore, these adjusted preference solutions are optimal to a certain degree of objectivity, and can be understood and accepted by experts in a feedback strategy. Zhang and Pedrycz (2018) built several goal programming models to manage consistency and consensus of intuitionistic multiplicative preference relations. Furthermore, the interactive consistency process and interactive consensus process based on the multi-stage models were also designed (Wu et al. 2019a). Fan et al. (2006) constructed a linear goal programming model to integrate the two different formats of preference relations and to achieve consensus. Zhang et al. (2018) proposed a consensus-oriented aggregation model to obtain a collective opinion with maximum consensus degree by minimizing the information deviation between individual and collective opinions. Wu et al. (2019b) constructed an optimization model to directly identify hesitant fuzzy linguistic preference values, which greatly improves the efficiency of the consensus reaching process. Zhang et al. (2019) developed a multi-stage optimization-based consensus reaching processes, which have better comprehensive consensus efficiency in different GDM settings.

Although different kinds of optimization methods and models have been developed to solve consensus issues for different preference relations. However, two key problems have not been adequately addressed.

Firstly, when feedbacking modification suggestions to the corresponding experts in consensus reaching process, the suggested preferences for improving consensus were usually represented in continuous forms (Xu et al. 2016). However, these preferences with continuous forms usually have no clear meanings considering that the original preference terms have been expressed by discrete terms (Kou and Lin 2014). Thus, it is very interesting and challenging to research the correspondence between the suggested preferences and the given original discrete terms. Additionally, some scholars also developed some models in which the suggested preferences were represented by discrete scales (Wu and Xu 2018). Therefore, motivated by Wu et al. (2019a) and to fully respect the experts' expression habits, we will focus on establishing some consensus optimization models by considering the suggested preferences represented in both the continuous scale and the discrete scale.

Secondly, most of the existing optimization models only focus on minimizing the size of change for preferences and do not consider the uniqueness of the obtained solutions. As far as we know, there usually are multiple optimal solutions when solving an optimization model, so it is unreasonable to consider only one aspect of them. Therefore, to provide more refined solutions to experts, we will develop a multi-stage consensus optimization model which consists of three objectives including minimizing the deviations of the modification magnitudes, minimizing the cardinal number of modifications while keeping the value of the first objective constant, and minimizing the number of experts who need to change their evaluations.

4.2.1 Consistency Index of DHLPR

When dealing with LPOs, it is necessary to transform all LPOs into DHLPRs and obtain the uniformed information when the original linguistic preferences are represented by LPOs. As we know, the most important and primary work is to ensure that all DHLPRs are consistent before making a decision with DHLPRs. Therefore, this subsection will develop transformation models which can obtain the DHLPRs with acceptably additive consistencies. Before establishing the transformation models, the definition of additive consistent DHLPR should be proposed firstly.

Definition 4.4 (Gou et al. 2020c). A DHLPR $\mathbb{R} = (r_{ij})_{m\times m} \subset A \times A$ can be called an additive consistent DHLPR if it satisfies

$$r_{ij} = r_{i\rho} + r_{\rho j} \quad (i, j, \rho = 1, 2, \ldots, m, i \neq j) \tag{4.4}$$

Then, a theorem can be developed as follows:

Theorem 4.2 (Gou et al. 2020c). *Let* $\mathbb{R} = (r_{ij})_{m\times m} \subset A \times A$ *be a DHLPR. If* $\overline{r}_{ij} = \frac{1}{m}(\bigoplus_{\rho=1}^{m}(r_{i\rho} + r_{\rho j}))$ *for* $i, j, \rho = 1, 2, \ldots, m, i \neq j$*, then* $\overline{\mathbb{R}} = (\overline{r}_{ij})_{m\times m} \subset A \times A$ *is an additive consistent DHLPR.*

Additionally, based on the transformation function f provided in Sect. 4.1, the definition of an additive consistency for a DHLPR also can be developed as follows:

Definition 4.5 (Gou et al. 2020c). A DHLPR $\mathbb{R} = (r_{ij})_{m\times m} \subset A \times A$ can be called an additive consistent DHLPR if it satisfies

$$f(r_{ij}) = f(r_{i\rho}) + f(r_{\rho j}) - 0.5 \quad (i, j, \rho = 1, 2, \ldots, m, i \neq j) \tag{4.5}$$

Similarly, a theorem can be obtained as follows:

Theorem 4.3 (Gou et al. 2020c). *Let* $\mathbb{R} = (r_{ij})_{m\times m} \subset A \times A$ *be a DHLPR. If* $f(\overline{r}_{ij}) = \frac{1}{m}(\bigoplus_{\rho=1}^{m}(f(r_{i\rho}) + f(r_{\rho j}) - 0.5))$ *for* $i, j, \rho = 1, 2, \ldots, m, i \neq j$*, then* $\overline{\mathbb{R}} = (\overline{r}_{ij})_{m\times m} \subset A \times A$ *is an additive consistent DHLPR.*

In actual decision-making processes, it is very important to check whether a DHLPR is of acceptable consistency. Based on Theorem 4.3, the consistency index for a DHLPR is defined as follows:

Definition 4.6 (Gou et al. 2020c). Let $\mathbb{R} = (r_{ij})_{m \times m} \subset A \times A$ be a DHLPR. Then the consistency index for $\mathbb{R}$ is obtained as:

$$CI(\mathbb{R}) = 1 - \frac{2}{3m(m-1)(m-2)} \sum_{i=1}^{m-1} \sum_{j=1, j \neq i}^{m} \sum_{\rho=1, \rho \neq i,j}^{m} \left| f(r_{ij}) - f(r_{i\rho}) - f(r_{\rho j}) + 0.5 \right| \tag{4.6}$$

Obviously, $CI(\mathbb{R}) = 1$ means that $f(r_{ij}) = f(r_{i\rho}) + f(r_{\rho j}) - 0.5$, which is equivalent to $f(r_{ij}) - 0.5 = (f(r_{i\rho}) - 0.5) + (f(r_{\rho j}) - 0.5)$, and this situation is called additive transitivity or additive consistency (Herrera-Viedma et al. 2004).

Additionally, there exists a mass of redundant computations in Eq. (4.6). For example, let $\delta_{ij\rho} = \left| f(r_{ij}) - f(r_{i\rho}) - f(r_{\rho j}) + 0.5 \right|$, then we can obtain that $\delta_{ij\rho} = \delta_{i\rho j} = \delta_{ji\rho} = \delta_{j\rho i} = \delta_{\rho ij} = \delta_{\rho ji}$. Therefore, Eq. (4.6) can be rewritten as:

$$\begin{aligned} CI(\mathbb{R}) &= 1 - \frac{4}{m(m-1)(m-2)} \sum_{1 \leq i < j < \rho \leq m}^{m-1} \left| f(r_{ij}) - f(r_{i\rho}) - f(r_{\rho j}) + 0.5 \right| \\ &= 1 - \frac{4}{m(m-1)(m-2)} \sum_{i=1}^{m-2} \sum_{j=i+1}^{m-1} \sum_{\rho=j+1}^{m} \left| f(r_{ij}) - f(r_{i\rho}) - f(r_{\rho j}) + 0.5 \right|. \end{aligned} \tag{4.7}$$

Based on Eq. (4.7), the DHLPR $\mathbb{R}$ is completely consistent when $CI(\mathbb{R}) = 1$. In actual decision-making processes, however, it is very difficult that the DHLPRs provided by experts are completely consistent. Therefore, in the beginning of decision-making, it is necessary to set a consistency threshold, denoted by $\overline{CI}$, which can be found in Table 3.14. If $CI(\mathbb{R}) < \overline{CI}$, then the experts need to revise their preference information according to the feedback suggestions.

4.2.2 *Transformation Models for Transforming LPOs into DHLPRs*

Based on these two kinds of LPOs discussed in Subsection 1.2.3, and the consistency index of DHLPR, Gou et al. (2020c) developed two transformation models to transform LPOs into DHLPRs on the basis of the additive consistent DHLPRs, respectively.

Firstly, for a $LPO = \left\{ A_{\sigma(1)} \overset{s_{t<o_k>}^{(\sigma(1),\sigma(2))}}{>} A_{\sigma(2)} \overset{s_{t<o_k>}^{(\sigma(2),\sigma(3))}}{>} \ldots \overset{s_{t<o_k>}^{(\sigma(m-1),\sigma(m))}}{>} A_{\sigma(m)} \right\}$, Gou et al. (2020c) transformed the first kind of LPO, the LPO in continuous form, into a completely consistent DHLPR $\mathbb{R} = (r_{ij})_{m \times m}$ on the basis of Definition 4.5.

To get all preference information of a DHLPR, we only need to calculate the elements of the upper triangular matrix. For a DHLPR $\mathbb{R} = (r_{ij})_{m \times m}$, the elements of main diagonal is $r_{ii} = s_{0<o_0>}$ $(i = 1, 2, \ldots, m)$. Additionally, we can also obtain the elements of $r_{\sigma(i)\sigma(i+1)} = s_{t<o_k>}^{(\sigma(i)\sigma(i+1))}$ $(i = 1, 2, \ldots, m-1)$ from the given LPO directly. Then we can only need to calculate the remaining elements of the upper triangular matrix and the number of them is $(m-1)(m-2)/2$.

Firstly, Gou et al. (2020c) developed an equation set including $(m-1)(m-2)/2$ equations, which can be used to obtain the remaining elements of the upper triangular matrix:

$$\begin{cases} r_{\sigma(1),\sigma(3)} = \frac{1}{m}\left(\overset{m}{\underset{\rho=1}{\oplus}} (r_{\sigma(1),\rho} + r_{\rho,\sigma(3)})\right); \ldots; r_{\sigma(1),\sigma(m)} = \frac{1}{m}\left(\overset{m}{\underset{\rho=1}{\oplus}} (r_{\sigma(1),\rho} + r_{\rho,\sigma(m)})\right) \\ r_{\sigma(2),\sigma(4)} = \frac{1}{m}\left(\overset{m}{\underset{\rho=1}{\oplus}} (r_{\sigma(2),\rho} + r_{\rho,\sigma(4)})\right); \ldots; r_{\sigma(2),\sigma(m)} = \frac{1}{m}\left(\overset{m}{\underset{\rho=1}{\oplus}} (r_{\sigma(2),\rho} + r_{\rho,\sigma(m)})\right) \\ \ldots \\ r_{\sigma(m-2),\sigma(m)} = \frac{1}{m}\left(\overset{m}{\underset{\rho=1}{\oplus}} (r_{\sigma(m-2),\rho} + r_{\rho,\sigma(m)})\right) \end{cases} \tag{4.8}$$

Similarly, based on Definition 4.5, the other kind of equation set can be developed to obtain the remaining elements of the upper triangular matrix:

$$\begin{cases} f(r_{\sigma(1),\sigma(3)}) = \frac{1}{m}\left(\overset{m}{\underset{\rho=1}{\oplus}} (f(r_{\sigma(1),\rho}) + f(r_{\rho,\sigma(3)}) - 0.5)\right); \\ \ldots; f(r_{\sigma(1),\sigma(m)}) = \frac{1}{m}\left(\overset{m}{\underset{\rho=1}{\oplus}} (f(r_{\sigma(1),\rho}) + f(r_{\rho,\sigma(m)}) - 0.5)\right) \\ f(r_{\sigma(2),\sigma(4)}) = \frac{1}{m}\left(\overset{m}{\underset{\rho=1}{\oplus}} (f(r_{\sigma(2),\rho}) + f(r_{\rho,\sigma(4)}) - 0.5)\right); \\ \ldots; f(r_{\sigma(2),\sigma(m)}) = \frac{1}{m}\left(\overset{m}{\underset{\rho=1}{\oplus}} (f(r_{\sigma(2),\rho}) + f(r_{\rho,\sigma(m)}) - 0.5)\right) \\ \ldots \\ f(r_{\sigma(m-2),\sigma(m)}) = \frac{1}{m}\left(\overset{m}{\underset{\rho=1}{\oplus}} (f(r_{\sigma(m-2),\rho}) + f(r_{\rho,\sigma(m)}) - 0.5)\right) \end{cases} \tag{4.9}$$

Obviously, in both models (4.8) and (4.9), some repeating elements $r_{ii}(i = 1, 2, \ldots, m)$ exist in every equation. Therefore, the models (4.8) and (4.9) can be rewritten as follows:

$$\left\{\begin{array}{l} r_{\sigma(1),\sigma(3)}=\frac{1}{m-2}\left(\mathop{\oplus}\limits_{\substack{\rho=1\\ \rho\neq\sigma(1)\\ \rho\neq\sigma(3)}}^{m}(r_{\sigma(1),\rho}+r_{\rho,\sigma(3)})\right);\ldots;r_{\sigma(1),\sigma(m)}=\frac{1}{m}\left(\mathop{\oplus}\limits_{\substack{\rho=1\\ \rho\neq\sigma(1)\\ \rho\neq\sigma(m)}}^{m}(r_{\sigma(1),\rho}+r_{\rho,\sigma(m)})\right)\\ r_{\sigma(2),\sigma(4)}=\frac{1}{m-2}\left(\mathop{\oplus}\limits_{\substack{\rho=1\\ \rho\neq\sigma(2)\\ \rho\neq\sigma(4)}}^{m}(r_{\sigma(2),\rho}+r_{\rho,\sigma(4)})\right);\ldots;r_{\sigma(2),\sigma(m)}=\frac{1}{m}\left(\mathop{\oplus}\limits_{\substack{\rho=1\\ \rho\neq\sigma(2)\\ \rho\neq\sigma(m)}}^{m}(r_{\sigma(2),\rho}+r_{\rho,\sigma(m)})\right)\\ \cdots\\ r_{\sigma(m-2),\sigma(m)}=\frac{1}{m-2}\left(\mathop{\oplus}\limits_{\substack{\rho=1\\ \rho\neq\sigma(m-2)\\ \rho\neq\sigma(m)}}^{m}(r_{\sigma(m-2),\rho}+r_{\rho,\sigma(m)})\right) \end{array}\right. \tag{4.10}$$

$$\left\{\begin{array}{l} f(r_{\sigma(1),\sigma(3)})=\frac{1}{m-2}\left(\mathop{\oplus}\limits_{\substack{\rho=1\\ \rho\neq\sigma(1)\\ \rho\neq\sigma(3)}}^{m}(f(r_{\sigma(1),\rho})+f(r_{\rho,\sigma(3)})-0.5)\right);\\ f(r_{\sigma(2),\sigma(4)})=\frac{1}{m-2}\left(\mathop{\oplus}\limits_{\substack{\rho=1\\ \rho\neq\sigma(2)\\ \rho\neq\sigma(4)}}^{m}(f(r_{\sigma(2),\rho})+f(r_{\rho,\sigma(4)})-0.5)\right);\\ \cdots\\ f(r_{\sigma(m-2),\sigma(m)})=\frac{1}{m-2}\left(\mathop{\oplus}\limits_{\substack{\rho=1\\ \rho\neq\sigma(m-2)\\ \rho\neq\sigma(m)}}^{m}(f(r_{\sigma(m-2),\rho})+f(r_{\rho,\sigma(m)})-0.5)\right) \end{array}\right. \tag{4.11}$$

Remark 4.3 (Gou et al. 2020c). When we transform the second kind of LPO in decentralized form, the fundamental is also to calculate the unknown elements with the existing linguistic preference information. Therefore, we can also utilize the model (4.10) or (4.11) to obtain the remaining elements of a DHLPR. Additionally, considering that the DHLPR is developed based on Definition 4.5, so the obtained DHLPR $\mathbb{R}$ is completely consistent.

Two examples can be set up to show the process of transformation models clearly:

Example 4.2 (Gou et al. 2020c). Let $S_O = \{s_{t<o_k>} | t = -4, \ldots, 4; \ k = -4, \ldots, 4\}$ be a DHLTS, $LPO' = \left\{A_2 \overset{s_{2<o_1>}}{>} A_3 \overset{s_{1<o_2>}}{>} A_1 \overset{s_{0<o_1>}}{>} A_4\right\}$ be a LPO. Then the incomplete DHLPR, denoted by $\mathbb{R}'$, is established as:

$$\mathbb{R}' = (r_{ij})_{4\times4} = \begin{pmatrix} s_{0<o_0>} & - & s_{-1<o_{-2}>} & s_{0<o_1>} \\ - & s_{0<o_0>} & s_{2<o_1>} & - \\ s_{1<o_2>} & s_{-2<o_{-1}>} & s_{0<o_0>} & - \\ s_{0<o_{-1}>} & - & - & s_{0<o_0>} \end{pmatrix}$$

Therefore, it is only necessary to calculate r_{12}, r_{24}, and r_{34}. Based on the model (4.10) or (4.11), two kinds of equation sets are listed as follows:

$$\begin{cases} f(r_{12}) = \frac{1}{2}\left(\underset{\rho=3,\rho=4}{\oplus} (f(r_{1\rho}) + f(r_{\rho 2}) - 0.5) \right) \\ f(r_{24}) = \frac{1}{2}\left(\underset{\rho=1,\rho=3}{\oplus} (f(r_{2\rho}) + f(r_{\rho 4}) - 0.5) \right) \\ f(r_{34}) = \frac{1}{2}\left(\underset{\rho=1,\rho=2}{\oplus} (f(r_{3\rho}) + f(r_{\rho 4}) - 0.5) \right) \end{cases} \text{and} \begin{cases} r_{12} = \frac{1}{2}\left(\underset{\rho=3,\rho=4}{\oplus} (r_{1\rho} + r_{\rho 2}) \right) \\ r_{24} = \frac{1}{2}\left(\underset{\rho=1,\rho=3}{\oplus} (r_{2\rho} + r_{\rho 4}) \right) \\ r_{34} = \frac{1}{2}\left(\underset{\rho=1,\rho=2}{\oplus} (r_{3\rho} + r_{\rho 4}) \right) \end{cases}.$$

Solving these two equation sets, we obtain $r_{12} = s_{-4<o_2>}$, $r_{24} = s_{4<o_0>}$, and $r_{34} = s_{1<o_2>}$. Then, the consistent DHLPR $\overline{\mathbb{R}}'$ is obtained as:

$$\overline{\mathbb{R}}' = \begin{pmatrix} s_{0<o_0>} & s_{-4<o_1>} & s_{-1<o_{-2}>} & s_{0<o_1>} \\ s_{4<o_{-1}>} & s_{0<o_0>} & s_{2<o_1>} & s_{4<o_0>} \\ s_{1<o_2>} & s_{-2<o_{-1}>} & s_{0<o_0>} & s_{1<o_3>} \\ s_{0<o_{-1}>} & s_{-4<o_0>} & s_{-1<o_{-3}>} & s_{0<o_0>} \end{pmatrix}.$$

Example 4.3 (Gou et al. 2020c). Let $S_O = \{s_{t<o_k>} \mid t = -4,\ldots,4;\ k = -4,\ldots,4\}$ be a DHLTS, $LPO'' = \left\{ A_2 \overset{s_{1<o_1>}}{>} A_3, A_2 \overset{s_{2<o_1>}}{>} A_1, A_1 \overset{s_{1<o_{-1}>}}{>} A_4 \right\}$ be a LPO. Then the incomplete DHLPR, denoted by $\mathbb{R}''$, is established as:

$$\mathbb{R}'' = (r_{ij})_{4\times4} = \begin{pmatrix} s_{0<o_0>} & s_{-2<o_{-1}>} & - & s_{1<o_{-1}>} \\ s_{2<o_1>} & s_{0<o_0>} & s_{1<o_1>} & - \\ - & s_{-1<o_{-1}>} & s_{0<o_0>} & - \\ s_{-1<o_1>} & - & - & s_{0<o_0>} \end{pmatrix}$$

Similarly, we need to calculate r_{13}, r_{24}, and r_{34} based on Eq. (4.10) or (4.11), and two kinds of the equation sets are listed as follows:

$$\begin{cases} r_{13} = \frac{1}{2}\left(\underset{\rho=2,\rho=4}{\oplus}(r_{1\rho} + r_{\rho 3})\right) \\ r_{24} = \frac{1}{2}\left(\underset{\rho=1,\rho=3}{\oplus}(r_{2\rho} + r_{\rho 4})\right) \\ r_{34} = \frac{1}{2}\left(\underset{\rho=1,\rho=2}{\oplus}(r_{3\rho} + r_{\rho 4})\right) \end{cases} \text{and} \begin{cases} f(r_{13}) = \frac{1}{2}\left(\underset{\rho=2,\rho=4}{\oplus}(f(r_{1\rho}) + f(r_{\rho 3}) - 0.5)\right) \\ f(r_{24}) = \frac{1}{2}\left(\underset{\rho=1,\rho=3}{\oplus}(f(r_{2\rho}) + f(r_{\rho 4}) - 0.5)\right) \\ f(r_{34}) = \frac{1}{2}\left(\underset{\rho=1,\rho=2}{\oplus}(f(r_{3\rho}) + f(r_{\rho 4}) - 0.5)\right) \end{cases}.$$

Solving these two equation sets, we obtain $r_{13} = s_{-1<o_0>}$, $r_{24} = s_{3<o_0>}$, and $r_{34} = s_{1<o_3>}$. Then, the consistent DHLPR $\overline{\mathbb{R}}''$ is obtained as:

$$\overline{\mathbb{R}}'' = (r_{ij})_{4\times 4} = \begin{pmatrix} s_{0<o_0>} & s_{-2<o_{-1}>} & s_{-1<o_0>} & s_{1<o_{-1}>} \\ s_{2<o_1>} & s_{0<o_0>} & s_{1<o_1>} & s_{3<o_0>} \\ s_{1<o_0>} & s_{-1<o_{-1}>} & s_{0<o_0>} & s_{1<o_3>} \\ s_{-1<o_1>} & s_{-3<o_0>} & s_{-1<o_{-3}>} & s_{0<o_0>} \end{pmatrix}.$$

4.2.3 Consensus Model for LPOs

A GDM problem with linguistic preference information can be described as follows: $A = \{A_1, A_2, \ldots, A_m\}$ is a set of alternatives, $E = \{e^1, e^2, \ldots, e^n\}$ is a set of experts, and experts' preference information is expressed by some LPOs $\{LPO^1, LPO^2, \ldots, LPO^n\}$.

After obtaining the DHLPRs of all experts with the completely acceptable consistencies, the next step will focus on reaching group consensus by developing a consensus reaching method. Therefore, this subsection will introduce a multi-stage consensus optimization model to reach group consensus firstly, and then an interactive consensus reaching algorithm is proposed on the basis of the proposed consensus model.

(a) **Group consensus measures**

Next, we mainly propose two group consensus measures to calculate the consensus degrees based on the distance between an individual DHLPR and the collective DHLPR, and the distance between an individual DHLPR and the remaining DHLPRs, respectively.

Firstly, the collective DHLPRs can be obtained by the following aggregating method:

Definition 4.7 (Gou et al. 2020c). Let $\{\mathbb{R}_1, \mathbb{R}_2, \ldots, \mathbb{R}_n\}$ be a set of individual DHLPRs, where $\mathbb{R}_a = (r_{ij}^a)_{m\times m} (a = 1, 2, \ldots, n)$, and the weight vector of them is $w = (w_1, w_2, \ldots, w_n)^T$. Then the collective DHLPR, denoted by $\mathbb{R}_c = (r_{ij}^c)_{m\times m}$, is obtained, where $r_{ij}^c = \overset{n}{\underset{a=1}{\oplus}} w_a r_{ij}^a$.

Then, the consensus degree of an individual DHLPR can be defined as follows:

Definition 4.8 (Gou et al. 2020c). Let $\{\mathbb{R}_1, \mathbb{R}_2, \ldots, \mathbb{R}_n\}$ be a set of individual DHLPRs transformed from the LPOs provided by the experts $e^a(a = 1, 2, \ldots, n)$, where $\mathbb{R}_a = (r_{ij}^a)_{m\times m}$, $\mathbb{R}_c = (r_{ij}^c)_{m\times m}$ be a collective DHLPR. Then, the consensus degree of an individual DHLPR $\mathbb{R}_a$ is calculated by

$$CD'(\mathbb{R}_a) = 1 - d(\mathbb{R}_a, \mathbb{R}_c) = 1 - \frac{2}{m(m-1)} \sum_{i=1}^{m-1} \sum_{j=i+1}^{m} \left| f(r_{ij}^a) - f(r_{ij}^c) \right| \quad (4.12)$$

Additionally, the group consensus degree is obtained by

$$CD' = \min_a CD'(\mathbb{R}_a) \quad (4.13)$$

Suppose that $\overline{CD}$ is a given consensus threshold, then the group consensus has been reached if the group consensus degree satisfies

$$CD' = \min_a CD'(\mathbb{R}_a) \geq \overline{CD} \quad (4.14)$$

Additionally, if we only consider the distance between any two different DHLPRs instead of the distance between an individual and collective DHLPRs, the other consensus degree of a DHLPR is defined as follows:

Definition 4.9 (Gou et al. 2020c). Let $\{\mathbb{R}_1, \mathbb{R}_2, \ldots, \mathbb{R}_n\}$ be a set of individual DHLPRs transformed from the LPOs provided by the experts $e^a(a = 1, 2, \ldots, n)$, and $\mathbb{R}_a = (r_{ij}^a)_{m\times m}$. Then, the consensus degree for an individual DHLPR $\mathbb{R}_a$ is calculated by

$$\begin{aligned} CD''(\mathbb{R}_a) &= 1 - \frac{1}{n-1} \sum_{b=1; b\neq a}^{n} d(\mathbb{R}_a, \mathbb{R}_b) \\ &= 1 - \frac{2}{m(m-1)(n-1)} \sum_{b=1; b\neq a}^{n} \sum_{i=1}^{m-1} \sum_{j=i+1}^{m} \left| f(r_{ij}^a) - f(r_{ij}^b) \right| \end{aligned} \quad (4.15)$$

Similarly, based on Eq. (4.15), the other kind of group consensus degree is obtained by

$$CD'' = \min_a CD''(\mathbb{R}_a) \quad (4.16)$$

Then, the group consensus has been reached if

$$CD'' = \min_a CD''(\mathbb{R}_a) \geq \overline{CD} \tag{4.17}$$

Remark 4.4 (Gou et al. 2020c). Obviously, both of two types of consensus measures are very closely related to each other. However, Definition 4.9 has more restrictive than Definition 4.8. Therefore, this subsection mainly uses the second group consensus degree CD'' in the consensus reaching process.

Additionally, the given consensus threshold value $\overline{CD}$ can be used to decide whether the consensus reaching process can be carried out. We usually set the consensus threshold to be smaller than 0.9 (Herrera-Viedma et al. 2005; Parreiras et al. 2012). If $CD \geq \overline{CD}$, then the consensus degree of all experts is sufficiently high and the consensus reaching process is over. Otherwise, we should make some changes about preference to improve the consensus degree and reach the given consensus threshold value. Furthermore, according to the practical situation of the decision-making problem, the value of the consensus threshold $\overline{CD}$ can be adjusted. If the demand for consensus is strict, then the threshold should be given a higher value; Otherwise, a lower threshold value should be provided.

(b) **Multi-stage consensus optimization model**

Considering that all the transformed DHLPRs are completely consistent, it is not necessary to improve the consistencies of them before starting the consensus reaching process. Gou et al. (2020c) developed a multi-stage consensus optimization model which can not only reach group consensus but also maintain the consistency of every DHLPR. This multi-stage optimization model consists of three objectives, the first one is to minimize the deviations of the modification magnitudes, the second one is to minimize the cardinal number of modifications while keeping the value of the first objective constant, and the third one is to minimize the number of experts who need to change their evaluations.

Stage 1. Minimizing the deviations of the modification magnitudes

In general, the experts more prefer to keep their preferences constant or only want to change their preferences as little as possible. Therefore, the first stage of the multi-stage consensus optimization model is to make the original DHLPR and the adjusted DHLPR as close as possible. According to this, we can develop a model to minimize the deviations of the modification magnitudes. Let $\mathbb{R}_a = (r_{ij}^a)_{m \times m}(a = 1, 2, \ldots, n)$ be the transformed DHLPRs provided by the experts $e^a(a = 1, 2, \ldots, n)$, and $\mathbb{R}_a^* = \left(r_{ij}^{*a}\right)_{m \times m}(a = 1, 2, \ldots, n)$ be the corresponding adjusted DHLPR. Then the objective can be given as:

$$\sum_{a=1}^{n} d\left(\mathbb{R}_a, \mathbb{R}_a^*\right) = \sum_{a=1}^{n} \sum_{i=1}^{m-1} \sum_{j=i+1}^{m} \left| f(r_{ij}^a) - f(r_{ij}^{*a}) \right| \tag{4.18}$$

Due to the fact that all individual DHLPRs should be of acceptable consistencies and the group consensus also needs to be reached, then the model in this stage is developed as follows:

$$\min \quad G_1 = \sum_{a=1}^{n} d(\mathbb{R}_a, \mathbb{R}_a^*) = \sum_{a=1}^{n}\sum_{i=1}^{m-1}\sum_{j=i+1}^{m} \left| f(r_{ij}^a) - f(r_{ij}^{*a}) \right|$$
$$s.t. \begin{cases} CI(\mathbb{R}_a^*) \geq \overline{CI}, & a = 1,2,\ldots,n, \quad (4.19\text{–}1) \\ CD(\mathbb{R}_a) \geq \overline{CD}, & a = 1,2,\ldots,n, \quad (4.19\text{–}2) \\ r_{ij}^{*a} \in \overline{S}_O \text{ or } r_{ij}^{*a} \in S_O, & i<j, a = 1,2,\ldots,n, \quad (4.19\text{–}3) \end{cases} \tag{4.19}$$

Based on Eqs. (4.7) and (4.15), the model (4.19) can be changed into

$$\min \quad G_1 = \sum_{a=1}^{n} d(\mathbb{R}_a, \mathbb{R}_a^*) = \sum_{a=1}^{n}\sum_{i=1}^{m-1}\sum_{j=i+1}^{m} \left| f(r_{ij}^a) - f(r_{ij}^{*a}) \right|$$
$$s.t. \begin{cases} \sum_{i=1}^{m-2}\sum_{j=i+1}^{m-1}\sum_{\rho=j+1}^{m} \left| f(r_{ij}^{*a}) - f(r_{i\rho}^{*a}) + f(r_{j\rho}^{*a}) - 0.5 \right| & a = 1,2,\ldots,n, \quad (4.20\text{–}1) \\ \quad \leq SCI, & \\ \sum_{b=1;b\neq a}^{n}\sum_{i=1}^{m-1}\sum_{j=i+1}^{m} \left| f(r_{ij}^a) - f(r_{ij}^{*a}) \right| \leq SCD, & a = 1,2,\ldots,n, \quad (4.20\text{–}2) \\ r_{ij}^{*a} \in \overline{S}_O \text{ or } r_{ij}^{*a} \in S_O, & i<j, a = 1,2,\ldots,n, \quad (4.20\text{–}3) \end{cases} \tag{4.20}$$

where $SCI = (1 - \overline{CI}) \times \frac{m(m-1)(m-2)}{4}$ and $SCD = (1 - \overline{CD}) \times \frac{m(m-1)(n-1)}{2}$.

By solving Eq. (4.19) or (4.20), we can obtain at least one optimal solution, which is denoted as G_1^*.

Stage 2. Minimizing the cardinal number of modifications

In the consensus reaching process, lots of existing methods need to change all the original preferences such as the automatic adjustment method (Gou et al. 2018; Liao et al. 2016; Zhang and Chen 2019). However, it will create a huge amount of work and make the consensus process complicated. Therefore, this stage focuses on developing a model to minimize the number of modifications and keep the optimal solution obtained in first stage.

Then, we need to verify whether the original preference changes, so a constraint can be set up:

$$\left| f(r_{ij}^a) - f(r_{ij}^{*a}) \right| \leq \Im \phi_{ij}^a \tag{4.21}$$

where $\Im$ is a sufficiently large number and $\phi_{ij}^a \in \{0, 1\}$. Clearly, if $\left| f(r_{ij}^a) - f(r_{ij}^{*a}) \right| = 0$ or $\phi_{ij}^a = 0$, then the original preference r_{ij}^a has not been changed; otherwise, the original preference r_{ij}^a has been changed.

Based on the optimal solution G_1^*, the model of the second stage is developed as follows:

$$\min \quad G_2 = \sum_{a=1}^{n} \sum_{i=1}^{m-1} \sum_{j=i+1}^{m} \phi_{ij}^{a}$$

$$s.t.\begin{cases} \sum_{a=1}^{n} \sum_{i=1}^{m-1} \sum_{j=i+1}^{m} \left| f(r_{ij}^{a}) - f(r_{ij}^{*a}) \right| = G_1^*, & (4.22\text{–}1) \\ \sum_{i=1}^{m-2} \sum_{j=i+1}^{m-1} \sum_{\rho=j+1}^{m} \left| f(r_{ij}^{*a}) - f(r_{i\rho}^{*a}) - f(r_{\rho j}^{*a}) + 0.5 \right| \le SCI, \quad a = 1,2,\ldots,n, & (4.22\text{–}2) \\ \sum_{b=1;b\neq a}^{n} \sum_{i=1}^{m-1} \sum_{j=i+1}^{m} \left| f(r_{ij}^{*a}) - f(r_{ij}^{*b}) \right| \le SCD, \quad a = 1,2,\ldots,n, & (4.22\text{–}3) \\ \left| f(r_{ij}^{a}) - f(r_{ij}^{*a}) \right| \le \Im \phi_{ij}^{a}, \quad i<j, a = 1,2,\ldots,n, & (4.22\text{–}4) \\ \phi_{ij}^{a} \in \{0,1\}, \quad i<j, a = 1,2,\ldots,n, & (4.22\text{–}5) \\ r_{ij}^{*a} \in \overline{S}_O \ or \ r_{ij}^{*a} \in S_O, \quad i<j, a = 1,2,\ldots,n, & (4.22\text{–}6) \end{cases} \tag{4.22}$$

where $SCI = (1 - \overline{CI}) \times \frac{m(m-1)(m-2)}{4}$, $SCD = (1 - \overline{CD}) \times \frac{m(m-1)(n-1)}{2}$, and $\Im$ is a sufficiently large number.

Similarly, solving Eq. (4.22), we can also obtain at least one optimal solution, which is denoted as G_2^*.

Stage 3. Minimizing the number of experts who need to change their preferences

Considering that the consensus will be reached faster if we can minimize the number of experts who need to change their preferences within reasonable limits, so this stage mainly focuses on this goal. To verify whether an expert e^a changes his/her preference information, a constraint can be set up

$$\sum_{i=1}^{m-1} \sum_{j=i+1}^{m} \left| f(r_{ij}^{a}) - f(r_{ij}^{*a}) \right| \le \Im \varphi^{a} \tag{4.23}$$

where $\Im$ is a sufficiently large number and $\varphi^a \in \{0,1\}$. Therefore, if $\sum_{i=1}^{m-1} \sum_{j=i+1}^{m} \left| f(r_{ij}^{a}) - f(r_{ij}^{*a}) \right| = 0$ or $\varphi^z = 0$, then the expert e^a does not change his preference information; otherwise the expert e^a has changed some preference information.

Based on the optimal solution G_2^* obtained in Eq. (4.22), the model of the third stage can be established as follows:

$$\min \quad G_3 = \sum_{a=1}^{n} \varphi^a$$

$$s.t.\begin{cases} \sum_{a=1}^{n}\sum_{i=1}^{m-1}\sum_{j=i+1}^{m} \phi_{ij}^a = G_2^*, & & (4.24\text{–}1) \\ \sum_{a=1}^{n}\sum_{i=1}^{m-1}\sum_{j=i+1}^{m} \left|f(r_{ij}^a) - f(r_{ij}^{*a})\right| = G_1^*, & & (4.24\text{–}2) \\ \sum_{i=1}^{m-1}\sum_{j=i+1}^{m} \left|f(r_{ij}^a) - f(r_{ij}^{*a})\right| \le \Im\varphi^a, & a = 1,2,\ldots,n, & (4.24\text{–}3) \\ \varphi^a \in \{0,1\}, & a = 1,2,\ldots,n, & (4.24\text{–}4) \\ \sum_{i=1}^{m-2}\sum_{j=i+1}^{m-1}\sum_{\rho=j+1}^{m} \left|f(r_{ij}^{*a}) - f(r_{i\rho}^{*a}) - f(r_{\rho j}^{*a}) + 0.5\right| \le SCI, & a = 1,2,\ldots,n, & (4.24\text{–}5) \\ \sum_{b=1;b\neq a}^{n}\sum_{i=1}^{m-1}\sum_{j=i+1}^{m} \left|f(r_{ij}^{*a}) - f(r_{ij}^{*b})\right| \le SCD, & a = 1,2,\ldots,n, & (4.24\text{–}6) \\ \left|f(r_{ij}^a) - f(r_{ij}^{*a})\right| \le \Im\phi_{ij}^a, & i<j,\ a = 1,2,\ldots,n, & (4.24\text{–}7) \\ \phi_{ij}^a \in \{0,1\}, & i<j,\ z = 1,2,\ldots,n, & (4.24\text{–}8) \\ r_{ij}^{*a} \in \overline{S}_O \ or\ r_{ij}^{*a} \in S_O, & i<j, a = 1,2,\ldots,n, & (4.24\text{–}9) \end{cases} \tag{4.24}$$

Similarly, solving the model (4.24), at least one optimal solution can be obtained, denoted as G_3^*.

When we obtain the optimal adjusted DHLPRs $\mathbb{R}_a^* = (r_{ij}^{*a})_{m\times m}$ $(a = 1,2,\ldots,n)$, the next step is to identify the set of experts who should change their preferences. Moreover, it is necessary to identify the positions that should be changed for each expert e^a. The identification rules can be developed as follows:

(1) Based on the optimal solutions obtained in the model (4.24), the set of experts who need to change their preferences can be obtained:

$$ES = \left\{ e^a \left| \sum_{i=1}^{m-1}\sum_{j=i+1}^{m} \left|f(r_{ij}^a) - f(r_{ij}^{*a})\right| \neq 0, a = 1,2,\ldots,n \right.\right\} \tag{4.25}$$

(2) For every expert $e^a \in ES$, we have

$$P(e^a) = \left\{(i,j)\left|\left|f(r_{ij}^a) - f(r_{ij}^{*a})\right| > 0,\ i,j = 1,2,\ldots,m, i<j\right.\right\} \tag{4.26}$$

where the positions $(i,j)(i,j = 1,2,\ldots,m, i<j)$ that should be changed for the expert e^a can be identified.

Remark 4.5 (Gou et al. 2020c). After identifying the expert $e^a \in ES$ and the positions (i,j) by Eqs. (4.25) and (4.26), we need to feedback the adjustment suggestions to the corresponding experts. In this process, an interval value can be developed as follows:

$$IV = \left[\min\{r_{ij}^{a}, r_{ij}^{*a}\}, \max\{r_{ij}^{a}, r_{ij}^{*a}\}\right], \quad a = 1, 2, \ldots, n.$$

Then, the experts can provide their adjusted preferences according to the suggested interval values.

(c) **Interactive consensus reaching algorithm with LPOs**

Next, we mainly develop an interactive consensus reaching algorithm with LPOs in GDM. This algorithm consists of two parts. One is to transform the LPOs into the corresponding DHLPRs $\mathbb{R}_a = (r_{ij}^{a})_{m\times m}(a = 1, 2, \ldots, n)$ with complete consistencies, and we do not need to check the consistency of each DHLPR anymore. The other one is to reach group consensus on the basis of the multi-stage consensus optimization model discussed above.

The consensus reaching process is a dynamic decision-making process. Firstly, based on Definition 4.9, we can obtain the consensus degree $CD''(R_a)$ of each DHLPR and decide which one needs to adjust his/her preference. Secondly, we need to use the multi-stage consensus optimization model to obtain the optimal solution. In this process, we can develop an identification rules which can identify the expert who has to adjust his preference and the positions that should be adjusted. Additionally, we can also get the suggestions by combining the original DHLPR and the optimal solution, and then feedback the suggestions to the expert. Finally, we update the preference based on the adjustments provided by the expert.

Algorithm 4.2 (Gou et al. 2020c). Multi-stage interactive consensus reaching algorithm with LPOs

Input: The original LPOs $\{LPO^1, LPO^2, \ldots, LPO^n\}$, the given consistency threshold value $\overline{CI}$, the given consensus threshold value $\overline{CD}$, and the maximum iteration number $\mathbb{Z}_{\max}$.

Output: The improved DHLPRs $\{*\mathbb{R}_1, *\mathbb{R}_2, \ldots, *\mathbb{R}_n\}$, the final reached consensus degree for each expert e^a, $CD(*\mathbb{R}_a)$ $(a = 1, 2, \ldots, n)$, and the iteration number $\mathbb{Z}$.

Step 1. Based on Eq. (4.10) or (4.11), we transform each LPO to the corresponding DHLPR with complete consistency, $\mathbb{R}_a = (r_{ij}^{a})_{m\times m}(a = 1, 2, \ldots, n)$.

Step 2. Let $\mathbb{Z} = 1$, and $R_a^{(\mathbb{Z})} = (r_{ij}^{a,(\mathbb{Z})})_{m\times m} = (r_{ij}^{a})_{m\times m}(a = 1, 2, \ldots, n)$.

Step 3. Based on Eq. (4.15), we obtain the consensus degree for each DHLPR $\mathbb{R}_a^{(\mathbb{Z})}$, $CD''(\mathbb{R}_a^{(\mathbb{Z})})$ $(a = 1, 2, \ldots, n)$. If $CD''(\mathbb{R}_{a''}^{(\mathbb{Z})}) \geq \overline{CD}$ or $\mathbb{Z} > \mathbb{Z}_{\max}$, then go to Step 6; otherwise, go to Step 4.

Step 4. Calculate the optimal adjusted DHLPR $\mathbb{R}_a^{*(\mathbb{Z})} = (r_{ij}^{*a,(\mathbb{Z})})_{m\times m}(a = 1, 2, \ldots, n)$ on the basis of the proposed multi-stage consensus optimization models.

Step 5. The identification rules proposed in Eqs. (4.25) and (4.26) can be used to identify the set of experts who need to change their preferences, denoted as $ES^{(\mathbb{Z})}$, and the position $P(e^a)$ that should be changed for the expert e^a. $\mathbb{R}_a^{*(\mathbb{Z})}(a = 1,2,\ldots,n)$ can be regarded as a decision aid and fed back to the corresponding expert. Let $R_a^{(\mathbb{Z}+1)} = (r_{ij}^{a,(\mathbb{Z}+1)})_{m\times m}(a = 1,2,\ldots,n)$ be the adjusted DHLPRs provided by experts. We can send the following suggestions to experts:

$$r_{ij}^{a,(\mathbb{Z}+1)} \in [\min\{r_{ij}^{a,(\mathbb{Z})}, r_{ij}^{*a,(\mathbb{Z})}\}, \max\{r_{ij}^{a,(\mathbb{Z})}, r_{ij}^{*a,(\mathbb{Z})}\}], \quad a = 1,2,\ldots,n \tag{4.27}$$

Then, the experts are advised to change their preferences. Let $\mathbb{Z} = \mathbb{Z}+1$, and go back to Step 2.

Step 6. Let $*\mathbb{R}_a = \mathbb{R}_a^{(\mathbb{Z})}(a = 1,2,\ldots,n)$. Based on Definition 3.7, we obtain the collective DHLPR $*\mathbb{R}_c = (*r_{ij}^c)_{m\times m}$. Then, we calculate the synthetical value of each alternative $SV(A_i) = \sum_{j=1}^{m} f(*r_{ij}^c)$. Finally, we obtain the rank of all alternatives by ranking the synthetical values in descending order.

Step 7. End.

One figure is drawn to show the multi-stage interactive consensus reaching algorithm with LPOs (Fig. 4.2).

4.2.4 Application of Multi-stage Interactive Consensus Reaching Algorithm with LPOs

This subsection mainly sets up two numerical examples to show the proposed transformation methods and the multi-stage interactive consensus reaching algorithm. Then, some comparative analyses with some existing methods are made to show the advantages and effectiveness of the proposed methods.

(1) **Numerical examples about transformation methods and comparative analyses**

Example 4.4 (Gou et al. 2020c). Let $S_O = \{s_{t<o_k>} | t = -4,\ldots,4; k = -4,\ldots,4\}$. Suppose that two experts provide their LPOs for five alternatives $\{A_1, A_2,\ldots,A_5\}$ as follows:

$$LPO^1 = \left\{A_5 \overset{s_{0<o_2>}}{>} A_2 \overset{s_{2<o_1>}}{>} A_3 \overset{s_{1<o_{-1}>}}{>} A_1 \overset{s_{0<o_1>}}{>} A_4\right\},$$

$$LPO^2 = \left\{A_2 \overset{s_{1<o_1>}}{>} A_3, A_2 \overset{s_{2<o_1>}}{>} A_1, A_1 \overset{s_{1<o_{-1}>}}{>} A_4, A_5 \overset{s_{3<o_{-1}>}}{>} A_4\right\}.$$

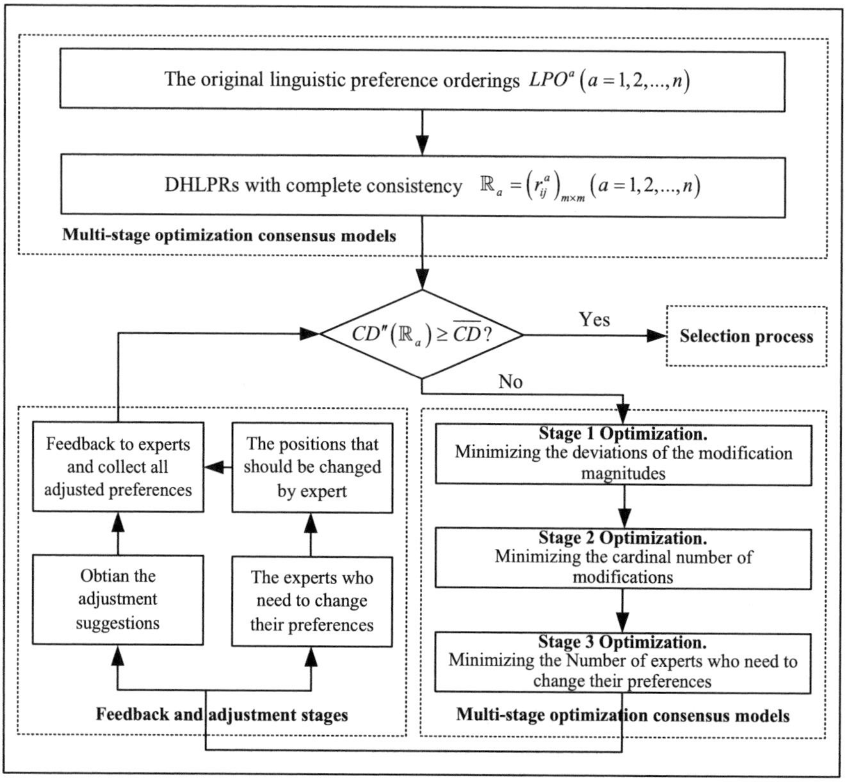

Fig. 4.2 Multi-stage interactive group consensus optimization process with LPOs

Firstly, based on Eqs. (4.10) or (4.11), the LPOs can be transformed into the corresponding DHLPRs $\mathbb{R}^1=(r_{ij}^1)_{5\times5}$ and $\mathbb{R}^2=(r_{ij}^2)_{5\times5}$, respectively. Then, based on the transformation function f, two DHLPRs are transformed into the corresponding fuzzy preference relations (FPRs) (Kacprzyk and Roubens 1988) $P^1=(p_{ij}^1)_{5\times5}$ and $P^2=(p_{ij}^2)_{5\times5}$, where $p_{ij}^k+p_{ji}^k=1$ and $p_{ii}^k=0.5\ (i,j=1,2,\ldots,5;k=1,2)$.

Additionally, the concept of classical preference ordering (Chiclana et al. 1998) can be shown as follows: An expert e^k provides his/her preference on A as a preference ordering, $O^k=\{o_1^k,o_2^k,\ldots,o_m^k\}(k=1,2,\ldots,n)$, where $o_i^k(i=1,2,\ldots,m)$ denote the positional orders of the alternatives $A_i(i=1,2,\ldots,m)$. Then, Chiclana et al. (1998) developed two functions, denoted as g^1 and g^2, to transform preference ordering O^k into the FPR $P^k=(p_{ij}^k)_{m\times m}$:

$$(1)\ g^1:p_{ij}^k=g^1(o_j^k-o_i^k)=\begin{cases}1 & if\ o_j^k>o_i^k\\ 0 & if\ o_i^k<o_j^k\end{cases}\quad(i\neq j);$$

(2) $g^2 : p_{ij}^k = g^2(o_j^k - o_i^k) = \begin{cases} 1 & \text{if } o_j^k > o_i^k \\ \frac{1}{2} & \text{if } o_j^k = o_i^k \\ 0 & \text{if } o_j^k < o_i^k \end{cases}$;

(3) By defining the concept of utility value $u_i = \frac{m - o_i^k}{m-1}$, Dombi (1995) utilized a function g^3 to transform O^k into the FPR $P^k = (p_{ij}^k)_{m \times m}$, where $p_{ij}^k = g^3(u_i - u_j) = \frac{1}{2}(1 + u_i - u_j)$;

(4) By the utility values given by Tanino (1988), Chiclana et al. (1998) utilized a function g^4 to transform the O^k into the FPR P^k, where $p_{ij}^k = g^4(o_j^k - o_i^k) = \frac{1}{2}(1 + \frac{o_j^k}{m-1} - \frac{o_i^k}{m-1})$.

Based on the transformation function f and other transformation functions $g^k (k = 1, 2, 3, 4)$ (Chiclana et al. 1998; Dombi 1995), we can transform LPO^1 into four corresponding FPRs $P^{1k} = (p_{ij}^{1k})_{5 \times 5} \, (k = 1, 2, 3, 4)$. The transformation results are shown in Table 4.5.

Based on Table 4.5, some comparative analyses are summarized as follows:

(1) As we know, the DHLPRs transformed from LPOs are completely consistent. Additionally, in this transformation process, we do not lose any original linguistic information. Furthermore, based on the transformation function f, two DHLPRs are transformed into the corresponding FPRs with complete consistencies.

(2) For the transformation function g^1, it is obvious that the FPR P^{11} does not reflect the case when there is an indifference between two alternatives. Additionally, the transformed FPR only uses the elements 0 and 1 to represent the relationship between two alternatives, but it cannot reflect the degree (the DHLT) to which one alternative is superior to another. Finally, the transformation function g^1 cannot ensure that the transformed FPR P^{11} is with complete consistency.

(3) Even though the transformation function g^2 can reflect the case when there is an indifference between two alternatives, both it cannot reflect any kind of intensity of preference between alternatives when we compare the pair of alternatives, and cannot ensure that the transformed FPR P^{12} is with complete consistency.

(4) It is clear that both the transformation functions g^3 and g^4 are equal. Both of them use the utility values to reflect the degree (the DHLT) to which one alternative is superior to another, but the degree has a certain proportion and cannot reflect the real relationship of any two alternatives. Additionally, it also cannot ensure that the transformed FPRs P^{13} and P^{14} are with complete consistencies.

In conclusion, the transformation functions $g^k (k = 1, 2, 3, 4)$ have different shortcomings. The transformation function proposed in this subsection can not only ensure that the DHLPRs and FPRs are with complete consistencies, but also fully keep the original linguistic information (DHLTSs) unchanged.

Table 4.5 The transformation results based on different methods

Methods	Transformation results based on LPO^1	Transformation results based on LPO^2
The proposed method	$\mathbb{R}^1 = \begin{pmatrix} s_{0<o_{0}>} & s_{-3<o_{0}>} & s_{-1<o_{1}>} & s_{0<o_{1}>} & s_{-3<o_{-2}>} \\ s_{3<o_{0}>} & s_{0<o_{0}>} & s_{2<o_{1}>} & s_{3<o_{1}>} & s_{0<o_{-2}>} \\ s_{1<o_{-1}>} & s_{-2<o_{-1}>} & s_{0<o_{0}>} & s_{1<o_{0}>} & s_{-2<o_{-3}>} \\ s_{0<o_{-1}>} & s_{-3<o_{-1}>} & s_{-1<o_{0}>} & s_{0<o_{0}>} & s_{-3<o_{-3}>} \\ s_{3<o_{2}>} & s_{0<o_{2}>} & s_{2<o_{3}>} & s_{3<o_{3}>} & s_{0<o_{0}>} \end{pmatrix}$ $(DHLPRs \rightarrow FPRs) \downarrow$ $P^1 = \begin{pmatrix} 1/2 & 1/8 & 13/32 & 17/32 & 1/16 \\ 7/8 & 1/2 & 25/32 & 29/32 & 7/16 \\ 19/32 & 7/32 & 1/2 & 5/8 & 5/32 \\ 15/32 & 3/32 & 3/8 & 1/2 & 1/32 \\ 15/16 & 11/16 & 27/32 & 31/32 & 1/2 \end{pmatrix}$	$\mathbb{R}^2 = \begin{pmatrix} s_{0<o_{0}>} & s_{-2<o_{-1}>} & s_{-1<o_{0}>} & s_{1<o_{-1}>} & s_{-2<o_{0}>} \\ s_{1<o_{1}>} & s_{0<o_{0}>} & s_{1<o_{1}>} & s_{3<o_{0}>} & s_{0<o_{1}>} \\ s_{1<o_{0}>} & s_{-1<o_{-1}>} & s_{0<o_{0}>} & s_{2<o_{-1}>} & s_{-1<o_{0}>} \\ s_{-1<o_{1}>} & s_{-3<o_{0}>} & s_{-2<o_{1}>} & s_{0<o_{0}>} & s_{-3<o_{1}>} \\ s_{2<o_{0}>} & s_{0<o_{-1}>} & s_{1<o_{0}>} & s_{3<o_{-1}>} & s_{0<o_{0}>} \end{pmatrix}$ $(DHLPRs \rightarrow FPRs) \downarrow$ $P^2 = \begin{pmatrix} 1/2 & 7/32 & 3/8 & 19/32 & 1/4 \\ 25/32 & 1/2 & 21/32 & 7/8 & 17/32 \\ 5/8 & 11/32 & 1/2 & 23/32 & 3/8 \\ 13/32 & 1/8 & 9/32 & 1/2 & 5/32 \\ 3/4 & 15/32 & 5/8 & 27/32 & 1/2 \end{pmatrix}$
Chiclana et al. (1998): g^1	$P^{11} = \begin{pmatrix} - & 0 & 0 & 1 & 0 \\ 1 & - & 1 & 1 & 0 \\ 1 & 0 & - & 1 & 0 \\ 0 & 0 & 0 & - & 0 \\ 1 & 1 & 1 & 1 & - \end{pmatrix}$	–
Chiclana et al. (1998): g^2	$P^{12} = \begin{pmatrix} 1/2 & 0 & 0 & 1 & 0 \\ 1 & 1/2 & 1 & 1 & 0 \\ 1 & 0 & 1/2 & 1 & 0 \\ 0 & 0 & 0 & 1/2 & 0 \\ 1 & 1 & 1 & 1 & 1/2 \end{pmatrix}$	–
Dombi (1995): g^3 Chiclana et al. (1998): g^4	$P^{13} = P^{14} = \begin{pmatrix} 1/2 & 3/4 & 5/8 & 3/8 & 7/8 \\ 1/4 & 1/2 & 3/8 & 1/8 & 5/8 \\ 3/8 & 5/8 & 1/2 & 1/4 & 3/4 \\ 5/8 & 7/8 & 3/4 & 1/2 & 1 \\ 1/8 & 3/8 & 1/4 & 0 & 1/2 \end{pmatrix}$	–

(2) **Numerical examples about consensus model and comparative analyses**

Example 4.5 (Gou et al. 2020c). Suppose that a logistics company needs to choose the most suitable supplier from five competitive alternatives $\{A_1, A_2, \ldots, A_5\}$. To do so, four experts $\{e^1, e^2, e^3, e^4\}$ are invited to evaluate these five suppliers. These four experts provide their preferences with LPOs $LPO^a(a = 1, 2, 3, 4)$ based on DHLTS $S_O = \{s_{t<o_k>} | t = -4, \ldots, 4; k = -4, \ldots, 4\}$ with $S = \{s_{-4} = extremely\ bad, s_{-3} = very\ bad, s_{-2} = bad,\ s_{-1} = slightlybad, s_0 = medium, s_1 = slightly\ good,\ s_2 = good,\ s_3 = very\ good,\ s_4 = extremely\ good\}$ and $O = \{o_{-4} = far\ from, o_{-3} = scarcely, o_{-2} = only\ a\ little, o_{-1} = a\ little, o_0 = just\ right, o_1 = much,\ o_2 = very\ much, o_3 = extremely\ much, o_4 = entirely\}$.

$$LPO^1 = \left\{A_5 \overset{s_{0<o_2>}}{>} A_2 \overset{s_{1<o_1>}}{>} A_3 \overset{s_{1<o_{-1}>}}{>} A_1 \overset{s_{0<o_1>}}{>} A_4\right\}$$

$$LPO^2 = \left\{A_2 \overset{s_{1<o_1>}}{>} A_3, A_2 \overset{s_{2<o_1>}}{>} A_1, A_1 \overset{s_{1<o_{-1}>}}{>} A_4, A_5 \overset{s_{2<o_1>}}{>} A_4\right\}$$

$$LPO^3 = \left\{A_3 \overset{s_{0<o_1>}}{>} A_5 \overset{s_{2<o_0>}}{>} A_2 \overset{s_{1<o_{-2}>}}{>} A_4 \overset{s_{0<o_3>}}{>} A_1\right\}$$

$$LPO^4 = \left\{A_1 \overset{s_{2<o_1>}}{>} A_3, A_2 \overset{s_{1<o_1>}}{>} A_3, A_5 \overset{s_{1<o_{-1}>}}{>} A_1, A_2 \overset{s_{1<o_{-2}>}}{>} A_4\right\}$$

Obviously, this is an actual GDM problem. Based on Table 3.14, we have $\overline{CI} = 0.8256$, and according to Remark 4.4, we can let $\overline{CD} = 0.85$ and $\mathbb{Z}_{\max} = 5$. But in the multi-stage consensus optimization model, to obtain the efficient intervals and send experts more helpful suggestions, we set $\overline{CI} = 0.9$ and $\overline{CD} = 0.9$.

Using Algorithm 4.2, the consensus reaching process is shown as follows:

Step 1. Transform all LPOs $LPO^a(a = 1, 2, 3, 4)$ into the DHLPRs $\mathbb{R}_a = (r_{ij}^a)_{5\times 5}\ (a = 1, 2, 3, 4)$:

$$\mathbb{R}_1 = \begin{pmatrix} s_{0<o_0>} & s_{-2<o_0>} & s_{-1<o_1>} & s_{0<o_1>} & s_{-2<o_{-2}>} \\ s_{2<o_0>} & s_{0<o_0>} & s_{1<o_1>} & s_{2<o_1>} & s_{0<o_{-2}>} \\ s_{1<o_{-1}>} & s_{-1<o_{-1}>} & s_{0<o_0>} & s_{1<o_0>} & s_{-1<o_{-3}>} \\ s_{0<o_{-1}>} & s_{-2<o_{-1}>} & s_{-1<o_0>} & s_{0<o_0>} & s_{-2<o_{-3}>} \\ s_{2<o_2>} & s_{0<o_2>} & s_{1<o_3>} & s_{2<o_3>} & s_{0<o_0>} \end{pmatrix}$$

$$\mathbb{R}_2 = \begin{pmatrix} s_{0<o_0>} & s_{-2<o_{-1}>} & s_{-1<o_0>} & s_{1<o_{-1}>} & s_{-1<o_{-2}>} \\ s_{2<o_1>} & s_{0<o_0>} & s_{1<o_1>} & s_{3<o_0>} & s_{1<o_{-1}>} \\ s_{1<o_0>} & s_{-1<o_{-1}>} & s_{0<o_0>} & s_{2<o_{-1}>} & s_{0<o_{-2}>} \\ s_{-1<o_1>} & s_{-3<o_0>} & s_{-2<o_1>} & s_{0<o_0>} & s_{-2<o_{-1}>} \\ s_{1<o_2>} & s_{-1<o_1>} & s_{0<o_2>} & s_{2<o_1>} & s_{0<o_0>} \end{pmatrix}$$

$$\mathbb{R}_3 = \begin{pmatrix} s_{0<o_0>} & s_{-1<o_{-1}>} & s_{-3<o_{-2}>} & s_{0<o_{-3}>} & s_{-3<o_{-1}>} \\ s_{1<o_1>} & s_{0<o_0>} & s_{-2<o_{-1}>} & s_{1<o_{-2}>} & s_{-2<o_0>} \\ s_{3<o_2>} & s_{2<o_1>} & s_{0<o_0>} & s_{3<o_{-1}>} & s_{0<o_1>} \\ s_{0<o_3>} & s_{-1<o_2>} & s_{-3<o_1>} & s_{0<o_0>} & s_{-3<o_2>} \\ s_{3<o_1>} & s_{2<o_0>} & s_{0<o_{-1}>} & s_{3<o_{-2}>} & s_{0<o_0>} \end{pmatrix}$$

$$\mathbb{R}_4 = \begin{pmatrix} s_{0<o_0>} & s_{1<o_0>} & s_{2<o_1>} & s_{2<o_{-2}>} & s_{-1<o_1>} \\ s_{-1<o_0>} & s_{0<o_0>} & s_{1<o_1>} & s_{1<o_{-2}>} & s_{-2<o_1>} \\ s_{-2<o_{-1}>} & s_{-1<o_{-1}>} & s_{0<o_0>} & s_{0<o_{-3}>} & s_{-3<o_0>} \\ s_{-2<o_2>} & s_{-1<o_2>} & s_{0<o_3>} & s_{0<o_0>} & s_{-1<o_1>} \\ s_{1<o_{-1}>} & s_{2<o_{-1}>} & s_{3<o_0>} & s_{1<o_{-1}>} & s_{0<o_0>} \end{pmatrix}$$

Step 2. Let $\mathbb{Z} = 1$, and $R_a^{(1)} = (r_{ij}^{a,(1)})_{5\times5} = (r_{ij}^a)_{5\times5}$ $(a = 1, 2, 3, 4)$.

Step 3. Based on Eq. (4.15), we can obtain the consensus degree $CD''(\mathbb{R}_a^{(1)})$ for each DHLPR $\mathbb{R}_a^{(1)}$, which is shown in Table 4.6. Additionally, considering that all DHLPRs are completely consistent, so the consistency indices of them are $CI(\mathbb{R}_a^{(1)}) = 1(a = 1, 2, 3, 4)$.

Obviously, $CD'' = 0.7438 < 0.85$ and the consensus is not reached. Therefore, we utilize the multi-stage consensus optimization model to improve the group consensus degree. In this subsection, the optimization model is in continuous scale, that is $r_{ij}^{*a,(\mathbb{Z})} \in \bar{S}_O = \{s_{t<o_k>} | t = [-4, 4]; k = [-4, 4]\}$ $(a = 1, 2, 3, 4; i, j = 1, 2, \ldots, 5)$.

Interactive consensus reaching process:
Round 1.

Step 4^1. Let $G_q^{(\mathbb{Z})}$, $N_q^{(\mathbb{Z})}$, $NE_q^{(\mathbb{Z})}$ be the size of change, the modification number of the DHLTs in all DHLPRs, and the number of experts that have to change their preferences respectively at the stages $q(q = 1, 2, 3)$ in the round $\mathbb{Z}$. Using Eqs. (4.20), (4.22), and (4.24), we have

$$G_1^{(1)} = 3.0938, N_1^{(1)} = 21, NE_1^{(1)} = 4;$$
$$G_2^{(1)} = 3.0938, N_2^{(1)} = 20, NE_2^{(1)} = 3;$$
$$G_3^{(1)} = 3.0938, N_3^{(1)} = 20, NE_3^{(1)} = 3.$$

Table 4.6 The consistency indices and consensus degrees of DHLPRs $\mathbb{R}_a^{(1)}(a = 1, 2, 3, 4)$

	$\mathbb{R}_1^{(1)}$	$\mathbb{R}_2^{(1)}$	$\mathbb{R}_3^{(1)}$	$\mathbb{R}_4^{(1)}$
$CI(\mathbb{R}_a^{(1)})$	1	1	1	1
$CD''(\mathbb{R}_a^{(1)})$	0.8354	0.8188	0.7563	0.7438
CD''	0.7438			

Because $NE_2^{(1)} = NE_3^{(1)} = 3$, there exists no any further improvement in this improvement process. Then $\mathbb{R}_1^{*(1)} = \mathbb{R}_1$ and the optimal adjusted DHLPRs can be shown as follows:

$$\mathbb{R}_2^{*(1)} = \begin{pmatrix} s_{0<o_0>} & s_{-2<o_{-1}>} & s_{-1<o_0>} & s_{1<o_{-1}>} & s_{-1<o_{-2}>} \\ s_{2<o_1>} & s_{0<o_0>} & \underline{s_{0<o_{-1.05}>}} & \underline{s_{0<o_{-2.47}>}} & \underline{s_{0<o_{-2.98}>}} \\ s_{1<o_0>} & s_{0<o_{1.05}>} & s_{0<o_0>} & s_{2<o_{-1}>} & \underline{s_{0<o_{-3.51}>}} \\ s_{-1<o_1>} & s_{0<o_{2.47}>} & s_{-2<o_1>} & s_{0<o_0>} & s_{-2<o_{-1}>} \\ s_{1<o_2>} & s_{0<o_{2.98}>} & s_{0<o_{3.51}>} & s_{2<o_1>} & s_{0<o_0>} \end{pmatrix}$$

$$\mathbb{R}_3^{*(1)} = \begin{pmatrix} s_{0<o_0>} & \underline{s_{-1<o_{-2.06}>}} & \underline{s_{-2<o_{-1.3}>}} & \underline{s_{0<o_{-0.05}>}} & \underline{s_{-3<o_{1.94}>}} \\ s_{1<o_{2.06}>} & s_{0<o_0>} & \underline{s_{-2<o_{3.35}>}} & \underline{s_{1<o_{-1.78}>}} & \underline{s_{-2<o_{1.91}>}} \\ s_{2<o_{1.3}>} & s_{-2<o_{3.35}>} & s_{0<o_0>} & \underline{s_{2<o_{-1.61}>}} & \underline{s_{0<o_{-2.96}>}} \\ s_{0<o_{0.05}>} & s_{-1<o_{1.78}>} & s_{-2<o_{1.61}>} & s_{0<o_0>} & s_{-3<o_2>} \\ s_{3<o_{-1.94}>} & s_{2<o_{-1.91}>} & s_{0<o_{2.96}>} & s_{3<o_{-2}>} & s_{0<o_0>} \end{pmatrix}$$

$$\mathbb{R}_4^{*(1)} = \begin{pmatrix} s_{0<o_0>} & \underline{s_{-1<o_{-2.87}>}} & \underline{s_{0<o_{-2.99}>}} & \underline{s_{0<o_{1.01}>}} & \underline{s_{-1<o_{-2.7}>}} \\ s_{-1<o_{2.87}>} & s_{0<o_0>} & s_{1<o_1>} & s_{1<o_{-2}>} & s_{-2<o_1>} \\ s_{0<o_{2.99}>} & s_{-1<o_{-1}>} & s_{0<o_0>} & \underline{s_{1<o_0>}} & \underline{s_{-3<o_{2.16}>}} \\ s_{0<o_{-1.01}>} & s_{-1<o_2>} & s_{-1<o_0>} & s_{0<o_0>} & \underline{s_{-1<o_0>}} \\ s_{1<o_{2.7}>} & s_{2<o_{-1}>} & s_{3<o_{-2.16}>} & s_{1<o_0>} & s_{0<o_0>} \end{pmatrix}.$$

The consistency indices and consensus degrees of optimal adjusted DHLPRs $\mathbb{R}_a^{*(1)}(a = 1, 2, 3, 4)$ in this stage are shown in Table 4.7.

Then, the optimal adjusted DHLPRs $\mathbb{R}_a^{*(1)}(a = 1, 2, 3, 4)$ can be fed back to experts for improving the group consensus degrees.

Step 5$^{\mathbf{1}}$. With the help of the suggestions obtained above, the experts can decide how to adjust their preferences.

Based on the results obtained in the multi-stage optimization model, the experts e^2, e^3, and e^4 need to improve their preferences. Combining $\mathbb{R}_a^{(1)}(a = 2, 3, 4)$ and $\mathbb{R}_a^{*(1)}(a = 2, 3, 4)$, the suggestions can be listed as follows:

For expert e^2: $r_{23}^{2,(2)} \in [s_{0<o_{-1.05}>}, s_{1<o_1>}]$, $r_{24}^{2,(2)} \in [s_{0<o_{-2.47}>}, s_{3<o_0>}]$, $r_{25}^{2,(2)} \in [s_{0<o_{-2.98}>}, s_{1<o_{-1}>}]$ and $r_{35}^{2,(2)} \in [s_{0<o_{-3.51}>}, s_{0<o_{-2}>}]$.

For expert e^3: $r_{12}^{3,(2)} \in [s_{-1<o_{-2.06}>}, s_{-1<o_{-1}>}]$, $r_{13}^{3,(2)} \in [s_{-3<o_{-2}>}, s_{-2<o_{-1.3}>}]$, $r_{14}^{3,(2)} \in [s_{0<o_{-3}>}, s_{0<o_{-0.05}>}]$, $r_{15}^{3,(2)} \in [s_{-3<o_{-1}>}, s_{-3<o_{1.94}>}]$, $r_{23}^{3,(2)} \in [s_{-2<o_{-1}>},$

Table 4.7 The consistency indices and consensus degrees of optimal adjusted DHLPRs $\mathbb{R}_a^{*(1)}(a = 1, 2, 3, 4)$

	$\mathbb{R}_1^{*(1)}$	$\mathbb{R}_2^{*(1)}$	$\mathbb{R}_3^{*(1)}$	$\mathbb{R}_4^{*(1)}$
$CI(\mathbb{R}_a^{*(1)})$	1	0.9	0.9689	0.9
$CD''(\mathbb{R}_a^{*(1)})$	0.9	0.9	0.9	0.9
CD''	0.9			

$s_{-2<o_{3.35}>}]$, $r_{24}^{3,(2)} \in [s_{1<o_{-2}>}, s_{1<o_{-1.78}>}]$, $r_{25}^{3,(2)} \in [s_{-2<o_0>}, s_{-2<o_{1.91}>}]$, $r_{34}^{3,(2)} \in [s_{2<o_{-1.61}>}, s_{3<o_{-1}>}]$ and $r_{35}^{3,(2)} \in [s_{0<o_{-2.96}>}, s_{0<o_1>}]$.

For expert e^4: $r_{12}^{4,(2)} \in [s_{-1<o_{-2.87}>}, s_{1<o_0>}]$, $r_{13}^{4,(2)} \in [s_{0<o_{-2.99}>}, s_{2<o_1>}]$, $r_{14}^{4,(2)} \in [s_{0<o_{1.01}>}, s_{2<o_{-2}>}]$, $r_{15}^{4,(2)} \in [s_{-1<o_{-2.7}>}, s_{-1<o_1>}]$, $r_{34}^{4,(2)} \in [s_{0<o_{-3}>}, s_{1<o_0>}]$, $r_{35}^{4,(2)} \in [s_{-3<o_0>}, s_{-3<o_{2.16}>}]$ and $r_{45}^{4,(2)} \in [s_{-1<o_0>}, s_{-1<o_1>}]$.

Suppose that the experts provide their adjusted preferences as follows:

For expert e^2: $r_{23}^{2,(2)} = s_{0<o_{-2}>}$, $r_{24}^{2,(2)} = s_{0<o_0>}$, $r_{25}^{2,(2)} = s_{1<o_{-2}>}$, and $r_{35}^{2,(2)} = s_{0<o_{-3}>}$.

For expert e^3: $r_{12}^{3,(2)} = s_{-1<o_{-2}>}$, $r_{13}^{3,(2)} = s_{-3<o_1>}$, $r_{14}^{3,(2)} = s_{0<o_{-1}>}$, $r_{15}^{3,(2)} = s_{-3<o_1>}$, $r_{23}^{3,(2)} = s_{-2<o_2>}$, $r_{25}^{3,(2)} = s_{-2<o_1>}$, $r_{34}^{3,(2)} = s_{2<o_1>}$ and $r_{35}^{3,(2)} = s_{0<o_{-1}>}$.

For expert e^4: $r_{12}^{4,(2)} = s_{-1<o_1>}$, $r_{13}^{4,(2)} = s_{0<o_1>}$, $r_{14}^{4,(2)} = s_{0<o_3>}$, $r_{15}^{4,(2)} = s_{-1<o_{-1}>}$, $r_{34}^{4,(2)} = s_{0<o_3>}$ and $r_{35}^{4,(2)} = s_{-3<o_1>}$.

Round 2.

Step 4². The consistency indices and consensus degrees of the adjusted DHLPRs $\mathbb{R}_a^{(2)}(a = 1, 2, 3, 4)$ can be obtained as follows: (Table 4.8)

Obviously, $CD'' = 0.8208 < 0.85$ and the consensus is also not reached. Therefore, we still need to utilize the multi-stage consensus optimization model to improve the group consensus degree. Then, we have

$$G_1^{(2)} = 1.1063, N_1^{(2)} = 13, NE_1^{(2)} = 4;$$
$$G_2^{(2)} = 1.1063, N_2^{(2)} = 11, NE_2^{(2)} = 4;$$
$$G_3^{(2)} = 1.1063, N_3^{(2)} = 11, NE_3^{(2)} = 4.$$

The optimal adjusted DHLPRs are shown as follows:

$$\mathbb{R}_1^{*(2)} = \begin{pmatrix} s_{0<o_0>} & s_{-2<o_0>} & s_{-1<o_1>} & s_{0<o_1>} & s_{-2<o_{-2}>} \\ s_{2<o_0>} & s_{0<o_0>} & s_{1<o_1>} & \underline{s_{2<o_{-0.3}>}} & s_{0<o_{-2}>} \\ s_{1<o_{-1}>} & s_{-1<o_{-1}>} & s_{0<o_0>} & s_{1<o_0>} & s_{-1<o_{-3}>} \\ s_{0<o_{-1}>} & s_{-2<o_{-1}>} & s_{-1<o_0>} & s_{0<o_0>} & s_{-2<o_{-3}>} \\ s_{2<o_2>} & s_{0<o_2>} & s_{1<o_3>} & s_{2<o_3>} & s_{0<o_0>} \end{pmatrix}$$

Table 4.8 The consensus degrees of the adjusted DHLPRs $\mathbb{R}_a^{(2)}(a = 1, 2, 3, 4)$

	$\mathbb{R}_1^{(2)}$	$\mathbb{R}_2^{(2)}$	$\mathbb{R}_3^{(2)}$	$\mathbb{R}_4^{(2)}$
$CI(\mathbb{R}_a^{(2)})$	0.9539	0.9609	0.9654	0.9643
$CD''(\mathbb{R}_a^{(2)})$	0.8604	0.8625	0.8396	0.8208
CD''	0.8208			

$$\mathbb{R}_2^{*(2)} = \begin{pmatrix} s_{0<o_0>} & s_{-2<o_{-1}>} & s_{-1<o_0>} & s_{1<o_{-1}>} & s_{-1<o_{-2}>} \\ s_{2<o_1>} & s_{0<o_0>} & s_{0<o_{-2}>} & s_{0<o_0>} & \underline{s_{0<o_{1.7}>}} \\ s_{1<o_0>} & s_{0<o_2>} & s_{0<o_0>} & s_{2<o_{-1}>} & \underline{s_{-1<o_{-0.1}>}} \\ s_{-1<o_1>} & s_{0<o_0>} & s_{-2<o_1>} & s_{0<o_0>} & s_{-2<o_{-1}>} \\ s_{1<o_2>} & s_{0<o_{-1.7}>} & s_{1<o_{0.1}>} & s_{2<o_1>} & s_{0<o_0>} \end{pmatrix}$$

$$\mathbb{R}_3^{*(2)} = \begin{pmatrix} s_{0<o_0>} & s_{-1<o_{-2}>} & \underline{s_{-1<o_{-0.74}>}} & s_{0<o_{-1}>} & s_{-3<o_1>} \\ s_{1<o_2>} & s_{0<o_0>} & \underline{s_{0<o_{-2}>}} & s_{1<o_{-2}>} & s_{-2<o_1>} \\ s_{1<o_{0.74}>} & s_{0<o_2>} & s_{0<o_0>} & s_{2<o_1>} & \underline{s_{0<o_{-3.14}>}} \\ s_{0<o_1>} & s_{-1<o_2>} & s_{-2<o_{-1}>} & s_{0<o_0>} & s_{-3<o_2>} \\ s_{3<o_{-1}>} & s_{2<o_{-1}>} & s_{0<o_{3.14}>} & s_{3<o_{-2}>} & s_{0<o_0>} \end{pmatrix}$$

$$\mathbb{R}_4^{*(2)} = \begin{pmatrix} s_{0<o_0>} & \underline{s_{-1<o_{-1.98}>}} & \underline{s_{0<o_{-3}>}} & s_{0<o_3>} & \underline{s_{-1<o_{-1.99}>}} \\ s_{1<o_{1.98}>} & s_{0<o_0>} & s_{1<o_1>} & s_{1<o_{-2}>} & s_{-2<o_1>} \\ s_{0<o_3>} & s_{-1<o_{-1}>} & s_{0<o_0>} & \underline{s_{1<o_0>}} & s_{-3<o_1>} \\ s_{0<o_{-3}>} & s_{-1<o_2>} & s_{-1<o_0>} & s_{0<o_0>} & \underline{s_{-1<o_{-2.33}>}} \\ s_{1<o_{1.99}>} & s_{2<o_{-1}>} & s_{3<o_{-1}>} & s_{1<o_{2.33}>} & s_{0<o_0>} \end{pmatrix}.$$

The consistency indices and consensus degrees of all the optimal adjusted DHLPRs $\mathbb{R}_a^{*(2)}(a = 1,2,3,4)$ in this stage are obtained as follows: (Table 4.9)

Step 5². Based on the results obtained in multi-stage optimization model in this round, all the experts need to adjust their preferences, and the suggestions are:

For expert e^1: $r_{24}^{1,(3)} \in [s_{2<o_{-0.3}>}, s_{2<o_1>}]$.

For expert e^2: $r_{25}^{2,(3)} \in [s_{0<o_{1.7}>}, s_{1<o_{-2}>}]$ and $r_{35}^{2,(3)} \in [s_{-1<o_{-0.1}>}, s_{0<o_{-3}>}]$.

For expert e^3: $r_{13}^{3,(3)} \in [s_{-3<o_1>}, s_{-1<o_{-0.74}>}]$, $r_{23}^{3,(3)} \in [s_{-2<o_2>}, s_{0<o_{-2}>}]$ and $r_{35}^{3,(3)} \in [s_{0<o_{-3.14}>}, s_{0<o_{-1}>}]$.

For expert e^4: $r_{12}^{4,(3)} \in [s_{-1<o_{-1.98}>}, s_{1<o_1>}]$, $r_{13}^{4,(3)} \in [s_{0<o_{-3}>}, s_{0<o_1>}]$, $r_{15}^{4,(3)} \in [s_{-1<o_{-1.99}>}, s_{-1<o_{-1}>}]$, $r_{34}^{4,(3)} \in [s_{0<o_3>}, s_{1<o_0>}]$ and $r_{45}^{4,(3)} \in [s_{-1<o_{-2.33}>}, s_{-1<o_{-1}>}]$.

Suppose that the experts provide their adjusted preferences as follows:

For expert e^1: $r_{24}^{1,(3)} = s_{2<o_0>}$.

For expert e^2: $r_{35}^{2,(3)} = s_{-1<o_0>}$.

For expert e^3: $r_{13}^{3,(3)} = s_{-1<o_{-2}>}$, $r_{23}^{3,(3)} = s_{-1<o_0>}$ and $r_{35}^{3,(3)} = s_{0<o_{-3}>}$.

Table 4.9 The consistency levels and consensus levels of optimal adjusted DHLPRs $\mathbb{R}_a^{*(2)}(a = 1,2,3,4)$

	$\mathbb{R}_1^{*(2)}$	$\mathbb{R}_2^{*(2)}$	$\mathbb{R}_3^{*(2)}$	$\mathbb{R}_4^{*(2)}$
$CI(\mathbb{R}_a^{*(2)})$	0.9919	0.9	0.9445	0.9
$CD''(\mathbb{R}_a^{*(2)})$	0.9	0.9	0.9	0.9
CD''	0.9			

For expert e^4: $r_{12}^{4,(3)} = s_{-1<o_2>}$, $r_{13}^{4,(3)} = s_{0<o_{-1}>}$, $r_{34}^{4,(3)} = s_{1<o_0>}$ and $r_{45}^{4,(3)} = s_{-1<o_{-2}>}$.

Round 3.

The consensus degrees of the adjusted DHLPRs can be obtained as in Table 4.10.

In Table 4.10, there is $CD'' = 0.8729 > 0.85$, so all experts have reached the group consensus and the interactive consensus reaching process is terminated. Additionally, the consistency indices of all DHLPRs also satisfy the given threshold value.

Step 6. Finally, we can obtain the collective DHLPR and calculate the synthetical value of each of the alternatives $SV(A_1) = 1.9922$, $SV(A_2) = 2.7109$, $SV(A_3) = 2.5703$, $SV(A_4) = 1.8906$, $SV(A_5) = 3.3359$. Then we can obtain the ranking of all alternatives: $A_5 \succ A_2 \succ A_3 \succ A_1 \succ A_4$.

Step 7. End.

Next, we can utilize the existing methods to deal with this GDM problem:

(1) We can use the method introduced by He and Xu (2018) to deal with this GDM problem. Firstly, we need to transform LPO^1 and LPO^3 into the corresponding preference orderings by deleting the DHLTs. Additionally, it is difficult to obtain the corresponding preference orderings of LPO^2 and LPO^4 directly, but we can get them by the transformed DHLPRs. The final preference orderings of these four LPOs are shown as follows:

$$PO^1 = \{4, 2, 3, 5, 1\}, PO^2 = \{4, 1, 3, 5, 2\}, PO^3 = \{5, 3, 1, 4, 2\},$$
$$PO^4 = \{2, 3, 5, 4, 1\}$$

Based on the consensus measure proposed by He and Xu (2018), we can obtain the decision-making result as $A_5 \succ A_2 \succ A_3 \succ A_1 \succ A_4$ based on both the section process and consensus process.

(2) We can also utilize the method proposed by Chiclana et al. (1998) to deal with this GDM problem. Firstly, based on the transformation function f, four LPOs can be transformed to FPR $P^k (k = 1, 2, 3, 4)$. Then, we can obtain the collective FPR $P^c = (p_{ij}^c)_{5\times5}$, and compute the quantifier-guided dominance degree $QGDD_i$ and the quantifier guided non-dominance degree $QGNDD_i$ of each alternative A_i. The decision-making result is shown in Table 4.11.

Table 4.10 The group consensus index of the adjusted DHLPRs $\mathbb{R}_a^{(3)} (a = 1, 2, 3, 4)$

	$\mathbb{R}_1^{(3)}$	$\mathbb{R}_2^{(3)}$	$\mathbb{R}_3^{(3)}$	$\mathbb{R}_4^{(3)}$
$CI(\mathbb{R}_a^{(3)})$	0.9552	0.9617	0.9630	0.9659
$CD''(\mathbb{R}_a^{(3)})$	0.8875	0.8875	0.8813	0.8729
CD''	0.8729			

Table 4.11 The $QGDD_i$ and $QGNDD_i$ of all alternatives $A_i(i = 1, 2, \ldots, 5)$

	A_1	A_2	A_3	A_4	A_5	Rank
$QGDD_i$	0.3926	0.5684	0.5098	0.3359	0.6934	$A_5 \succ A_2 \succ A_3 \succ A_1 \succ A_4$
$QGNDD_i$	0.7578	0.9453	0.8984	0.6719	1	$A_5 \succ A_2 \succ A_3 \succ A_1 \succ A_4$
Final rank	$A_5 \succ A_2 \succ A_3 \succ A_1 \succ A_4$					

Based on multi-stage consensus optimization model and the existing methods (He and Xu 2018; Chiclana et al. 1998), some comparative analyses are summarized as follows:

(1) He and Xu (2018) discussed a consensus framework with three kinds of preference orderings including preference orderings (Chiclana et al. 1998), interval preference orderings (IPOs) (González-Pachón and Romero 2001), and hesitant preference orderings set (HPOS) (He and Xu 2018); Chiclana et al. (1998) discussed some transformation methods to transform preference orderings into FPRs. However, the firstly and most important gap of all the preference orderings proposed by He and Xu (2018) and Chiclana et al. (1998) do not exist any relationship between two adjacent alternatives, which leads to incomplete evaluation information. Additionally, He and Xu (2018) and Chiclana et al. (1998) only discussed some preference orderings in continuous forms, but ignored the preference orderings in decentralized forms. Furthermore, because all the preference orderings proposed by He and Xu (2018) and Chiclana et al. (1998) only consider the orderings among all alternatives and ignore the relationship between two adjacent alternatives, so the transformation methods given by Chiclana et al. (1998) only utilize the preference ordering to obtain the corresponding FPRs.
To overcome the gaps discussed above, we propose two kinds of LPOs. Firstly, the LPOs utilize the DHLTs to represent the relationships between two adjacent alternatives. Secondly, by fully considering the behaviors of the experts, we define two different LPOs in continuous forms and decentralized forms, respectively. Thirdly, combining the DHLTs and different forms of LPOs, the transformed DHLPRs will be more complete.

(2) In the transformation process, the transformation function proposed in this subsection cannot only ensure that the DHLPRs and FPRs are with complete consistencies, but also fully consider the original linguistic information (DHLTSs). However, the transformation functions $g^k(k = 1, 2, 3, 4)$ discussed by Chiclana et al. (1998) cannot ensure that the transformed FPRs are of complete consistencies. Furthermore, even though some functions of them use the utility values to reflect the degree to which one alternative is superior to another, but the degree has a certain proportion and cannot reflect the real relationship of any two alternatives.

(3) In the consensus reaching process, firstly, we do not need to improve the consistencies of the transformed DHLPRs considering that all the transformed

DHLPRs are completely consistent. Secondly, the multi-stage consensus optimization model proposed in this subsection consists of three objectives: minimizing the deviations of the modification magnitudes, minimizing the cardinal number of modifications while keeping the value of the first objective constant, and minimizing the number of experts who need to change their evaluations. Therefore, this model can be used to achieve consensus using minimal changes in the size of the change, the number of modifications, and the number of individuals who need to revise their preferences.

The method proposed by He and Xu (2018) only discusses how to identify the expert with the smallest consensus degree and proposed an automatic adjustment method to improve the consensus degree. Therefore, the multi-stage consensus optimization model proposed in this subsection is more efficient and targeted.

4.3 Group Consensus Decision-Making Methods with Self-confident DHLPRs

In this subsection, by proposing a new concept of Self-confident DHLPRs, we mainly research the experts' weights-determining method, the consensus reaching process and the simulation experiment:

(1) In different decision-making areas, the experts maybe various and each of them usually has different specialized knowledge or influence. Therefore, it is very important to determine the experts' weights in GDM. Up to now, there has existed a large number of weight-determining methods including the dynamic weights-determining approach based on the goal programming model (Liu et al. 2019), the optimization model (Park et al. 2011), and the AHP method (Ramanathan and Ganesh 1994), etc. However, these methods only utilize evaluation values to calculate the experts' weights and obtain only one kind of objective weights. Therefore, they have a common weakness, i.e., the subjective weights and the other objective weights are lost. To overcome this shortcoming, this subsection fully considers all kinds of information and obtains the weight vector of experts including the subjective weights and two kinds of objective weights. Firstly, the experts can evaluate themselves where the evaluation values can be regarded as their subjective weights; Additionally, each expert can be evaluated by the remaining experts and one kind of objective weights are obtained; Moreover, the evaluation matrix provided by each expert can be utilized to calculate the other kind of objective weights. Finally, the synthetic weights of experts can be obtained by combining all of these three weights.

(2) In the processes of GDM with preference relations, the element of priority vector reflects the importance degree of the corresponding alternative, and the difference between the individual priority vector and the collective priority

vector represents the proximity degree of an expert's preference and group's preference (Saaty 1980). Therefore, obtaining the individual priority vector and the collective priority vector are very important to reach consensus and make a decision. Based on this, in the consensus reaching process, this subsection develops two models to calculate the individual priority vector of each expert and the collective priority vector of all experts. These two priority vectors can not only be used to judge whether all experts reach consensus, but also be used to obtain the ranking of all alternatives.

(3) We hope that the consensus can be reached as soon as possible, and the number of iterations is as small as possible. In this regard, three comparison criteria are proposed to reflect the consensus efficiency of the proposed DHSM-based consensus model, including the number of iterations, the consensus success ratio and the distance between the original and adjusted preferences. Motivated by the analyses above, a simulation experiment is devised to testify the proposed DHSM-based consensus model by comparing it with two other consensus reaching models: One is the DHLPR without the self-confident degrees; the other is that the self-confident degrees are not changed in the consensus reaching process.

4.3.1 Self-confident DHLPR

In Sect. 1.2.4, Gou and Xu (2019) developed the concept of DHLPR. However, it can be seen from Definition 1.8 that the DHLPR only contains double hierarchy linguistic information, which can express the complex linguistic information completely. Additionally, it cannot reflect the self-confident degrees of the experts given for all DHLTs. In recent years, Liu et al. (2017) proposed lots of preference relations with self-confidences including multiplicative preference relation with self-confidence, additive preference relation with self-confidence and ordinal 2-tuple linguistic preference relation with self-confidence. This research orientation is very important to reflect the comprehensive decision information of the experts by adding the self-confident degree to each element of preference relation. Based on which, Gou et al. (2020b) defined the self-confident DHLPR.

In actual decision-making processes with preference relations, the priority vector is a useful tool to obtain the ranking of alternatives. In this subsection, as an important tool to calculate the priority vector from DHLPR, a least squares approach needs to be developed firstly. Then, considering that adding the self-confident degree to each element of preference relation can reflect the comprehensive decision information of the experts, and motivated by the preference relations with self-confidences proposed by Liu et al. (2017), a novel concept of self-confident DHLPR is proposed by adding the self-confident degrees to the basic elements of DHLPRs.

(a) **A least squares approach**

As we know, the least squares approach is one of the important methods to derive a priority vector from preference relations such as multiplicative preference relations (Saaty 1980) and subjective preference relations (Crawford and Williams 1985). In this subsection, we develop a least squares approach to calculate the priority vector of a DHLPR, which can be used as a basis for ranking alternatives and obtaining the consensus degree in consensus reaching process.

Let $\omega = (\omega_1, \omega_2, \ldots, \omega_m)^T$ be the priority vector of the DHLPR $\mathbb{R} = (r_{ij})_{m \times m}$, where $\omega_i > 0$ $(i = 1, 2, \ldots, m)$ and $\sum_{i=1}^{m} \omega_i = 1$. The priority vector can be used to characterize a consistent DHLPR, i.e., the DHLPR $\mathbb{R} = (r_{ij})_{m \times m}$ satisfies $f(r_{ij}) = f(r_{ik}) - f(r_{jk}) + 0.5, \ \forall\, i, j, k$. Therefore, for a consistent DHLPR, there exists

$$f(r_{ij}) = \omega_i - \omega_j + 0.5, \quad i, j = 1, 2, \ldots, m \tag{4.28}$$

In general, the DHLPR is not always consistent. The error between the preference element r_{ij} and the corresponding consistent preference element can be obtained by the following formula:

$$\varepsilon_{ij} = 0.5 \times (\omega_i - \omega_j) + 0.5 - f(r_{ij}), \quad i, j = 1, 2, \ldots, m \tag{4.29}$$

where the adjustment coefficient 0.5 is used to ensure that the range of ε_{ij} belongs to [−1, 1].

In general, the priority vector will be more reasonable if the DHLPR is more consistent. In other words, the smaller the error ε_{ij} between the preference element and the corresponding consistent preference element is, the more reasonable the priority vector should be. Therefore, based on Eq. (4.29), the priority vector $\omega = (\omega_1, \omega_2, \ldots, \omega_m)^T$ can be obtained by establishing and solving the following least squares model:

$$\begin{aligned} & \min \sum_{i=1}^{m} \sum_{i<j}^{m} (0.5 \times (\omega_i - \omega_j) + 0.5 - f(r_{ij}))^2 \\ & s.t. \begin{cases} \sum\limits_{i=1}^{m} \omega_i = 1, \\ w_i > 0, \ i = 1, 2, \ldots, m \end{cases} \end{aligned} \tag{4.30}$$

Table 4.12 The detailed information about the 7-point numerical set

Numerical value	Semantic meaning
1	Extremely low self-confident
2	Very low self-confident
3	Low self-confident
4	Medium self-confident
5	High self-confident
6	Very high self-confident
7	Extremely high self-confident

(b) **Self-confident DHLPR**

When providing the preference values, the experts may give self-confident degrees to those values to reflect their attitudes. Therefore, adding self-confident degrees to the DHLPR makes the complete evaluations. Let $S^{SL} = \{1, 2, \ldots, N\}$ be a numerical set used by the experts to express their self-confident degrees over the provided preference values in the DHLPR, where the experts have N ratings (Friedman and Amoo 1999). Without loss of generality, this subsection utilizes 7-point numerical set to express the experts' self-confident degrees, and we can list the meaning of each element in numerical set in Table 4.12.

Motivated by the DHLPR and the concept of self-confident degree, given a fixed set of alternatives $A = \{A_1, A_2, \ldots, A_m\}$, the self-confident DHLPR can be defined as follows:

Definition 4.10 (Gou et al. 2020b). A matrix $\Re = ((r_{ij}, sl_{ij}))_{m \times m}$ is called a self-confident DHLPR when its basic elements have the following two parts: (1) the first part, $r_{ij} \in S_O$, expresses the preference value of the alternative A_i over A_j; (2) the second part, $sl_{ij} \in S^{SL}$, expresses the self-confident degree with respect to the preference value r_{ij}. The following conditions are assumed: $r_{ij} \oplus r_{ji} = s_{0<o_0>}$, $r_{ii} = s_{0<o_0>}$, $sl_{ij} = sl_{ji}$, and $sl_{ii} = N$ for $i, j = 1, 2, \ldots, m$.

Example 4.6 (Gou et al. 2020b). Let $S_O = \{s_{t<o_k>} | t = -4, \ldots, 4;\ k = -4, \ldots, 4\}$ be a DHLTS, $S^{SL} = \{1, 2, \ldots, 7\}$ be a 7-point numerical set to express the experts' self-confident degrees, and $A = \{A_1, A_2, A_3\}$ be a set of three alternatives. Then, an expert can provide his/her self-confident DHLPR, denoted by $\Re$, as follows:

$$\Re = \begin{pmatrix} (s_{0<o_0>}, 7) & (s_{2<o_{-1}>}, 2) & (s_{1<o_2>}, 5) \\ (s_{-2<o_1>}, 2) & (s_{0<o_0>}, 7) & (s_{-1<o_1>}, 4) \\ (s_{-1<o_{-2}>}, 5) & (s_{1<o_{-1}>}, 4) & (s_{0<o_0>}, 7) \end{pmatrix}$$

In this self-confident DHLPR $\Re$, $r_{13} = s_{1<o_2>}$ expresses that the preference value of the alternative A_1 over A_3 is $s_{1<o_2>}$, and its concrete meaning can be obtained by the DHLTS S_O. $sl_{13} = 5$ expresses that the self-confident degree with respect to r_{13} is 5. Similarly, the remaining elements can also be explained.

4.3.2 Weights-Determining Methods

In GDM, when we obtain the self-confident DHLPRs provided by the experts, the next step is to ensure whether all experts reach consensus. Therefore, a suitable consensus reaching model for self-confident DHLPRs needs to be developed. Additionally, if the experts do not reach consensus, the consensus reaching model also needs to consist of two aspects' adjustments. One is to check whether the expert's consensus degree is lower than the overall consensus degree and to adjust the double hierarchy linguistic preference information, the other one is to adjust the self-confident degrees by increasing the self-confident degrees of the expert whose consensus degree is higher than the overall consensus degree and decreasing the self-confident degrees of the expert whose consensus degree is lower than the overall consensus degree. Based on these, this subsection mainly proposes an iteration-based consensus reaching model for self-confident DHLPRs. The proposed consensus reaching model is a double hierarchy linguistic preference values and self-confident degrees modifying-based consensus model, which can be abbreviated as a DHSM-based consensus model.

Firstly, the consensus reaching problem with self-confident DHLPRs can be described as follows:

Let $S_O = \{s_{t<o_k>} | t = -\tau, \ldots, -1, 0, 1, \ldots, \tau;\ k = -\varsigma, \ldots, -1, 0, 1, \ldots, \varsigma\}$ be a DHLTS, and $S^{SL} = \{1, 2, \ldots, N\}$ be a N-point numerical set to express the experts' self-confident degrees. For a GDM problem under double hierarchy linguistic environment, a set of experts $\{e^1, e^2, \ldots, e^n\}(n \geq 2)$ are invited to evaluate a set of alternatives $\{A_1, A_2, \ldots, A_m\}(m \geq 2)$. Let $w = (w_1, w_2, \ldots, w_n)^T$ be the weight vector of the experts, where $w_k \geq 0$ $(k = 1, 2, \ldots, n)$ expresses the weight of the expert e^k, and $\sum_{k=1}^{n} \omega_k = 1$. Let $\Re^k = ((r_{ij}^k, sl_{ij}^k))_{m \times m}$ be a self-confident DHLPR provided by the expert e^k, where r_{ij}^k expresses the preference value of the alternative A_i over A_j, and $sl_{ij}^k \in S^{SL}$ expresses the self-confident degree with respect to the preference value r_{ij}^k.

Then, as we know, the existing weights-determining methods can only determine objective weights of experts by making some operations on evaluation information. However, the subjective information cannot be neglected if we want to consider all useful information. Additionally, the objective weight information of one expert may contain another part, i.e., the evaluation information provided by the remaining experts. Therefore, the synthetic weight of each expert can be obtained by taking into account subjective weights and objective weights simultaneously:

(1) Subjective weights: The expert's own evaluation information, which can be obtained by each expert directly and denoted by $\kappa_k^S (k = 1, 2, \ldots, n)$. Then the

subjective weight w_k^S of each expert can be obtained by normalizing the evaluation information as: $w_k^S = \kappa_k^S / \sum_{k=1}^{n} \kappa_k^S$.

(2) Objective weights: One is obtained from all of the provided self-confident DHLPRs $\Re^k = ((r_{ij}^k, sl_{ij}^k))_{m\times m}(k = 1, 2, \ldots, n)$, denoted by $w_k^{O_1}$; the other is provided by the remaining experts, denoted by $w_k^{O_2}$.

To calculate the first objective weight $w_k^{O_1}$, a distance measure between any two self-confident DHLPRs needs to be defined.

Definition 4.11 (Gou et al. 2020b). Let $\Re^k = ((r_{ij}^k, sl_{ij}^k))_{m\times m}(k = 1, 2)$ be two self-confident DHLPRs. The distance between them can be obtained by

$$d_{12} = d(\Re^1, \Re^2) = \sqrt{\frac{2}{m(m-1)} \sum_{i=1}^{m} \sum_{i<j}^{m} (f(r_{ij}^1) \times sl_{ij}^1 - f(r_{ij}^2) \times sl_{ij}^2)^2} \quad (4.31)$$

Based on Eq. (4.31), all the distance between any two pairs of experts can be established as a matrix:

$$D = (d_{zk})_{n\times n} = \begin{array}{c} \\ e^1 \\ e^2 \\ \vdots \\ e^n \end{array} \begin{array}{c} \begin{array}{cccc} e^1 & e^2 & \cdots & e^n \end{array} \\ \begin{pmatrix} 0 & d_{12} & \cdots & d_{1n} \\ d_{21} & 0 & \cdots & d_{2n} \\ \vdots & \vdots & \ddots & \vdots \\ d_{n1} & d_{n2} & \cdots & 0 \end{pmatrix} \end{array}$$

Then, one kind of the objective weight of each expert can be obtained by

$$w_k^{O_1} = \sum_{z=1}^{n} d_{zk} \Big/ \sum_{k=1}^{n} \sum_{z=1}^{n} d_{zk} \quad (4.32)$$

In addition, in GDM, each expert can provide the evaluations for others according to how well he/she knows them. The evaluations can compose of a mutual evaluation matrix $\Upsilon = (\lambda_{zk})_{n\times n}$ as follows:

$$\Upsilon = (\lambda_{zk})_{n\times n} = \begin{array}{c} \\ e^1 \\ e^2 \\ \vdots \\ e^n \end{array} \begin{array}{c} \begin{array}{cccc} e^1 & e^2 & \cdots & e^n \end{array} \\ \begin{pmatrix} - & \lambda_{12} & \cdots & \lambda_{1n} \\ \lambda_{21} & - & \cdots & \lambda_{2n} \\ \vdots & \vdots & \ddots & \vdots \\ \lambda_{n1} & \lambda_{n2} & \cdots & - \end{pmatrix} \end{array}$$

where the element $\lambda_{zk} \in [0, 1]$ expresses one expert's evaluation value for other expert. The larger the value of λ_{zk} is, the higher the evaluation is. Combining all these evaluation values of each column, we can obtain the importance degree of each expert, i.e.,

$$\lambda_k = \frac{1}{n-1} \sum_{\substack{z=1 \\ z \neq k}}^{n} \lambda_{zk} \tag{4.33}$$

Furthermore, we can normalize them and the other kind of objective weight of each expert can be obtained by

$$w_k^{O_2} = \lambda_k \Big/ \sum_{k=1}^{n} \lambda_k \tag{4.34}$$

Combining these three kinds of weights of the experts, the synthetic weight of each expert can be obtained by

$$w_k = \alpha w_k^S + \beta w_k^{O_1} + \gamma w_k^{O_2}, \quad k = 1, 2, \ldots, n \tag{4.35}$$

where α, β, γ are three adjustment parameters and satisfy $\alpha, \beta, \gamma \geq 0$ and $\alpha + \beta + \gamma = 1$. In general, these parameters can be given by the moderator in the GDM process directly.

An example can be set up to explain the weight-determining process:

Example 4.7 (Gou et al. 2020b). Let $S_O = \{s_{t<o_k>} | t = -4, \ldots, 4;\ k = -4, \ldots, 4\}$ be a DHLTS, three experts $\{e^1, e^2, e^3\}$ are invited to evaluate three alternatives $\{A_1, A_2, A_3\}$. Suppose that

(1) The experts' subjective weight vector is $\kappa^S = (0.8, 0.9, 0.8)^T$,
(2) Their evaluations $\Re^k (k = 1, 2, 3)$ are

$$\Re^1 = \begin{pmatrix} (s_{0<o_0>}, 7) & (s_{2<o_{-1}>}, 2) & (s_{1<o_2>}, 4) \\ (s_{-2<o_1>}, 2) & (s_{0<o_0>}, 7) & (s_{-1<o_1>}, 5) \\ (s_{-1<o_{-2}>}, 4) & (s_{1<o_{-1}>}, 5) & (s_{0<o_0>}, 7) \end{pmatrix}$$

$$\Re^2 = \begin{pmatrix} (s_{0<o_0>}, 7) & (s_{1<o_1>}, 2) & (s_{1<o_{-2}>}, 5) \\ (s_{-1<o_{-1}>}, 2) & (s_{0<o_0>}, 7) & (s_{0<o_1>}, 4) \\ (s_{-1<o_2>}, 5) & (s_{0<o_{-1}>}, 4) & (s_{0<o_0>}, 7) \end{pmatrix}$$

$$\Re^3 = \begin{pmatrix} (s_{0<o_0>}, 7) & (s_{2<o_1>}, 4) & (s_{2<o_1>}, 3) \\ (s_{-2<o_{-1}>}, 4) & (s_{0<o_0>}, 7) & (s_{-1<o_0>}, 4) \\ (s_{-2<o_{-1}>}, 3) & (s_{1<o_0>}, 4) & (s_{0<o_0>}, 7) \end{pmatrix}$$

(3) Each expert can provide the evaluations for others, then a mutual evaluation matrix $\Upsilon = (\lambda_{zk})_{3\times 3}$ is established as follows:

$$\Upsilon = (\lambda_{zk})_{3\times 3} = \begin{matrix} & e^1 & e^2 & e^3 \\ e^1 & - & 0.8 & 0.7 \\ e^2 & 0.8 & - & 0.7 \\ e^3 & 0.6 & 0.5 & - \end{matrix}$$

Then, the weights of the experts can be calculated by:

(1) By normalizing the experts' subjective weight vector, the subjective weight vector is $w^S = (0.32, 0.36, 0.32)^T$;
(2) Based on Eqs. (4.31) and (4.32), the first objective weight vector is obtained as $w^{O_1} = (0.25, 0.27, 0.48)^T$;
(3) Based on Eqs. (4.33) and (4.34), the second objective weight vector is obtained as $w^{O_2} = (0.34, 0.32, 0.34)^T$.

Based on Eq. (4.35), and let $\alpha = 0.4$, $\beta = 0.3$, and $\gamma = 0.3$, then the synthetic weight vector is obtained as $w = (0.31, 0.32, 0.37)^T$.

4.3.3 Consensus Measure and Adjustment Mechanism with Self-confident DHLPRs

In the consensus reaching process, a large number of consensus models have been developed, including the consensus models with different preference expression structures (Chen et al. 2015), the consensus models based on consistency and consensus measures (Gou et al. 2018), the multiple stages optimization consensus models (Gou et al. 2020c), the consensus models under dynamic contexts (Dong et al. 2017), and the consensus models considering the behaviors/attitudes of the experts (Wu et al. 2017), etc. However, the self-confident DHLPR has not been considered in the existing consensus models due to the complexity of this kind of decision-making environment. To fill this research gap, Gou et al. (2020b) proposed a consensus measure and consensus adjustment mechanism. In this subsection, based on Model (4.30), two models are established to obtain the individual and collective priority vectors of self-confident DHLPRs, respectively. Additionally, the consensus measure is developed by the obtained individual and collective priority vectors.

(a) **Consensus measures**

- **Model for obtaining the individual priority vector**

Motivated by Model (4.30), an extended least squares model is set up to calculate the individual priority vector of the expert e^k:

$$\begin{aligned} & \min \sum_{i=1}^{m} \sum_{i<j}^{m} sl_{ij}^k \times (0.5 \times (\omega_i^k - \omega_j^k) + 0.5 - f(r_{ij}^k))^2 \\ & s.t. \begin{cases} \sum_{i=1}^{m} \omega_i^k = 1, \\ \omega_i^k > 0, \ i = 1, 2, \ldots, m \end{cases} \end{aligned} \tag{4.36}$$

The self-confident degree sl_{ij}^k in Model (4.36) reflects the magnification of error degree, and the larger the value of sl_{ij}^k is, the higher the magnification of error degree should be. Solving Model (4.36), the optimum result of the individual priority vector $\omega^k = (\omega_1^k, \omega_2^k, \ldots, \omega_m^k)^T$ can be obtained.

- **Model for obtaining the collective priority vector**

Let $w = (w_1, w_2, \ldots, w_n)^T$ be the weight vector of all experts. The collective priority vector $\omega^c = (\omega_1^c, \omega_2^c, \ldots, \omega_m^c)^T$ can be obtained by solving Model (4.37):

$$\begin{aligned} & \min \sum_{k=1}^{n} w_k \sum_{i=1}^{m} \sum_{i<j}^{m} sl_{ij}^k \times (0.5 \times (\omega_i^c - \omega_j^c) + 0.5 - f(r_{ij}^k))^2 \\ & s.t. \begin{cases} \sum_{i=1}^{m} \omega_i^c = 1, \\ \omega_i^c > 0, \ i = 1, 2, \ldots, m \end{cases} \end{aligned} \tag{4.37}$$

Generally, the consensus measure for the GDM problems is obtained by calculating the distance or similarity degree between individual and collective priority vectors. Gou et al. (2020b) developed a consensus measure by measuring the similarity degree between individual and collective priority vectors:

(1) Individual consensus degree

Let $\omega^k = (\omega_1^k, \omega_2^k, \ldots, \omega_m^k)^T$ and $\omega^c = (\omega_1^c, \omega_2^c, \ldots, \omega_m^c)^T$ be the individual and collective priority vectors obtained from Model (4.36) and Model (4.37), respectively. Then, the individual consensus degree $CD(e^k)$ of the expert e^k is defined as:

$$CD(e^k) = 1 - \sqrt{\frac{1}{m} \sum_{i=1}^{m} (\omega_i^k - \omega_i^c)^2} \tag{4.38}$$

(2) Collective consensus degree

Based on the individual consensus degrees $CD(e^k)(k = 1, 2, \ldots, n)$, the collective consensus degree among all experts is measured by

$$CD = \frac{1}{n}\sum_{k=1}^{n} CD(e^k) \tag{4.39}$$

Obviously, $CD(e^k) \in [0, 1]$ and $CD \in [0, 1]$. The bigger the value of CD is, the higher the collective consensus degree among all experts should be. Specially, $CD = 1$ means that all experts have full consensus with the collective opinion. In the consensus reaching process for a GDM, a consensus threshold $\xi \in [0, 1]$ is usually provided in advance. If $CD \geq \xi$, then all experts' collective consensus is reached. Otherwise, the corresponding individual opinion should be adjusted.

(b) **Feedback adjustment mechanism**

In the consensus reaching process, the experts need to adjust preference information if all experts' collective consensus is not reached. To communicate fully with the experts and get their feedbacks, this subsection develops a feedback adjustment mechanism to help the experts adjust their preferences and improve the collective consensus degree. In general, the feedback adjustment mechanism consists of two consensus rules:

(1) Identification rule (IR). The IR can be used to identify the experts who should adjust their preferences to improve the collective consensus degree.
(2) Direction rule (DR). The DR can be used to find out the adjustment direction and guide the experts to improve their preferences.

Based on these two rules, we establish a feedback adjustment mechanism, which can be used to identify and adjust both the preferences and self-confident degrees. The feedback adjustment mechanism is descripted as follows:

Step 1. Identify preferences and self-confident degrees that need to be adjusted.

Based on Eqs. (4.38) and (4.39), we can find out that the experts whose individual consensus degrees are lower than the given consensus threshold $\xi \in [0, 1]$, i.e., $ECD = \{e^k | CD(e^k) < \xi\}$. Then we transform the collective priority vector $\omega^c = (\omega_1^c, \omega_2^c, \ldots, \omega_m^c)^T$ into the DHLPR and obtain the collective DHLPR $W^c = (\omega_{ij}^c)_{m\times m} = (f^{-1}(0.5 \times (\omega_i^c - \omega_j^c) + 0.5))_{m\times m}$. Let $\Phi^k = (\phi_{ij}^k)_{m\times m}$ be the error matrix with respect to the self-confident DHLPR $\Re^k = ((r_{ij}^k, sl_{ij}^k))_{m\times m}$ of the expert $e^k \in ECD$, and

$$\phi_{ij}^k = (f(r_{ij}^k) - f(\omega_{ij}^c))^2 \tag{4.40}$$

where ϕ_{ij}^k reflects the error degree of the preference r_{ij}^k. The larger the value of ϕ_{ij}^k is, the higher the error degree of the preference r_{ij}^k should be. Then, we can identify the experts whose preferences with error degrees are larger than the upper error threshold $\overline{\phi}(\overline{\phi} \geq 0)$:

$$EL^k = \{(i,j)|\phi_{ij}^k > \overline{\phi},\ i<j\} \tag{4.41}$$

Similarity, we can also identify the experts whose preferences with error degrees are smaller than the lower error threshold $\underline{\phi}\left(\underline{\phi} \geq 0\right)$:

$$EM^k = \{(i,j)|\phi_{ij}^k < \underline{\phi},\ i<j\} \tag{4.42}$$

Step 2. Based on the DR, we can determine the adjustment suggestions to improve the preferences and self-confident degrees.

Let $\widetilde{\Re^k} = ((\widetilde{r_{ij}^k}, \widetilde{sl_{ij}^k}))_{m\times m}$ be the adjusted self-confident DHLPR with respect to the preference relation $\Re^k = ((r_{ij}^k, sl_{ij}^k))_{m\times m}$ of the expert e^k. Based on the collective DHLPR $W^c = (\omega_{ij}^c)_{m\times m}$, the adjustment method can be developed by the following DRs:

(1) For each $(i,j) \in EL^k$, we can feedback it to the expert e^k and suggest him/her to adjust the preference r_{ij}^k based on the following rule:

$$\widetilde{r_{ij}^k} \in [\min(r_{ij}^k, \omega_{ij}^c), \max(r_{ij}^k, \omega_{ij}^c)] \tag{4.43}$$

Based on Eq. (4.43), it is clear that the preference r_{ij}^k hinders the consensus reaching process because it makes a bigger error degree between the preference r_{ij}^k and the collective preference ω_{ij}^c. Therefore, it is necessary to decrease the self-confident degree sl_{ij}^k of the expert e^k. Then we provide the following suggestion to the expert e^k:

$$\widetilde{sl_{ij}^k} \in [1, sl_{ij}^k] \tag{4.44}$$

(2) For each $(i,j) \in EM^k$, the preference r_{ij}^k promotes the consensus reaching process because the preference r_{ij}^k is close to the collective preference w_{ij}^c. Therefore, it is necessary to increase the self-confident degree sl_{ij}^k of the expert e^k. Then we provide the following suggestions to the expert e^k:

$$\widetilde{r_{ij}^k} = r_{ij}^k \text{ and } \widetilde{sl_{ij}^k} \in [sl_{ij}^k, N] \tag{4.45}$$

(3) For the remaining preference $(i,j) \notin EL^k$ and $(i,j) \notin EM^k$, we suggest that it is not necessary to adjust the preference of the expert e^k, i.e., $\widetilde{r_{ij}^k} = r_{ij}^k$ and $\widetilde{sl_{ij}^k} = sl_{ij}^k$.

For the remaining self-confident DHLPR of the experts whose consensus degrees are bigger than the given consensus threshold ξ, i.e., $e^k \notin ECD$, we have $\Re^k = \widetilde{\Re^k}$.

(c) **The DHSM-based consensus model**

Based on the discussion above, the DHSM-based consensus model can be established. Firstly, based on Model (4.36) and Model (4.37), the individual and collective priority vectors can be obtained respectively based on the self-confident DHLPRs provided by all experts. Then, the individual consensus degrees and the collective consensus degree are calculated by Eqs. (4.38) and (4.39), respectively. If the collective consensus degree is lower than the given threshold value and the maximum number of iterations (i.e., $\mathbb{Z}_{\max} \geq 1$) is not reached, then we can provide the suggestions obtained by the feedback adjustment mechanism to the experts for repairing their self-confident DHLPRs.

Next, an algorithm can be developed to show the DHSM-based consensus model:

Algorithm 4.3 (Gou et al. 2020b). DHSM-based consensus model

Input: The self-confident DHLPRs $\Re^k = (r_{ij}^k, sl_{ij}^k)_{m\times m}$ $(k = 1,2,\ldots,n)$, the subjective evaluation values $\kappa_k^S (k = 1,2,\ldots,n)$, the mutual evaluation matrix $\Upsilon = (\lambda_{zk})_{n\times n}$, ξ, $\overline{\phi}$, $\underline{\phi}$, and $\mathbb{Z}_{\max} \geq 1$.

Output: The adjusted self-confident DHLPR $\widetilde{\Re^k} = ((\widetilde{r_{ij}^k}, \widetilde{sl_{ij}^k}))_{m\times m} (k = 1,2,\ldots,n)$, the collective priority vector $\omega^c = (\omega_1^c, \omega_2^c, \ldots, \omega_m^c)^T$, and the number of iterations $\mathbb{Z}$.

Step 1. Let $\mathbb{Z} = 0$, $\Re^{k(\mathbb{Z})} = ((r_{ij}^{k(\mathbb{Z})}, sl_{ij}^{k(\mathbb{Z})}))_{m\times m} = ((r_{ij}^k, sl_{ij}^k))_{m\times m}$.

Step 2. Based on Eqs. (4.31)-(4.35), experts' weight vector $w^{k(\mathbb{Z})} = (w_1^{(\mathbb{Z})}, w_2^{(\mathbb{Z})}, \ldots, w_m^{(\mathbb{Z})})^T$ is obtained.

Step 3. By Model (4.36) and Model (4.37), the individual priority vectors $\omega^{k(\mathbb{Z})} = (\omega_1^{k(\mathbb{Z})}, \omega_2^{k(\mathbb{Z})}, \ldots, \omega_m^{k(\mathbb{Z})})^T (k = 1,2,\ldots,n)$ of the experts and the collective priority vector $\omega^{c(\mathbb{Z})} = (\omega_1^{c(\mathbb{Z})}, \omega_2^{c(\mathbb{Z})}, \ldots, \omega_m^{c(\mathbb{Z})})^T$ are obtained.

Step 4. Based on Eqs. (4.38) and (4.39), the individual consensus degrees $CD(e^k)^{(\mathbb{Z})}$ $(k = 1,2,\ldots,n)$ and the collective consensus degree $CD^{(\mathbb{Z})}$ are obtained. If $CD^{(\mathbb{Z})} \geq \xi$ and $\mathbb{Z} \geq \mathbb{Z}_{\max}$, then go to Step 6; Otherwise, go to the next step.

Step 5. Identify the experts with $CD(e^k)^{(\mathbb{Z})} < \xi$, and calculate the collective DHLPR $W^{c(\mathbb{Z})} = (\omega_{ij}^{c(\mathbb{Z})})_{m\times m} = (f^{-1}(0.5 \times (\omega_i^{c(\mathbb{Z})} - \omega_j^{c(\mathbb{Z})}) + 0.5))_{m\times m}$ and the error matrix $\Phi^{k(\mathbb{Z})} = (\phi_{ij}^{k(\mathbb{Z})})_{m\times m}$ based on Eq. (4.40). Then, the self-confident

$\Re^{k(\mathbb{Z}+1)} = ((r_{ij}^{k(\mathbb{Z}+1)}, sl_{ij}^{k(\mathbb{Z}+1)}))_{m\times m}$ of the expert e^k is established. The expert is advised to adjust the preference information and the self-confident degree as follows:

(1) For $(i,j) \in EL^k$, $r_{ij}^{k(\mathbb{Z})} \in [\min(r_{ij}^{k(\mathbb{Z})}, \omega_{ij}^{c(\mathbb{Z})}), \max(r_{ij}^{k(\mathbb{Z})}, \omega_{ij}^{c(\mathbb{Z})})]$ and $sl_{ij}^{k(\mathbb{Z}+1)} \in [1, sl_{ij}^{k(\mathbb{Z})}]$.
(2) For $(i,j) \in EM^k$, $r_{ij}^{k(\mathbb{Z}+1)} = r_{ij}^{k(\mathbb{Z})}$ and $sl_{ij}^{k(\mathbb{Z}+1)} \in [sl_{ij}^{k(\mathbb{Z})}, N]$.
(3) For $(i,j) \notin EL^k$ and $(i,j) \notin EM^k$, $r_{ij}^{k(\mathbb{Z}+1)} = r_{ij}^{k(\mathbb{Z})}$ and $sl_{ij}^{k(\mathbb{Z}+1)} = sl_{ij}^{k(\mathbb{Z})}$.

Besides, if $CD(e^k)^{(\mathbb{Z})} \geq \xi$, then we have $\Re^{k(\mathbb{Z}+1)} = \Re^{k(\mathbb{Z})}$.
Let $\mathbb{Z} = \mathbb{Z}+1$. Go back to Step 2.

Step 6. Let $\widetilde{\Re^k} = \Re^{k(\mathbb{Z})}(k = 1, 2, \ldots, n)$ and $\omega^c = \omega^{c(\mathbb{Z})}$. Output $\widetilde{\Re^k}$, ω^c and $\mathbb{Z}$. The final ranking of alternatives can be obtained on the basis of ω^c.

Step 7. End.

To understand Algorithm 4.3 clearly, the DHSM-based consensus model for the GDM problems with self-confident DHLPRs can be descripted in Fig. 4.3.

Remark 4.6 (Gou et al. 2020b). Figure 4.3 mainly consists of four parts:

(1) Weights' determining process. This part can obtain all experts' weights, then the weights can be used to calculate the collective priority vector.
(2) Priority vectors' calculated process. In this part, the main work is to calculate the individual priority vectors of experts and the collective priority vector.
(3) Consensus process. If the consensus level is acceptable or the number of iterations reaches the given maxrounds, then go to the selection process. If not,

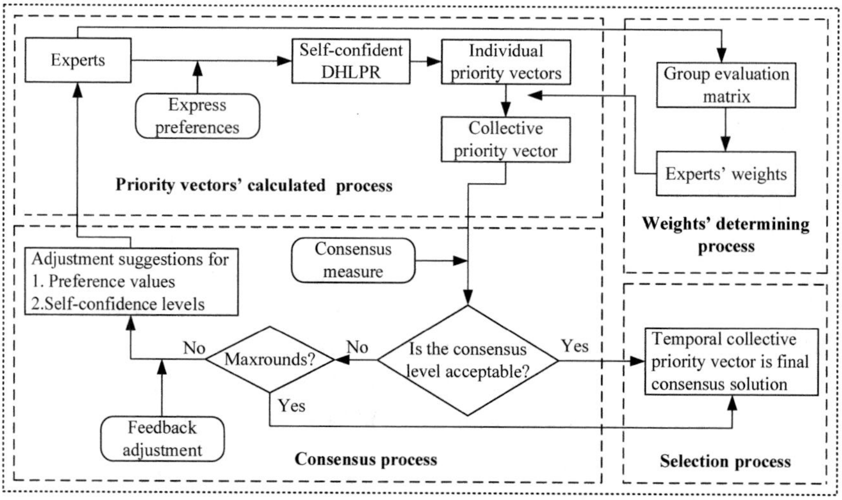

Fig. 4.3 The DHSM-based consensus model for GDM problems with self-confident DHLPRs

then the experts should adjust their preferences until the group consensus is reached or the number of iterations reaches the given maxrounds.

(4) Selection process. Based on the temporal collective priority vector, the ranking of alternatives is obtained.

4.3.4 Application of the DHSM-Based Consensus Model for GDM Problems with Self-confident DHLPRs

In this subsection, we apply the proposed DHSM-based consensus model to deal with a practical GDM problem concerning the selection of optimal Telemedicine technology.

Example 4.8 (Gou et al. 2020b). Broadly speaking, Telemedicine refers to the long-distance diagnosis, treatment and consultation of the wounded and sick in remote areas, islands or ships with poor medical conditions by taking advantages of the medical technology and medical equipment of large hospitals or specialized medical centers through computer technology, remote sensing, telemetry and remote control technology. In recent years, China has received attention and development of Telemedicine. First of all, Telemedicine mitigates to some extent the unbalanced status of China's expert resources and population distribution; The second is to alleviate the problem of high patient referral rate and high cost in remote areas. To promote the development of Telemedicine, a lot of hospitals have carried out the pilot work of telemedicine.

Now a city in China decides to choose an optimal hospital in the field of Telemedicine from four alternatives $A = \{A_1, A_2, A_3, A_4\}$. Four experts $E = \{e^1, e^2, e^3, e^4\}$ are invited to form a group to provide their self-confident DHLPRs over A, where $S^{SL} = \{1, 2, \ldots, 7\}$ is a 7-point numerical set to express the experts' self-confident degrees, and $S_O = \{s_{t<o_k>} | t = -4, \ldots, 4;\ k = -4, \ldots, 4\}$ is a DHLTS with $S = \{s_{-4} = extremely\ bad, s_{-3} = very\ bad, s_{-2} = bad, s_{-1} = slightly$ $\{bad, s_0 = equal, s_1 = slightly\ good, s_2 = good, s_3 = very\ \ good, s_4 = extremely$ $good$ and $O = \{o_{-4} = far\ from, o_{-3} = scarcely, o_{-2} = only\ a\ littlle, o_{-1} = a\ little,$ $o_0 = just\ \ right, o_1 = much, o_2 = very\ much, o_3 = extremely\ much, o_4 = entirely\}$.

The self-confident DHLPRs $\Re^k = ((r^k_{ij}, sl^k_{ij}))_{4\times4} (k = 1, 2, 3, 4)$ provided by the experts are shown as follows:

$$\Re^1 = \begin{pmatrix} (s_{0<o_0>}, 7) & (s_{-1<o_{-1}>}, 4) & (s_{1<o_{-2}>}, 3) & (s_{2<o_{-3}>}, 5) \\ (s_{1<o_1>}, 4) & (s_{0<o_0>}, 7) & (s_{3<o_{-1}>}, 2) & (s_{2<o_3>}, 6) \\ (s_{-1<o_2>}, 3) & (s_{-3<o_1>}, 2) & (s_{0<o_0>}, 7) & (s_{-1<o_2>}, 2) \\ (s_{-2<o_3>}, 5) & (s_{-3<o_{-3}>}, 6) & (s_{1<o_{-3}>}, 2) & (s_{0<o_0>}, 7) \end{pmatrix}$$

$$\Re^2 = \begin{pmatrix} (s_{0<o_0>},7) & (s_{-1<o_1>},3) & (s_{0<o_{-2}>},2) & (s_{-2<o_2>},5) \\ (s_{1<o_{-1}>},3) & (s_{0<o_0>},7) & (s_{2<o_2>},6) & (s_{2<o_3>},3) \\ (s_{-3<o_2>},2) & (s_{-2<o_{-2}>},6) & (s_{0<o_0>},7) & (s_{-1<o_3>},4) \\ (s_{-3<o_{-2}>},5) & (s_{-1<o_{-3}>},3) & (s_{1<o_{-3}>},4) & (s_{0<o_0>},7) \end{pmatrix}$$

$$\Re^3 = \begin{pmatrix} (s_{0<o_0>},7) & (s_{-1<o_1>},4) & (s_{1<o_2>},1) & (s_{1<o_{-2}>},5) \\ (s_{0<o_{-1}>},4) & (s_{0<o_0>},7) & (s_{2<o_0>},3) & (s_{2<o_2>},5) \\ (s_{-1<o_{-2}>},1) & (s_{-2<o_0>},3) & (s_{0<o_0>},7) & (s_{-1<o_{-3}>},2) \\ (s_{2<o_2>},5) & (s_{2<o_{-2}>},5) & (s_{-3<o_3>},2) & (s_{0<o_0>},7) \end{pmatrix}$$

$$\Re^4 = \begin{pmatrix} (s_{0<o_0>},7) & (s_{-1<o_1>},4) & (s_{1<o_1>},3) & (s_{2<o_{-1}>},5) \\ (s_{1<o_{-1}>},4) & (s_{0<o_0>},7) & (s_{1<o_{-1}>},2) & (s_{2<o_1>},4) \\ (s_{-3<o_{-1}>},3) & (s_{-1<o_1>},2) & (s_{0<o_0>},7) & (s_{-2<o_1>},5) \\ (s_{-2<o_{-3}>},5) & (s_{-2<o_{-1}>},4) & (s_{2<o_{-1}>},5) & (s_{0<o_0>},7) \end{pmatrix}$$

Additionally, each expert's own evaluation information is $\kappa_k^S = (0.8, 0.9, 0.7, 0.8)^T$. Furthermore, each expert can provide the evaluations for others, which establish a mutual evaluation matrix $\Upsilon = (\lambda_{zk})_{4\times4}$ ($\lambda_{zk} \in [0,1]$) as follows:

$$\Upsilon = (\lambda_{zk})_{4\times4} = \begin{matrix} & e^1 & e^2 & e^3 & e^4 \\ e^1 & - & 0.5 & 0.7 & 0.5 \\ e^2 & 0.6 & - & 0.6 & 0.7 \\ e^3 & 0.7 & 0.6 & - & 0.5 \\ e^4 & 0.5 & 0.6 & 0.9 & - \end{matrix}$$

For this GDM problem, the given consensus threshold is $\xi = 0.95$, the upper and lower error thresholds are $\overline{\phi} = 0.15$ and $\underline{\phi} = 0.05$, and the maximum number of iterations $\mathbb{Z}_{\max} = 5$.

Step 1. Let $\mathbb{Z} = 0$, $\Re^{k(0)} = ((r_{ij}^{k(0)}, sl_{ij}^{k(0)}))_{4\times4} = ((r_{ij}^k, sl_{ij}^k))_{4\times4} (k = 1,2,3,4)$.

Step 2. Let $\alpha = 0.2$, $\beta = 0.5$, and $\gamma = 0.3$. Based on Eqs. (4.31)–(4.35), the experts' weight vector is obtained as $w^{(0)} = (0.24, 0.30, 0.23, 0.24)^T$.

Step 3. The individual priority vectors of all experts can be calculated from the self-confident DHLPRs by Model (4.36):

$$w^{1(0)} = (0.1311, 0.5357, 0.3332, 0)^T, w^{2(0)} = (0.3268, 0.5143, 0.0768, 0.0820)^T$$
$$w^{3(0)} = (0.2865, 0.5326, 0.1810, 0)^T, w^{4(0)} = (0.2808, 0.4683, 0.2058, 0.0451)^T.$$

Based on Model (4.37), the collective priority vector can be obtained as:

$$w^{c(0)} = (0.2920, 0.5334, 0.1746, 0)^T$$

Step 4. Based on Eqs. (4.38) and (4.39), we can calculate the individual consensus degrees $CD(e^k)^{(0)}$ $(k = 1, 2, \ldots, n)$ and the collective consensus degree $CD^{(0)}$, which are shown in Table 4.13.

Clearly, $CD^{(0)} = 0.9433 < \xi = 0.95$. This indicates that the consensus degree among all experts is not acceptable. Therefore, it is necessary to improve the consensus degree among all experts by the feedback adjustment mechanism.

Step 5. Based on Table 4.13, the individual consensus degrees of the experts $\{e^1, e^2\}$ are lower than the given consensus threshold, i.e., $ECD^{(0)} = \{e^1, e^2\}$. Then, we transform the collective priority vector $w^{c(0)}$ and obtain the collective DHLPR:

$$W^{c(0)} = \left(\omega_{ij}^{c(0)}\right)_{4\times4} \begin{pmatrix} s_0\!<\!o_0\!> & s_{-1}\!<\!o_{0.14}\!> & s_0\!<\!o_{1.88}\!> & s_1\!<\!o_{0.67}\!> \\ s_1\!<\!o_{-0.14}\!> & s_0\!<\!o_0\!> & s_1\!<\!o_{1.74}\!> & s_2\!<\!o_{0.53}\!> \\ s_0\!<\!o_{-1.88}\!> & s_{-1}\!<\!o_{-1.74}\!> & s_0\!<\!o_0\!> & s_0\!<\!o_{2.79}\!> \\ s_{-1}\!<\!o_{-0.67}\!> & s_{-2}\!<\!o_{-0.53}\!> & s_0\!<\!o_{-2.79}\!> & s_0\!<\!o_0\!> \end{pmatrix}$$

The error matrix $\Phi^{k(0)} = (\phi_{ij}^{k(0)})_{4\times4}$ of each expert $e^k \in ECD^{(0)} = \{e^1, e^2\}$ can be generated by Eq. (4.40):

$$\Phi^{1(0)} = \begin{pmatrix} 0 & 0.0355 & 0.0038 & 0.0103 \\ 0.0355 & 0 & 0.1643 & 0.0770 \\ 0.0038 & 0.1643 & 0 & 0.1498 \\ 0.0103 & 0.0770 & 0.1498 & 0 \end{pmatrix},$$

$$\Phi^{2(0)} = \begin{pmatrix} 0 & 0.0270 & 0.1212 & 0.3335 \\ 0.0270 & 0 & 0.1331 & 0.0770 \\ 0.1212 & 0.1331 & 0 & 0.1186 \\ 0.3335 & 0.0770 & 0.1186 & 0 \end{pmatrix}.$$

The upper error threshold is $\overline{\phi} = 0.15$. Based on Eq. (4.41), we can identify the experts whose preference information with error degrees are larger than $\overline{\phi}$:

$$EL^{1(0)} = \{(2,3)\} \text{ and } EL^{2(0)} = \{(1,4)\}$$

Table 4.13 The individual and collective consensus degrees

	e^1	e^2	e^3	e^4
$CD(e^k)^{(0)}$	0.8870	0.9332	0.9958	0.9571
$CD^{(0)}$	0.9433			

Similarly, the lower error threshold is $\underline{\phi} = 0.05$. Based on Eq. (4.42), we can identify the experts whose preference information with error degrees are lower than $\underline{\phi}$:

$$EM^{1(0)} = \{(1,2),(1,3),(1,4)\} \text{ and } EM^{2(0)} = \{(1,2)\}$$

Let $\Re^{k(1)} = ((r_{ij}^{k(1)}, sl_{ij}^{k(1)}))_{4\times 4}$ be the adjusted self-confident DHLPRs of the experts $e^k(1,2)$, respectively. Then,

1. When constructing $\Re^{1(1)} = ((r_{ij}^{1(1)}, sl_{ij}^{1(1)}))_{4\times 4}$, we advise that

(1) $r_{23}^{1(1)} \in [s_{1<o_{1.74}>}, s_{3<o_{-1}>}]$ and $sl_{23}^{1(1)} \in [1,2]$.

(2) $r_{12}^{1(1)} = s_{-1<o_{-1}>}$ and $sl_{12}^{1(1)} \in [4,7]$; $r_{13}^{1(1)} = s_{1<o_{-2}>}$ and $sl_{13}^{1(1)} \in [3,7]$; $r_{14}^{1(1)} = s_{2<o_{-3}>}$ and $sl_{14}^{1(1)} \in [5,7]$.

The adjusted results can be shown as: For $i<j$, the remain unidentified elements of $\Re^2$ are not changed. For $i>j$, $r_{ji}^{k(1)} \oplus r_{ij}^{k(1)} = s_{0<o_0>}$ and $sl_{ji}^{k(1)} = sl_{ij}^{k(1)}$ are used.

2. When constructing $\Re^{2(1)} = ((r_{ij}^{2(1)}, sl_{ij}^{2(1)}))_{4\times 4}$, there are:

(1) $r_{14}^{2(1)} \in [s_{-2<o_2>}, s_{1<o_{0.67}>}]$ and $sl_{14}^{2(1)} \in [1,5]$;

(2) $r_{12}^{2(1)} = s_{-1<o_1>}$ and $sl_{12}^{2(1)} \in [3,7]$.

The adjusted results can be shown as: For $i<j$, the remaining unidentified elements of $\Re^2$ are not changed. For $i>j$, $r_{ji}^{k(1)} \oplus r_{ij}^{k(1)} = s_{0<o_0>}$ and $sl_{ji}^{k(1)} = sl_{ij}^{k(1)}$ are used.

Then, we can provide these adjustment suggestions to the experts e^1 and e^2, and they are advised to adjust the preference information and the self-confident degrees. Without loss of generality, they provide the adjusted self-confident DHLPRs according to the adjustment suggestions as:

$$\Re^{1(1)} = \begin{pmatrix} (s_{0<o_0>},7) & (s_{-1<o_{-1}>},5) & (s_{1<o_{-2}>},4) & (s_{2<o_{-3}>},6) \\ (s_{1<o_1>},4) & (s_{0<o_0>},7) & (s_{2<o_{-2}>},1) & (s_{2<o_3>},6) \\ (s_{-1<o_2>},3) & (s_{-3<o_1>},2) & (s_{0<o_0>},7) & (s_{-1<o_2>},2) \\ (s_{-2<o_3>},5) & (s_{-3<o_{-3}>},6) & (s_{1<o_{-3}>},2) & (s_{0<o_0>},7) \end{pmatrix}$$

$$\Re^{2} = \begin{pmatrix} (s_{0<o_0>},7) & (s_{-1<o_1>},4) & (s_{0<o_{-2}>},2) & (s_{1<o_0>},2) \\ (s_{1<o_{-1}>},3) & (s_{0<o_0>},7) & (s_{2<o_2>},6) & (s_{2<o_3>},3) \\ (s_{-3<o_2>},2) & (s_{-2<o_{-2}>},6) & (s_{0<o_0>},7) & (s_{-1<o_3>},4) \\ (s_{-3<o_{-2}>},5) & (s_{-1<o_{-3}>},3) & (s_{1<o_{-3}>},4) & (s_{0<o_0>},7) \end{pmatrix}$$

Then, go back to Step 2. Firstly, the weight vector of experts is obtained as $w^{(1)} = (0.25, 0.29, 0.23, 0.24)^T$. Then, by Model (4.36), the individual priority vectors of all experts can be calculated from the self-confident DHLPRs:

Table 4.14 The individual and collective consensus degrees

	e^1	e^2	e^3	e^4
$CD(e^k)^{(1)}$	0.9057	0.9635	0.9700	0.9623
$CD^{(1)}$	0.9504			

$$\omega^{1(1)} = (0.1396, 0.4849, 0.3755, 0)^T, \omega^{2(1)} = (0.2887, 0.5417, 0.1696, 0)^T,$$
$$\omega^{3(1)} = (0.2865, 0.5326, 0.1810, 0)^T, \omega^{4(1)} = (0.2808, 0.4683, 0.2058, 0.0451)^T.$$

Based on Model (4.37), the collective priority vector can be obtained as:

$$\omega^{c(1)} = (0.2538, 0.5173, 0.2290, 0)^T.$$

The individual consensus degrees $CD(e^k)^{(1)}$ $(k = 1, 2, 3, 4)$ and collective consensus degree $CD^{(1)}$ can be obtained and shown in Table 4.14.

Clearly, $CD^{(1)} = 0.9504 > \xi = 0.95$. This indicates that all experts reach consensus. Based on the collective priority vector $\omega^{c(1)}$, the final ranking of alternatives can be obtained as $A_2 \succ A_1 \succ A_3 \succ A_4$.

4.3.5 Comparative Analyses and Simulation

This subsection mainly makes some comparative analyses about the validity of the proposed consensus reaching method by setting up a simulation experiment.

(a) **Comparison objects**

The comparative analyses mainly consist of three objects:

(1) The proposed DHSM-based consensus model.
(2) Without considering the adjustments of the self-confident degrees, we can obtain a double hierarchy linguistic preference values and self-confidence degrees-unchanged modifying-based (DHSUM-based) consensus model. Therefore, another new algorithm about the DHSUM-based consensus model can be established if all self-confident degrees remain unchanged in the consensus reaching process.
(3) Without the self-confident degrees, the self-confident DHLPR can be transformed into DHLPR. That is to say, the DHLPR implies that the expert is fully self-confident of his/her evaluation information. In other words, the self-confident degrees of all evaluations in a DHLPR are the same, and they can be omitted for notation simplification. Therefore, the DHLPR can be regarded as a special case of the self-confident DHLPR. Here, a consensus reaching method of DHLPR, named as double hierarchy linguistic preference values modifying-based (DHM-based) consensus model can be established by

omitting the adjustment of self-confident degrees in Algorithm 4.3. Then the DHM-based consensus model can be established by deleting all self-confidence degrees and their adjustment rules in Algorithm 4.3.

Then, we can compare the DHSM-based consensus model, the DHSUM-based consensus model and the DHM-based consensus model in the next simulation experiment.

(b) **Comparison criteria**

In the consensus reaching process, it is expected that the consensus can be reached as soon as possible, and the adjustment numbers are as small as possible. Therefore, three comparison criteria are proposed to reflect the consensus efficiency of the consensus reaching models.

(1) It is unavoidable that a number of consensus rounds will be needed to reach the given consensus threshold. Therefore, the number of iterations (denoted as $\mathbb{Z}$) in the consensus reaching process reflects the consensus efficiency, and can be regarded as the first criterion in comparison.
(2) The consensus success ratio (denoted as P) expresses the ratio of reaching the consensus within the range of the allowed number of iterations $\mathbb{Z} < \mathbb{Z}_{\max}$.
(3) The distance between the original and the adjusted preference information is denoted as AD. Suppose that $\Re^k = ((r_{ij}^k, sl_{ij}^k))_{m\times m}$ and $\widetilde{\Re^k} = ((\widetilde{r_{ij}^k}, \widetilde{sl_{ij}^k}))_{m\times m}$ are the original and adjusted preferences, respectively. Then, the overall distance of the experts $E = \{e^1, e^2, \ldots, e^n\}$ can be obtained by

$$AD = \sum_{k=1}^{n} \sqrt{\frac{2}{m(m-1)} \sum_{\substack{i,j=1 \\ i<j}}^{m} \left(r_{ij}^k - \widetilde{r_{ij}^k}\right)^2} \tag{4.46}$$

Obviously, $AD \in [0, 1]$. The smaller the value of AD, the less information loss of the preference relation.

(c) **Simulation and comparison results**

Let $\dot{\mathbb{Z}}_1$, $\dot{\mathbb{Z}}_2$ and $\dot{\mathbb{Z}}_3$ be the numbers of iterations of the DHSM-based consensus model, the DHSUM-based consensus model and the DHM-based consensus model, respectively; $\dot{P}_1$, $\dot{P}_2$ and $\dot{P}_3$ be the consensus success ratios of the DHSM-based consensus model, the DHSUM-based consensus model and the DHM-based consensus model, respectively; $\dot{AD}_1$, $\dot{AD}_2$ and $\dot{AD}_3$ be the adjustment distances of the DHSM-based consensus model, the DHSUM-based consensus model and the DHM-based consensus model, respectively. Then the simulation model can be described as follows:

Algorithm 4.4 (Gou et al. 2020b) Simulation model

Input: m, n, S^{SL}, ω, ξ, $\overline{\phi}$, $\underline{\phi}$, and $\mathbb{Z}_{\max}$.

Output: $\dot{\mathbb{Z}}_1$, $\dot{\mathbb{Z}}_2$, $\dot{\mathbb{Z}}_3$, $\dot{P}_1$, $\dot{P}_2$, $\dot{P}_3$, $A\dot{D}_1$, $A\dot{D}_2$ and $A\dot{D}_3$.

Step 1. Generate n self-confident DHLPRs $\Re^k = ((r_{ij}^k, sl_{ij}^k))_{m\times m}(k = 1, 2, \ldots, n)$.

Step 2. Utilize the DHSM-based consensus model (Algorithm 4.3) to deal with all self-confident DHLPRs $\Re^k$ to obtain the adjusted self-confident DHLPRs $\widetilde{\Re^{k,1}} = ((\widetilde{r_{ij}^{k,1}}, \widetilde{sl_{ij}^{k,1}}))_{m\times m}(k = 1, 2, \ldots, n)$ and the number of consensus rounds $\dot{\mathbb{Z}}_1$.

Step 3. Utilize the DHSUM-based consensus model (Algorithm 4.3) to deal with all self-confident DHLPRs $\Re^k$ to obtain the adjusted self-confident DHLPRs $\widetilde{\Re^{k,2}} = ((\widetilde{r_{ij}^{k,2}}, \widetilde{sl_{ij}^{k,2}}))_{m\times m}(k = 1, 2, \ldots, n)$ and the number of consensus rounds $\dot{\mathbb{Z}}_2$.

Step 4. Transform all self-confident DHLPRs $\Re^k(k = 1, 2, \ldots, n)$ to the DHLPRs $\mathbb{R}^k = (r_{ij}^k)_{m\times m}(k = 1, 2, \ldots, n)$ by deleting the self-confident degrees, and utilize the DHM-based consensus model to deal with all DHLPRs $\mathbb{R}^k$ to obtain the adjusted DHLPRs $\widetilde{\mathbb{R}^k}$ $(k = 1, 2, \ldots, n)$ and the number of consensus rounds $\dot{\mathbb{Z}}_3$.

Step 5. Calculate the overall distance between $\Re^k$ and $\widetilde{\Re^{k,1}}$ $(k = 1, 2, \ldots, n)$ and obtain $A\dot{D}_1$; Calculate the overall distance between $\Re^k$ and $\widetilde{\Re^{k,2}}$ $(k = 1, 2, \ldots, n)$ and obtain $A\dot{D}_2$; calculate the overall distance between $\mathbb{R}^k$ and $\widetilde{\mathbb{R}^k}$ $(k = 1, 2, \ldots, n)$ and obtain $A\dot{D}_3$.

Step 6. If $CD^{(\mathbb{Z}_1)} > \xi$, then $\dot{P}_1 = 1$; otherwise, $\dot{P}_1 = 0$. If $CD^{(\mathbb{Z}_2)} > \xi$, then $\dot{P}_2 = 1$; otherwise, $\dot{P}_2 = 0$. If $CD^{(\mathbb{Z}_3)} > \xi$, then $\dot{P}_3 = 1$; otherwise, $\dot{P}_3 = 0$.

Step 7. Output $\dot{\mathbb{Z}}_1$, $\dot{\mathbb{Z}}_2$, $\dot{\mathbb{Z}}_3$, $\dot{P}_1$, $\dot{P}_2$, $\dot{P}_3$, $A\dot{D}_1$, $A\dot{D}_2$ and $A\dot{D}_3$.

Step 8. End.

In the simulation method, we set $S_O = \{s_{t<o_k>} | t = -4, \ldots, 4;\ k = -4, \ldots, 4\}$, $S^{SL} = \{1, 2, \ldots, 7\}$, $\xi = 0.82$, $\overline{\phi} = 0.2$, $\underline{\phi} = 0.05$, and $\mathbb{Z}_{\max} = 5$. The weights of all experts are equal. Then, based on different parameters m and n, we run the simulation method 1000 times to obtain the average values of $\dot{\mathbb{Z}}_1$, $\dot{\mathbb{Z}}_2$, $\dot{\mathbb{Z}}_3$, $\dot{P}_1$, $\dot{P}_2$, $\dot{P}_3$, $A\dot{D}_1$, $A\dot{D}_2$ and $A\dot{D}_3$, respectively. Specially, $\mathbb{Z}_1$, $\mathbb{Z}_2$ and $\mathbb{Z}_3$ reflect the average values of the number of iterations required to reach the consensus in the DHSM-based consensus model, the DHSUM-based consensus model and the DHM-based consensus model, respectively; P_1, P_2 and P_3 reflect the average values of the consensus success ratios of the DHSM-based consensus model, the DHSUM-based consensus model and the DHM-based consensus model, respectively; AD_1, AD_2 and AD_3 reflect the average values of the adjustment distances in the DHSM-based consensus model, the DHSUM-based consensus model and the DHM-based consensus model, respectively.

The simulation results can be shown in Figs. 4.4, 4.5 and 4.6.

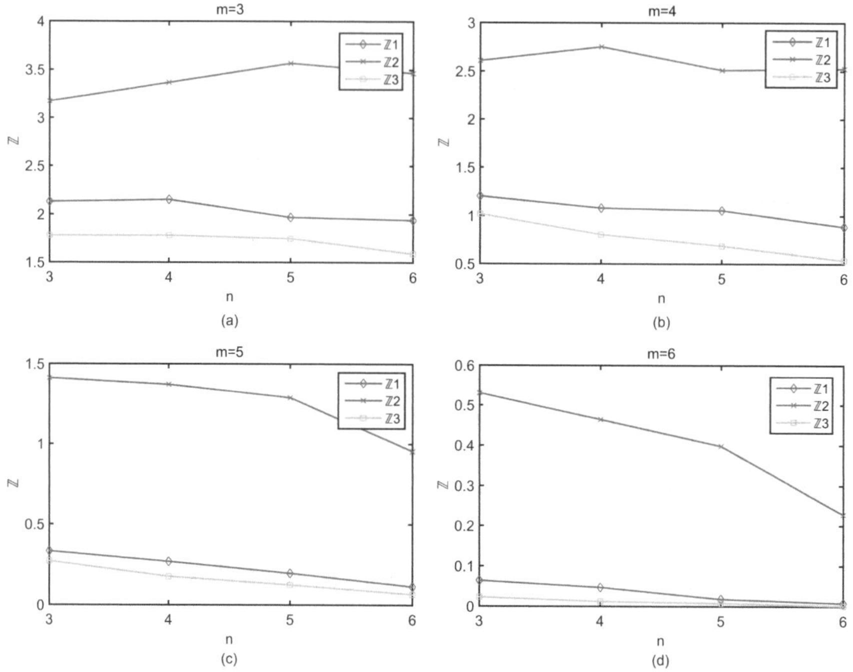

Fig. 4.4 The average values of $\mathbb{Z}_1$, $\mathbb{Z}_2$ and $\mathbb{Z}_3$ with different values of m and n

Based on Figs. 4.4, 4.5 and 4.6, some results and discussions can be summarized as follows:

(1) Figure 4.4 shows that the proposed DHSM-based consensus model needs less consensus round than the DHSUM-based consensus model with respect to different values of m and n, which means that the speed to reach consensus is accelerated by adjusting the self-confident degrees in the feedback consensus reaching process. Additionally, the proposed DHSM-based consensus model needs more consensus round than the DHM-based consensus model under different values of m and n. Considering that the DHLPR has no self-confident degree and is simpler than the self-confident DHLPR, it is obvious that the consensus reaching speed of the DHLPR must be faster than the self-confident DHLPR. Furthermore, the number of iterations gradually decreases as the number of experts increases as well as the dimension of the matrices increases.
(2) Figure 4.5 shows that compared with the DHSUM-based consensus model, the DHSM-based consensus model needs less adjustments with respect to different values of m and n, which means that the adjustments of self-confident degrees can decrease the loss of preference information. Similar to Fig. 4.4, without the self-confident degrees, the DHM-based consensus model based on the DHLPR will lose the least preference information compared with the DHSUM-based consensus model and the DHSM-based consensus model.

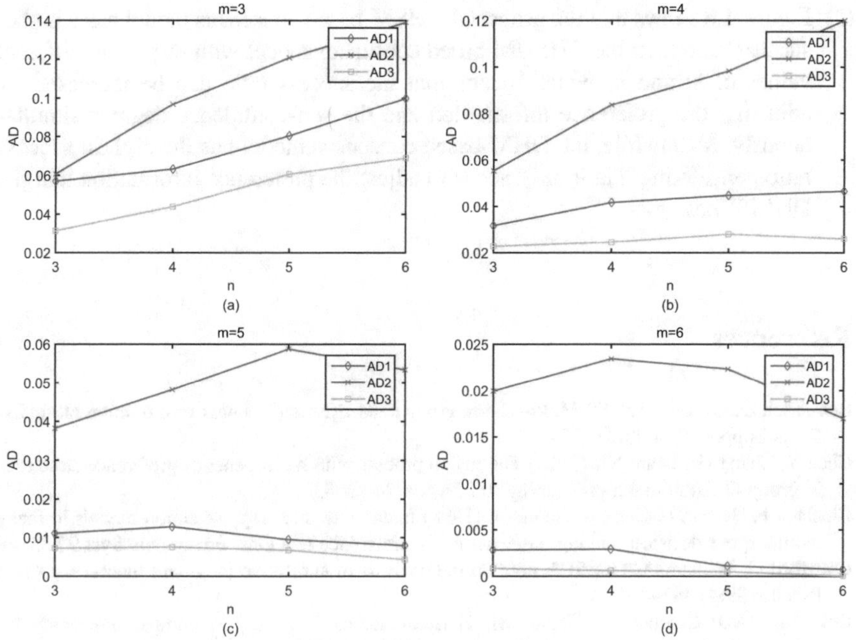

Fig. 4.5 The average values of AD_1, AD_2 and AD_3 with different values of m and n

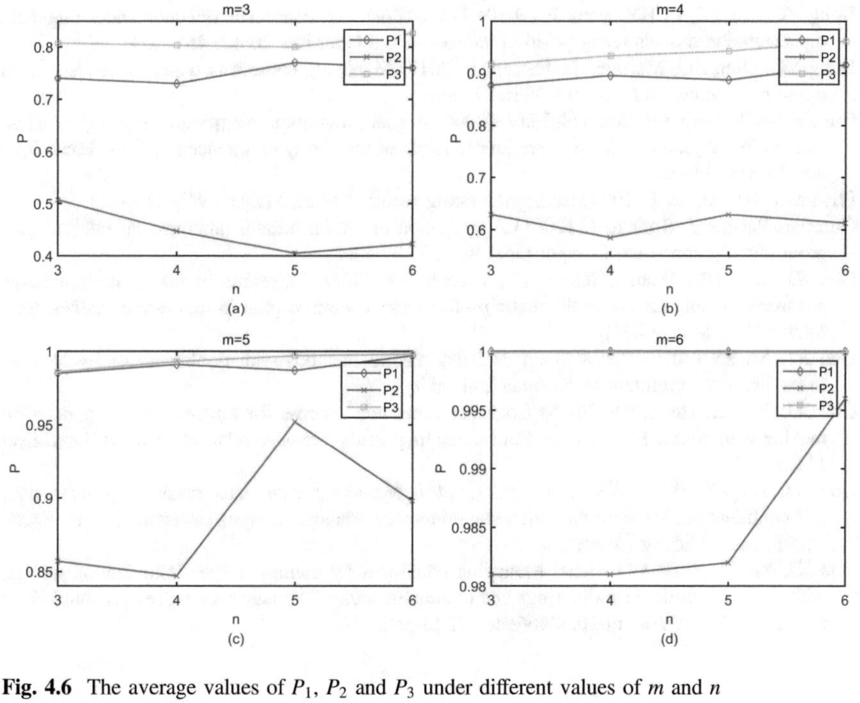

Fig. 4.6 The average values of P_1, P_2 and P_3 under different values of m and n

(3) Figure 4.6 shows that the proposed DHSM-based consensus model has a higher success ratio than the DHSUM-based consensus model with respect to different values of m and n, which means that the success ratio can be increased by adjusting the preference information and the self-confidence degrees simultaneously. Meanwhile, the DHM-based consensus model has the highest success ratio considering that it only needs to adjust the preference information that the DHLPR has.

References

Ben-Arieh D, Easton T (2007) Multi-criteria group consensus under linear cost opinion elasticity. Decis Support Syst 43(3):713–721

Chen X, Zhang HJ, Dong YC (2015) The fusion process with heterogeneous preference structures in group decision making: a survey. Inf Fusion 24:72–83

Chiclana F, Herrera F, Herrera-Viedma E (1998) Integrating three representation models in fuzzy multipurpose decision making based on fuzzy preference relations. Fuzzy Sets Syst 97:33–48

Crawford G, Williams C (1985) A note on the analysis of subjective judgment matrices. J Math Psychol 29(4):387–405

Del Moral MJ, Chiclana F, Tapia JM, Herrera-Viedma E (2018) A comparative study on consensus measures in group decision making. Int J Intell Syst 33(8):1624–1638

Dombi J (1995) A general framework for the utility-based and outranking methods. Fuzzy logic and soft computing. World Scientific, Singapore, pp 202–208

Dong QX, Cooper O (2016) A peer-to-peer dynamic adaptive consensus reaching model for the group AHP decision making. Eur J Oper Res 250(2):521–530

Dong YC, Xu YF, Li HY, Feng B (2010) The OWA-based consensus operator under linguistic representation models using position indexes. Eur J Oper Res 203(2):455–463

Dong YC, Ding ZG, Martínez L, Herrera F (2017) Managing consensus based on leadership in opinion dynamics. Inf Sci 397–398:187–205

Fan ZP, Ma J, Jiang YP, Sun YH, Ma L (2006) A goal programming approach to group decision making based on multiplicative preference relations and fuzzy preference relations. Eur J Oper Res 174(1):311–321

Friedman HH, Amoo T (1999) Rating the rating scales. J Mark Manage 9(3):114–123

González-Pachón J, Romero C (2001) Aggregation of partial ordinal rankings: an interval goal programming approach. Comput Oper Res 28:827–834

Gou XJ, Liao HC, Wang XX, Xu ZS, Herrera F (2020a) Consensus based on multiplicative consistent double hierarchy linguistic preferences: Venture capital in real estate market. Int J Strateg Prop Manag 42(1):1–23

Gou XJ, Xu ZS (2019) Research on decision-making methods with double hierarchy hesitant fuzzy linguistic preferences. Sichuan University

Gou XJ, Xu ZS, Herrera F (2018) Consensus reaching process for large-scale group decision making with double hierarchy hesitant fuzzy linguistic preference relations. Knowl Based Syst 157:20–33

Gou XJ, Xu ZS, Wang XX, Liao HC (2020b) Managing consensus reaching process with self-confident double hierarchy linguistic preference relations in group decision making. Fuzzy Optim Decis Making (Accept)

Gou XJ, Xu ZS, Zhou W (2020c) Managing consensus by multiple stages optimization models with linguistic preference orderings and double hierarchy linguistic preferences. Technol Econ Dev Econ. https://doi.org/10.3846/tede.2020.12013

He Y, Xu ZS (2018) A consensus framework with different preference ordering structures and its applications in human resource selection. Comput Ind Eng 118:80–88

Herrera-Viedma E, Herrera F, Chiclana F, Luque M (2004) Some issues on consistency of fuzzy preference relations. Eur J Oper Res 154(1):98–109

Herrera-Viedma E, Martínez L, Mata F, Chiclana F (2005) A consensus support system model for group decision-making problems with multigranular linguistic preference relations. IEEE Trans Fuzzy Syst 13(5):644–658

Kacprzyk J, Fedrizzi M (1988) A 'soft' measure of consensus in the setting of partial (fuzzy) preferences. Eur J Oper Res 34(3):316–325

Kacprzyk J, Roubens M (1988) Non-conventional preference relations in decision making. Springer, Berlin

Kamis NH, Chiclana F, Levesley J (2018) Preference similarity network structural equivalence clustering based consensus group decision making model. Appl Soft Comput 67:706–720

Karplus KP, Diederichs K (2012) Linking crystallographic model and data quality. Science 336 (6084):1030–1033

Kou G, Lin C (2014) A cosine maximization method for the priority vector derivation in AHP. Eur J Oper Res 235(1):225–232

Liao HC, Xu ZS, Zeng XJ, Xu DL (2016) An enhanced consensus reaching process in group decision making with intuitionistic fuzzy preference relations. Inf Sci 329:274–286

Liu WQ, Dong YC, Chiclana F, Cabrerizo FJ, Herrera-Viedma E (2017) Group decision making based on heterogeneous preference relations with self-confidence. Fuzzy Optim Decis Making 16(4):429–447

Liu JP, Song JM, Xu Q, Tao ZF, Chen HY (2019) Group decision making based on DEA cross-efficiency with intuitionistic fuzzy preference relations. Fuzzy Optim Decis Making 18:345–370

Ma LC (2016) A new group ranking approach for ordinal preferences based on group maximum consensus sequences. Eur J Oper Res 251(1):171–181

Meng FY, Tang J, Zhang SL (2019) Interval linguistic fuzzy decision making in perspective of preference relations. Technol Econ Dev Econ 25(5):998–1015

Morente-Molinera JA, Kou G, Pérez IJ, Samuylov K, Selamat A, Herrera-Viedma E (2018) A group decision making support system for the Web: How to work in environments with a high number of participants and alternatives. Appl Soft Comput 68:191–201

Morente-Molinera JA, Kou G, Samuylov K, Ureña R, Herrera-Viedma E (2019) Carrying out consensual group decision making processes under social networks using sentiment analysis over comparative expressions. Knowl Based Syst 165:335–345

Park JH, Cho HJ, Kwun YC (2011) Extension of the VIKOR method for group decision making with interval-valued intuitionistic fuzzy information. Fuzzy Optim Decis Making 10:233–253

Parreiras R, Ekel P, Bernardes F (2012) A dynamic consensus scheme based on a nonreciprocal fuzzy preference relation modeling. Inf Sci 211:1–17

Ramanathan R, Ganesh LS (1994) Group preference aggregation methods employed in AHP: an evaluation and an intrinsic process for deriving members' weightages. Eur J Oper Res 79 (2):249–265

Saaty TL (1980) The analytic hierarchy process. McGraw-Hill, New York

Tanino T (1988) Fuzzy preference relations in group decision making. Non-conventional preference relations in decision making. Springer, Berlin, pp 54–71

Wan SP, Wang F, Dong JY (2018) A group decision-making method considering both the group consensus and multiplicative consistency of interval-valued intuitionistic fuzzy preference relations. Inf Sci 466:109–128

Wu XL, Liao HC (2019) A consensus-based probabilistic linguistic gained and lost dominance score method. Eur J Oper Res 272(3):1017–1027

Wu ZB, Xu JP (2016) Managing consistency and consensus in group decision making with hesitant fuzzy linguistic preference relations. Omega 65(3):28–40

Wu ZB, Xu JP (2018) A consensus model for large-scale group decision making with hesitant fuzzy information and changeable clusters. Inf Fusion 41:217–231

Wu J, Chiclana F, Fujita H, Herrera-Viedma E (2017) A visual interaction consensus model for social network group decision making with trust propagation. Knowl Based Syst 122:39–50

Wu ZB, Jin BM, Xu JP (2018) Local feedback strategy for consensus building with probability-hesitant fuzzy preference relations. Appl Soft Comput 67:691–705

Wu ZB, Huang S, Xu JP (2019a) Multi-stage optimization models for individual consistency and group consensus with preference relations. Eur J Oper Res 275:182–194

Wu HY, Ren PJ, Xu ZS (2019b) Hesitant fuzzy linguistic consensus model based on trust-recommendation mechanism for hospital expert consultation. IEEE Trans Fuzzy Syst 27(1):2227–2241

Xu YJ, Herrera F, Wang HM (2016) A distance-based framework to deal with ordinal and additive inconsistencies for fuzzy reciprocal preference relations. Inf Sci 328:189–205

Xu YJ, Wen XW, Zhang WC (2018) A two-stage consensus method for large-scale multi-attribute group decision making with an application to earthquake shelter selection. Comput Ind Eng 116:113–129

Yu DJ, Xu ZS (2019) Intuitionistic fuzzy two-sided matching model and its application to personnel-position matching problems. J Oper Res Soc 71(2):312–321

Zhang ZM, Chen SM (2019) A consistency and consensus-based method for group decision making with hesitant fuzzy linguistic preference relations. Inf Sci 501:317–336

Zhang ZM, Pedrycz W (2018) Goal programming approaches to managing consistency and consensus for intuitionistic multiplicative preference relations in group decision making. IEEE Trans Fuzzy Syst 26(6):3261–3275

Zhang BW, Liang HM, Gao Y, Zhang GQ (2018) The optimization-based aggregation and consensus with minimum-cost in group decision making under incomplete linguistic distribution context. Knowl Based Syst 162:92–102

Zhang HJ, Dong YC, Francisco C, Yu S (2019) Consensus efficiency in group decision making: a comprehensive comparative study and its optimal design. Eur J Oper Res 275(2):580–598

Zhu B, Xu ZS (2018) Probability-hesitant fuzzy sets and the representation of preference relations. Technol Econ Dev Econ 24(3):1029–1040

Chapter 5
Large-Scale Group Consensus Decision-Making Methods with DHHFLPRs

With the rapid development of society and the increasingly complex economic environment, management and decision-making tasks are becoming more and more difficult. Meanwhile, with the progress of science and technology and the development of network environment, the communications between people are increasingly convenient. Therefore, large-scale group decision-making (LSGDM) has become the focus of decision-making problems. Generally, a GDM problem can be called LSGDM problem when the number of experts is more than 20 (Liu and Chen 2006). Now LSGDM are very commonly encountered in actual life, especially in the era of data (Labella 2018; Liu et al. 2014a, b, 2015a, b, c; Liu et al. 2016; Palomares et al. 2014a, b; Quesada et al. 2015; Wu and Liu 2016; Wu and Xu 2018; Xu et al. 2015; Zhang 2018; Zhang et al. 2018). Basically, large-scale problems in decision making, consensus reaching, voting, social choice etc. call for the use of sophisticated computational tools, analysis of computational efficiency of algorithm, etc. This is done by the so called computational social choice (Brandt et al. 2016; Chevalayre et al. 2007). These works are mathematically and algorithmically very sophisticated. Specially, in this chapter, the concept of a large-scale group consensus decision making related problem is not meant here in the sense of computational social choice and related areas (Brandt et al. 2016; Chevalayre et al. 2007). And the purpose of this chapter is to research the large-scale group consensus reaching methods based on DHHFLPRs. Firstly, because the clustering and the consensus reaching process are two important constituent parts, we will discuss the clustering method and the consensus reaching process in LSGDM with DHHFLPRs, propose a large-scale group consensus decision-making method and apply this method to the assessments of water resources in some cities of Sichuan province. Additionally, by constructing new clustering method and consensus model, and from the perspective of in-depth analyzing minority opinions and non-cooperative behaviors in LSGDM, this chapter puts forward a novel large-scale group consensus decision-making method based on DHHFLPRs, which is more in line with human cognition, and applies this method to the comprehensive assessments of the causes of haze formation.

X. Gou and Z. Xu, *Double Hierarchy Linguistic Term Set and Its Extensions*, Studies in Fuzziness and Soft Computing 396,
https://doi.org/10.1007/978-3-030-51320-7_5

5.1 Large-Scale Group Consensus Decision-Making Method with DHHFLPRs

LSGDM has been studied in some different fields and mainly includes two parts: consensus models and clustering methods. Firstly, some of the consensus models have been developed based on self-organizing maps (Palomares et al. 2014a), graphical monitoring tool (MENTOR) (Palomares et al. 2014b), expert weighting methodology (Quesada et al. 2015), and minimum adjustment cost feedback mechanism-based consensus model (Xu et al. 2012), etc. Additionally, two consensus models were built to deal with some LSGDM problems with non-cooperative behaviors and minority opinions (Xu et al. 2015), and individual concerns and satisfactions (Zhang et al. 2018), respectively. Furthermore, with the hesitant fuzzy information (Xia and Xu 2011), a consensus model for LSGDM was introduced, which is distinguished from previous studies about the obtained clusters and the feedback mechanism (Wu and Xu 2018). Zhang (2018) proposed a consistency-and consensus-based model based on probabilistic linguistic term sets (PLTSs) (Pang et al. 2016) under LSGDM. Secondly, amounts of clustering methods were developed including k-means clustering method (Wu and Xu 2018), fuzzy c-mean clustering method [9] (Palomares et al. 2014a), interval type-2 fuzzy equivalence clustering analysis (Wu and Liu 2016), the partial binary tree DEA-DA cyclic classification model (Liu et al. 2014a), and the hierarchical clustering approach (Zhu et al. 2016), etc. In this subsection, these two parts are also the contents which we need to focus on discussing.

The main work in this subsection is to discuss the clustering method and the consensus reaching process in LSGDM with DHHFLPRs, which can be summarized as follows:

(1) Based on the similarity measures of DHHFLTSs, we develop a clustering method for LSGDM based on information entropy theory, which can be understood very clearly by a dynamic clustering figure. By this method, the experts can be divided into several small groups. Additionally, we propose a weights-determining method, which can obtain the weight of each small group, the weights of the experts included in each small group, and the weights of all experts, respectively.
(2) We propose some consensus measures. A model is developed, which can precisely identify the alternatives, the pairs of alternatives and the experts that do not reach the consensus threshold, and then the moderator feeds these suggestions back to each small group and experts for modifying their preference information. This consensus measures can make the consensus degree improving process more targeted.
(3) Collecting the results obtained above, we establish a LSGDM model. It can be used to deal with LSGDM step by step. Moreover, a case study is set up to apply our model to deal with a practical LSGDM problem that is to evaluate Sichuan water resource management.

5.1.1 Similarity Degree-Based Clustering Algorithm

An LSGDM problem with DHHFLPRs can be described as: Let $A = \{A_1, A_2, \ldots, A_m\}$ be a set of alternatives, $E = \{e^1, e^2, \ldots, e^n\}$ be a set of experts, and $w = (w_1, w_2, \ldots, w_n)^T$ be the weight vector of experts with $0 \leq w_i \leq 1$ and $\sum_{i=1}^{n} w_i = 1$. Suppose that $\widetilde{H}^a_{S_O} = (h^a_{S_{O_{ij}}})_{m \times m}$ $(a = 1, 2, \ldots, n)$ be a DHHFLPR which indicates that the expert e^a gives his/her evaluations for all alternatives by making pairwise comparisons. Without loss of generality, we let $n \geq 20$ and $m \geq 3$.

Before we start work, the normalization of DHHFLPR is necessary. In order to not lose the original information, we can normalize DHHFLPR based on the linguistic expected value of DHHFLE proposed in Definition 3.1. Therefore, suppose that $\widetilde{H}_{S_O} = (h_{S_{O_{ij}}})_{m \times m}$ is a DHHFLPR, then we call $\widetilde{H}^N_{S_O} = (h^N_{S_{O_{ij}}})_{m \times m}$ a normalized DHHFLPR, which satisfies $h^N_{S_{O_{ij}}} = le(h_{S_{O_{ij}}})$, $le(h_{S_{O_{ij}}}) \oplus le(h_{S_{O_{ji}}}) = s_{0\langle o_0 \rangle}$, $le(h_{S_{O_{ii}}}) = s_{0\langle o_0 \rangle}$, and $\# le(h_{S_{O_{ij}}}) = \# le(h_{S_{O_{ji}}})$ $(i, j = 1, 2, \ldots, m)$.

In LSGDM, the discussions among the experts is very common. However, it will surely bring forth a huge amount of work and the communications among the experts also will not be smooth. To solve these problems, clustering is very necessary in the consensus reaching process because of a group with less experts is easier to discuss and improve preference information. Therefore, Gou et al. (2018) introduced how to cluster the experts in LSGDM on the basis of similarity measure. Firstly, based on the distance and similarity of DHHFLEs given in Chap. 3, the concept of similarity degree between two DHHFLEs can be defined as follows:

Definition 5.1 (Gou et al. 2018). Let $h^1_{S_O}$ and $h^2_{S_O}$ be two DHHFLEs, then the similarity degree between $h^1_{S_O}$ and $h^2_{S_O}$ is

$$sd(h_{S_{O_1}}, h_{S_{O_2}}) = 1 - d(h_{S_{O_1}}, h_{S_{O_2}}) = 1 - \left| f(le(h_{S_{O_1}})) - f(le(h_{S_{O_2}})) \right| \tag{5.1}$$

where f is the equivalent transformation function. Clearly, $sd(h_{S_{O_1}}, h_{S_{O_2}}) \in [0, 1]$, and the $sd(h_{S_{O_1}}, h_{S_{O_2}})$ is closer to 1, the more similar between $h^1_{S_O}$ and $h^2_{S_O}$ will be.

Then a similarity matrix $SM^{ab} = (sm^{ab}_{ij})_{m \times m}$ $(a, b = 1, 2, \ldots, n)$ for each pair of experts (e^a, e^b) can be established:

$$SM^{ab} = \begin{pmatrix} sm^{ab}_{11} & sm^{ab}_{12} & \cdots & sm^{ab}_{1m} \\ sm^{ab}_{21} & sm^{ab}_{22} & \cdots & sm^{ab}_{2m} \\ \vdots & \vdots & \ddots & \vdots \\ sm^{ab}_{m1} & sm^{ab}_{m2} & \cdots & sm^{ab}_{mm} \end{pmatrix} \tag{5.2}$$

where sm^{ab}_{ij} expresses the similarity degree between e^a and e^b in the position (i, j) and

$$sm_{ij}^{ab} = sd(h_{S_{O_{ij}}}^{a}, h_{S_{O_{ij}}}^{b}) \tag{5.3}$$

In general, the higher similarity degree two experts have, the greater possibility they belong to the same group. Therefore, a similarity degree-based clustering method can be developed as follows:

Algorithm 5.1 (Gou et al. 2018). **Similarity degree-based clustering algorithm**

Step 1. Establish the overall similarity matrix. Based on Eqs. (5.1) and (5.2), we can obtain a similarity matrix $SM^{ab} = (sm_{ij}^{ab})_{m \times m}(a, b = 1, 2, \ldots, n)$ associated with each pair of experts (e^a, e^b), and then aggregate all the similarity matrices and obtain the overall similarity matrix $OSM = (osm^{ab})_{n \times n}$, where

$$osm^{ab} = \frac{2}{m(m-1)} \sum_{i=1}^{m} \sum_{i<j}^{m} sm_{ij}^{ab} \tag{5.4}$$

Step 2. Choose the classification threshold. Ranking all the different elements of the upper triangular matrix of OSM (except the diagonal elements) following the order from big to small, denoted by $\eta_1 > \eta_2 > \cdots > \eta_p > \cdots > \eta_q$, where $q \leq n(n-1)/2$. Let $\eta = \eta_p$, obviously, $\eta \in [0, 1]$.

Step 3. Determine the optimal classification threshold $\eta*$. Let C_p be the rate of threshold change, obtained by

$$C_p = \frac{\eta_{p-1} - \eta_p}{n_p - n_{p-1}} \tag{5.5}$$

where η_{p-1} and η_p are the $(p-1)$-th and pth classification threshold, respectively; n_p and n_{p-1} are the number of the p-th and $(p-1)$-th classification, respectively. If $n_p = n$, then the operation is over. If

$$C_\mu = \max_p \{C_p\} \tag{5.6}$$

then we call the μth classification threshold the optimal classification threshold, namely, $\eta* = \eta_\mu$.

Step 4. Determine the classification result. Firstly, we collect all pairs of experts (e^a, e^b) into an overall group where $osm^{ab} \geq \eta*$, denoted by $B_1, B_2, \ldots, B_\zeta$, and then combine the elements of overall group into a group if they satisfy $B_{\zeta_i} \cap B_{\zeta_j} \neq \varnothing$

$(\zeta_i \neq \zeta_j;\ \zeta_i, \zeta_j = 1, 2, \ldots, \zeta)$. When $B_{\zeta_i} \cap B_{\zeta_j} = \varnothing$, then we can obtain the classification result of the large-scale group members, denoted as $B_\imath(\imath = 1, 2, \ldots, \Upsilon)$.

Step 5. End.

Remark 5.1 (Gou et al. 2018). This clustering is mainly based on the similarity degree between any two experts, which means that two experts can be deemed as a cluster if they have a high enough similarity degree. From Steps 1–3, we can obtain all pairs of experts (e^a, e^b), which can be collected into an overall group, denoted by $B_1, B_2, \ldots, B_\zeta$. For example, if $B_{\zeta_i} \cap B_{\zeta_j} \neq \varnothing$, then it is obvious that these two pairs of experts have same expert. Therefore, all experts are included in B_{ζ_i} and B_{ζ_j}. Similarly, we can obtain the final clustering result.

The similarity degree-based clustering process in LSGDM can be described in Fig. 5.1.

5.1.2 Double Hierarchy Information Entropy-Based Weights-Determining Method

At present, there exist a lot of weights-determining methods in decision-making, such as the dynamic weights-determining approach based on the intuitionistic fuzzy Bayesian network (Hao et al. 2018), the two-layer weights-determining method (Liu et al. 2015b), the AHP method (Ramanathan and Ganesh 1994), the Delphi methods (Linstone and Turoff 1975), the entropy-based method (Xu et al. 2012), the TOPSIS-based methods (Yue 2011), the projection method (Yue 2012), and the combined weighting methods (Hsu and Chen 1996; Lee 2000), etc. In this subsection, we also need to develop a weight-determining method for LSGDM. Based on the clustering result discussed in Subsection 5.1.1, a double hierarchy information entropy-based weights-determining method can be developed. This method can obtain three kinds of weights information including the weight of each group, the weights of the experts included in each group, and the weights of all experts. The process of this method can be shown as follows:

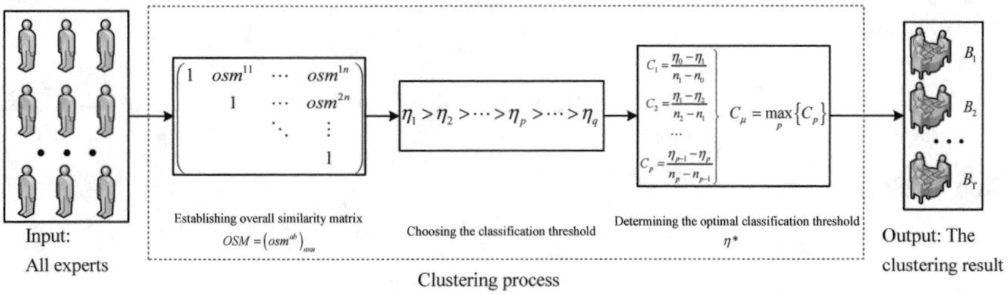

Fig. 5.1 The similarity degree-based clustering process in LSGDM

Algorithm 5.2 (Gou et al. 2018). **Double hierarchy information entropy-based weights-determining method**

Step 1. Determine the weight of each group mostly based on the number of experts. Suppose that the experts $e^1, e^2, \ldots, e^n$ are divided into T groups, and the t-th group contains φ_t experts, then the weight of each group ω_t can be obtained by

$$\omega_t = \frac{\varphi_t^2}{\sum_{t=1}^{T} \varphi_t^2}, \quad t = 1, 2, \ldots, T \tag{5.7}$$

Step 2. Utilize information entropy theory to determine the weights of experts included in each group. The first step is to obtain every expert's ordering vector $U^a = (u_1^a, u_2^a, \ldots, u_m^a)$ $(a = 1, 2, \ldots, n)$ for all alternatives, which can be calculated by

$$u_i^a = \frac{\sum_{j=1}^{m} f(le(h_{S_{O_{ij}}}^a))}{\sum_{i=1}^{m} \sum_{j=1}^{m} f(le(h_{S_{O_{ij}}}^a))}, \quad i = 1, 2, \ldots, m \tag{5.8}$$

Then the information entropy of the expert e^a can be obtained by

$$IE(U^a) = -\frac{1}{\log_2 m} \cdot \sum_{i=1}^{m} u_i^a \log_2 u_i^a \tag{5.9}$$

Information entropy indicates the uncertainty degree and the randomness of evaluation information. Therefore, the smaller the information entropy is, the bigger the certainty degree will be, which means that the corresponding expert plays a significant role and it is necessary to give him/her a bigger weight. Therefore, let $\overline{\omega}_t^a$ be the weight of the ath expert included in the t-th group, then

$$\overline{\omega}_t^a = \frac{(IE(U^a))^{-1}}{\sum_{a=1}^{\varphi_t} (IE(U^a))^{-1}} \tag{5.10}$$

Step 3. Obtain the weight of every expert by combining these two weight information:

$$w_a = \omega_t \cdot \overline{\omega}_t^a \tag{5.11}$$

Step 4. End.

5.1.3 Large-Scale Group Consensus Decision-Making Method with DHHFLPRs

Firstly, the fundamental of consensus reaching process in LSGDM can be shown in Fig. 5.2.

From Fig. 5.2, there exist four main issues in consensus reaching process:

(1) How to calculate the overall consensus degree.
(2) How to identify the alternatives, the part of alternatives, and the experts that need to improve preference relations.
(3) How to discuss and improve the preference relation in each group.
(4) How to determine some necessary parameters.

For the first issue, some consensus degrees can be developed to solve it. At the beginning, we aggregate all similarity matrices $SM^{ab} = (sm_{ij}^{ab})_{m\times m}(a, b = 1, 2, \ldots, n)$ associated with each pair of experts (e^a, e^b) and establish a consensus matrix $CM = (cm_{ij})_{m\times m}$ based on the similarity degrees, where

$$cm_{ij} = \frac{2}{n(n-1)} \sum_{a=1}^{n-1} \sum_{b=a+1}^{n} sm_{ij}^{ab}, \quad i, j = 1, 2, \ldots, m \tag{5.12}$$

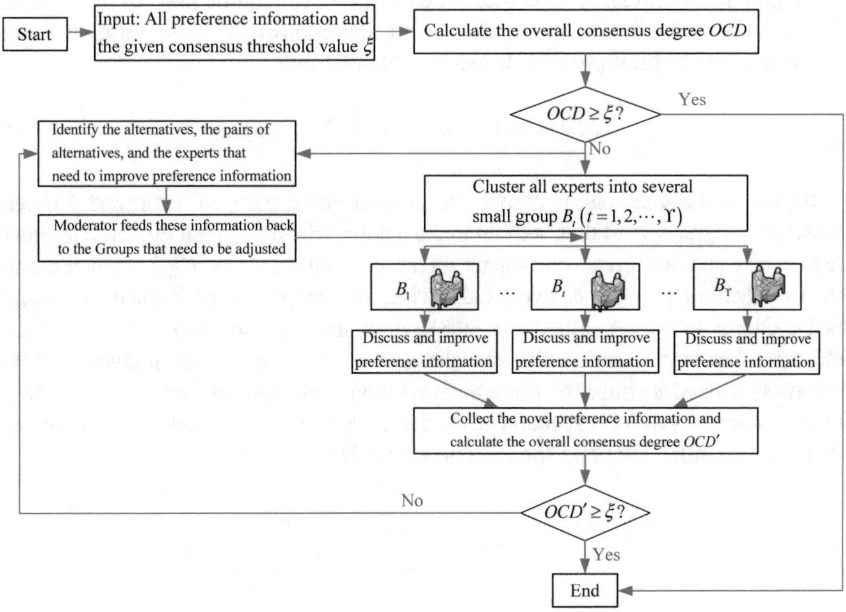

Fig. 5.2 The fundamental of consensus reaching process in LSGDM

Next, we determine the consensus degrees of all experts based on the following three parts:

(1) Consensus degree for each pair of alternatives. Considering each element cm_{ij} included in the consensus matrix $CM = (cm_{ij})_{m \times m}$ means the consensus level among all experts for the pair of alternatives (A_i, A_j), so we can use it to express the consensus degree for (A_i, A_j), denoted as $cdpa_{ij}$ and

$$cdpa_{ij} = cm_{ij} \ (i,j = 1,2,\ldots,m) \tag{5.13}$$

Obviously, the bigger the value of $cdpa_{ij}$ is, the greater agreement among all experts on the pair of alternatives (A_i, A_j) will be. Therefore, we can utilize this measure to obtain which position has a poor consensus level.

(2) Consensus degree for each alternative. By aggregating all elements included in each row of consensus matrix $CM = (cm_{ij})_{m \times m}$, the consensus degree for every alternative A_i, denoted by cda_i, can be developed to measure the consensus level among all experts for this alternative:

$$cda_i = \frac{1}{m-1} \sum_{j=1, i \neq j}^{m} cdpa_{ij} \ (j = 1,2,\ldots,m) \tag{5.14}$$

(3) Overall consensus degree for all preference relations. the overall consensus degree for all preference relations, denoted by ocd, can be used to measure the total consensus level among all experts and control the progress of the consensus researching process. It can be obtained by

$$ocd = \min_i \{cda_i\} \ (i = 1,2,\ldots,m) \tag{5.15}$$

Based on the discussions above, we propose three parts to determine different consensus degrees. And then we can make a comparison between overall consensus degree ocd and the given consensus threshold value ξ. If $ocd \geq \xi$, then the consensus reaching process is over; Otherwise, two steps are performed simultaneously: One is to cluster all experts into several small groups based on Sect. 5.1.1, and the other one is to identify the alternatives, the part of alternatives, and the experts that need to improve preference relations and how to improve them. Next we only need to solve the second issues, the method includes two kinds of rules: the identification rules (IR) and the direction rules (DR).

(a) **The identification rules (IR)**

The identification rules are mainly used to identify the alternatives, the pairs of alternatives and the experts that do not reach the given consensus threshold.

(I) Identify the alternatives ($IR-1$): Let AL be the set of alternatives in which the consensus degree cda_i is lower than the given consensus threshold value ξ. Then we can identify the alternatives based on

$$AL = \{A_i | cda_i < \xi, \ \ i = 1, 2, \ldots, m\} \tag{5.16}$$

Obviously, AL is a set and it may contain many alternatives. However, if we only want to change one alternative in each consensus reaching process, then the set AL can be developed as:

$$AL = \{A_i | \min_i \{cda_i < \xi, \ \ i = 1, 2, \ldots, m\}\} \tag{5.17}$$

(II) Identify the pairs of alternatives ($IR-2$): For any alternative $A_i \in AL$, this rule is utilized to identify which pair of alternatives (A_i, A_j) needs to be improved. These pairs of alternatives are named as a set PAL_i and can be obtained by

$$PAL_i = \{(A_i, A_j) | A_i \in AL \wedge cdpa_{ij} < \xi\} \tag{5.18}$$

Obviously, combining $IR-1$ and $IR-2$, we can determine which position needs to be changed.

(III) Identify the experts ($IR-3$): The experts who need to improve their preference relations can be decided by making some discussions among all experts in each group. Additionally, the next method can also be used as a reference for each group:

Let DM_{ij} be a set of experts who should change their preference information. Then we can calculate the distance between any of the experts $e^a (a = 1, 2, \ldots, n)$ and all the others e^b $(a \neq b)$ at the position (A_i, A_j) based on the formula below:

$$d_{ij}^a = \sum_{b=1, b \neq a}^{n} (1 - sd(h_{S_{O_{ij}}}^a, h_{S_{O_{ij}}}^b)) = n - 1 - \sum_{b=1, b \neq a}^{n} sm_{ij}^{ab} \tag{5.19}$$

The expert DM_{ij} who should change preference at (A_i, A_j) can be determined based on

$$DM_{ij} = \{e^* | (A_i, A_j) \in PAL_i \wedge d^*_{ij} = \max_a \{d^a_{ij}\}\} \tag{5.20}$$

Combining $IR-1$, $IR-2$ and $IR-3$, it is very easy to determine which expert and his/her position needs to be changed. Suppose that an expert $e^* \in DM_{ij}$, and he/she needs to change preference information $h^*_{S_{O_{ij}}}$, then a set can be set up to express these elements:

$$\Delta = \{(*, (i,j)) | e^* \in DM_{ij} \wedge (A_i, A_j) \in PAL_i\} \tag{5.21}$$

(b) **The direction rules (DR)**

These rules are utilized to send suggestions to each group and tell them how to increase the consensus level in the next round. Firstly, the moderator needs to set up a target and gives it to each group, and then each group can discuss how to change their preferences in the position (A_i, A_j). The target can be obtained by referencing the aggregation information of all experts' preferences.

Based on Eq. (3.24), the group DHHFLPRs $\widetilde{H}^c_{S_O} = (h^c_{S_{O_{ij}}})_{m \times m}$ can be established, and we call $h^c_{S_{O_{ij}}}$ the group preference element. Then the direction rules can be designed as follows:

(I) $DR-1$: If $h^*_{S_{O_{ij}}} < h^c_{S_{O_{ij}}}$, then the expert e^* should increase his/her evaluation associated with the pair of alternatives (A_i, A_j).
(II) $DR-2$: If $h^*_{S_{O_{ij}}} > h^c_{S_{O_{ij}}}$, then the expert e^* should decrease his/her evaluation associated with the pair of alternatives (A_i, A_j).

When the experts know how to change their evaluations associated with the pair of alternatives (A_i, A_j), the next problem is to decide the extent of the change. Suppose that $(h^*_{S_{O_{ij}}})^{(\mathbb{Z}+1)}$ and $(h^*_{S_{O_{ij}}})^{(\mathbb{Z})}$ are the $\mathbb{Z}+1$th and $\mathbb{Z}$th round preferences of the expert e^*, respectively. Then, the general range is

$$\left(h^*_{S_{O_{ij}}}\right)^{(\mathbb{Z}+1)} \in \left[\min\left\{\left(h^*_{S_{O_{ij}}}\right)^{(\mathbb{Z})}, \left(h^c_{S_{O_{ij}}}\right)^{(\mathbb{Z})}\right\}, \max\left\{\left(h^*_{S_{O_{ij}}}\right)^{(\mathbb{Z})}, \left(h^c_{S_{O_{ij}}}\right)^{(\mathbb{Z})}\right\}\right] \tag{5.22}$$

In fact, we can always find a parameter $\lambda \in (0,1)$, Eq. (5.22) is equivalent with

$$\left(h^*_{S_{O_{ij}}}\right)^{(\mathbb{Z}+1)} = \lambda\left(h^*_{S_{O_{ij}}}\right)^{(\mathbb{Z})} \oplus (1-\lambda)\left(h^c_{S_{O_{ij}}}\right)^{(\mathbb{Z}+1)} \tag{5.23}$$

Remark 5.2 (Gou et al. 2018). For the consensus reaching process of this subsection, the solutions of these two issues only are the references for each group. For the third issue, each group can be free to discuss and decide how to improve the preference information. Therefore, each expert who needs to improve their preference information has two choices: Change or no to change. For the first one, this group can discuss how to improve the preference information based on Eq. (5.22). But for the second one, we also have two choices: Delete this expert or change his/her preference information based on Eq. (5.23) randomly.

Theorem 5.1 (Gou et al. 2018). For any alternative A_i, if its related preference information needs to be changed, and the identification rules and the direction rules have been applied, then

$$(cda_i)^{(\mathbb{Z}+1)} > (cda_i)^{(\mathbb{Z})} \tag{5.24}$$

Proof To prove $(cda_i)^{(\mathbb{Z}+1)} > (cda_i)^{(\mathbb{Z})}$, it is equivalent to prove $(cm_{ij})^{(\mathbb{Z}+1)} > (cm_{ij})^{(\mathbb{Z})}$ and

$$\frac{2}{n(n-1)}\sum_{a=1}^{n-1}\sum_{b=a+1}^{n}\left(sm_{ij}^{ab}\right)^{(\mathbb{Z}+1)} > \frac{2}{n(n-1)}\sum_{a=1}^{n-1}\sum_{b=a+1}^{n}\left(sm_{ij}^{ab}\right)^{(\mathbb{Z})} \tag{5.25}$$

Based on Eqs. (5.1) and (5.3), Eq. (5.25) can be rewritten as:

$$\frac{2}{n(n-1)}\sum_{a=1}^{n-1}\sum_{b=a+1}^{n} sd\left(le\left(h_{S_{O_{ij}}}^{a}\right)^{(\mathbb{Z}+1)}, le\left(h_{S_{O_{ij}}}^{b}\right)^{(\mathbb{Z}+1)}\right) > \frac{2}{n(n-1)}\sum_{a=1}^{n-1}\sum_{b=a+1}^{n} sd\left(le\left(h_{S_{O_{ij}}}^{a}\right)^{(\mathbb{Z})}, le\left(h_{S_{O_{ij}}}^{b}\right)^{(\mathbb{Z})}\right) \tag{5.26}$$

which is equal to

$$\frac{2}{n(n-1)}\sum_{a=1}^{n-1}\sum_{b=a+1}^{n} d\left(le\left(h_{S_{O_{ij}}}^{a}\right)^{(\mathbb{Z}+1)}, le\left(h_{S_{O_{ij}}}^{b}\right)^{(\mathbb{Z}+1)}\right) < \frac{2}{n(n-1)}\sum_{a=1}^{n-1}\sum_{b=a+1}^{n} d\left(le\left(h_{S_{O_{ij}}}^{a}\right)^{(\mathbb{Z})}, le\left(h_{S_{O_{ij}}}^{b}\right)^{(\mathbb{Z})}\right) \tag{5.27}$$

Without a loss of generality, let e^1 be the expert who needs to change his /her preference for the part (A_i, A_j), then Eq. (5.27) can be developed into

$$\begin{aligned} & d\left(le\left(h_{S_{O_{ij}}}^{1}\right)^{(\mathbb{Z}+1)}, le\left(h_{S_{O_{ij}}}^{2}\right)^{(\mathbb{Z}+1)}\right) + d\left(le\left(h_{S_{O_{ij}}}^{1}\right)^{(\mathbb{Z}+1)}, le\left(h_{S_{O_{ij}}}^{2}\right)^{(\mathbb{Z}+1)}\right) + \cdots + d\left(le\left(h_{S_{O_{ij}}}^{1}\right)^{(\mathbb{Z}+1)}, le\left(h_{S_{O_{ij}}}^{n}\right)^{(\mathbb{Z}+1)}\right) \\ & < d\left(le\left(h_{S_{O_{ij}}}^{1}\right)^{(\mathbb{Z})}, le\left(h_{S_{O_{ij}}}^{2}\right)^{(\mathbb{Z})}\right) + d\left(le\left(h_{S_{O_{ij}}}^{1}\right)^{(\mathbb{Z})}, le\left(h_{S_{O_{ij}}}^{2}\right)^{(\mathbb{Z})}\right) + \cdots + d\left(le(h_{S_{O_{ij}}}^{1})^{(\mathbb{Z})}, le\left(h_{S_{O_{ij}}}^{n}\right)^{(\mathbb{Z})}\right) \end{aligned} \tag{5.28}$$

Based on Eq. (5.23), we have

$$
\begin{aligned}
& d\left(le\left(h_{S_{O_{ij}}}^{1}\right)^{(\mathbb{Z}+1)}, le\left(h_{S_{O_{ij}}}^{2}\right)^{(\mathbb{Z}+1)}\right) \\
& = \left| f\left(\lambda le\left(h_{S_{O_{ij}}}^{1}\right)^{(\mathbb{Z})} + (1-\lambda) le\left(h_{S_{O_{ij}}}^{c}\right)^{(\mathbb{Z})}\right) - f\left(\lambda le\left(h_{S_{O_{ij}}}^{2}\right)^{(\mathbb{Z})} + (1-\lambda) le\left(h_{S_{O_{ij}}}^{2}\right)^{(\mathbb{Z})}\right) \right| \\
& = \left| \lambda f\left(le\left(h_{S_{O_{ij}}}^{1}\right)^{(\mathbb{Z})} - le\left(h_{S_{O_{ij}}}^{2}\right)^{(\mathbb{Z})}\right) \right| \\
& \quad + \frac{(1-\lambda)}{n}\left(f\left(le\left(h_{S_{O_{ij}}}^{1}\right)^{(\mathbb{Z})} - le\left(h_{S_{O_{ij}}}^{2}\right)^{(\mathbb{Z})}\right) + f\left(le\left(h_{S_{O_{ij}}}^{2}\right)^{(\mathbb{Z})} - le\left(h_{S_{O_{ij}}}^{2}\right)^{(\mathbb{Z})}\right) - f\left(le\left(h_{S_{O_{ij}}}^{m}\right)^{(\mathbb{Z})} - le\left(h_{S_{O_{ij}}}^{2}\right)^{(\mathbb{Z})}\right)\right) \\
& \quad < \lambda d\left(le\left(h_{S_{O_{ij}}}^{1}\right)^{(\mathbb{Z})}, le\left(h_{S_{O_{ij}}}^{2}\right)^{(\mathbb{Z})}\right) + \frac{1-\lambda}{n}\left(d\left(le\left(h_{S_{O_{ij}}}^{1}\right)^{(\mathbb{Z})}, le\left(h_{S_{O_{ij}}}^{2}\right)^{(\mathbb{Z})}\right) + d\left(le\left(h_{S_{O_{ij}}}^{n}\right)^{(\mathbb{Z})}, le\left(h_{S_{O_{ij}}}^{2}\right)^{(\mathbb{Z})}\right)\right)
\end{aligned}
$$

Because the consensus degree between e^1 and e^2 are smallest, then we obtain

$$
d\left(le\left(h_{S_{O_{ij}}}^{1}\right)^{(\mathbb{Z}+1)}, le\left(h_{S_{O_{ij}}}^{2}\right)^{(\mathbb{Z}+1)}\right) < \left(\lambda + \frac{(1-\lambda)(n-1)}{n}\right) d\left(le\left(h_{S_{O_{ij}}}^{1}\right)^{(\mathbb{Z})}, le\left(h_{S_{O_{ij}}}^{2}\right)^{(\mathbb{Z})}\right) \tag{5.29}
$$

Based on Eq. (5.29), we have

$$
\begin{aligned}
& \left(\lambda + \frac{(1-\lambda)(n-1)}{n}\right) d\left(le\left(h_{S_{O_{ij}}}^{1}\right)^{(\mathbb{Z})}, le\left(h_{S_{O_{ij}}}^{2}\right)^{(\mathbb{Z})}\right) = \left(1 + \frac{\lambda - 1}{n}\right) d\left(le\left(h_{S_{O_{ij}}}^{1}\right)^{(\mathbb{Z})}, le\left(h_{S_{O_{ij}}}^{2}\right)^{(\mathbb{Z})}\right) \\
& < d\left(le\left(h_{S_{O_{ij}}}^{1}\right)^{(\mathbb{Z})}, le\left(h_{S_{O_{ij}}}^{2}\right)^{(\mathbb{Z})}\right)
\end{aligned}
$$

Therefore, we have $d(le(h_{S_{O_{ij}}}^{1})^{(\mathbb{Z}+1)}, le(h_{S_{O_{ij}}}^{2})^{(\mathbb{Z}+1)}) < d(le(h_{S_{O_{ij}}}^{1})^{(\mathbb{Z})}, le(h_{S_{O_{ij}}}^{2})^{(\mathbb{Z})})$. Similarly, there are

$$
\begin{aligned}
d\left(le\left(h_{S_{O_{ij}}}^{1}\right)^{(\mathbb{Z}+1)}, le\left(h_{S_{O_{ij}}}^{3}\right)^{(\mathbb{Z}+1)}\right) &< d\left(le\left(h_{S_{O_{ij}}}^{1}\right)^{(\mathbb{Z})}, le\left(h_{S_{O_{ij}}}^{3}\right)^{(\mathbb{Z})}\right) \\
d\left(le\left(h_{S_{O_{ij}}}^{1}\right)^{(\mathbb{Z}+1)}, le\left(h_{S_{O_{ij}}}^{4}\right)^{(\mathbb{Z}+1)}\right) &< d\left(le\left(h_{S_{O_{ij}}}^{1}\right)^{(\mathbb{Z})}, le\left(h_{S_{O_{ij}}}^{4}\right)^{(\mathbb{Z})}\right) \\
&\vdots \\
d\left(le\left(h_{S_{O_{ij}}}^{1}\right)^{(\mathbb{Z}+1)}, le\left(h_{S_{O_{ij}}}^{n}\right)^{(\mathbb{Z}+1)}\right) &< d\left(le\left(h_{S_{O_{ij}}}^{1}\right)^{(\mathbb{Z})}, le\left(h_{S_{O_{ij}}}^{n}\right)^{(\mathbb{Z})}\right)
\end{aligned}
$$

Add up all these inequalities, we can obtain the Eq. (5.28), then $(cda_i)^{(\mathbb{Z}+1)} > (cda_i)^{(\mathbb{Z})}$. Which completes the proof of Theorem 5.1. ■

For the final issue, some parameters need to be determined such as the given consensus threshold value ξ, and the number of iteration, denoted by CT. Xu et al. (2015) analyzed these two parameters and obtained that it is reasonable to set ξ to fall within the interval [0.7386, 0.85], and the maximum number of iterations may belong to [0, 6]. In this subsection, we can also determine two kinds of parameters in these two intervals respectively. However, both of them only are the references and the final values of them must be combined with the practical decision-making problem.

In an LSGDM, a moderator is invited to give the revision suggestions to the experts and guide them to modify their preference information. Then based on the above discussion, an LSGDM model with DHHFLPRs can be shown as follows:

Algorithm 5.3 (Gou et al. 2018). **LSGDM consensus model with DHHFLPRs**

Step 1. Check whether all experts' preference information reaches the given consensus threshold ξ based on Eqs. (5.13)–(5.15). If so, go to Step 5, else go to Step 2.
Step 2. Cluster all experts into several categories based on Algorithm 5.1 (This step only happens in the first time of consensus reaching process). Calculate all experts' weights and obtain the group DHHFLPR based on Eq. (3.24). Then go to Step 3..
Step 3. Identify the alternatives, the pairs of alternatives and the experts that need to improve their consensus degrees on the basis of Eqs. (5.16)–(5.21). The moderator feeds the above two kinds of information to all groups, then every group conducts a discussion. Every group can discuss and change the corresponding preference information based on Remark 5.2. Then go to Step 4.
Step 4. Collect all modified evaluation information of each group and go back to Step 1.
Step 5. Calculate all experts' weights and obtain the final group DHHFLPR $*\widetilde{H}^c_{S_O} = (*h^c_{S_{O_{ij}}})_{m\times m}$. Then we obtain the synthetical value $SV(A_i) = \sum_{j=1}^{m} E(*h^c_{S_{O_{ij}}})$ $(i = 1, 2, \ldots, m)$ of each alternative and the ranking order.
Step 6. End.

This LSGDM model with DHHFLPRs can be shown in Fig. 5.3.

5.1.4 Application of Large-Scale Group Consensus Decision-Making Method with DHHFLPRs

Next, we apply the LSGDM model with DHHFLPRs into a practical LSGDM problem about water resource management.

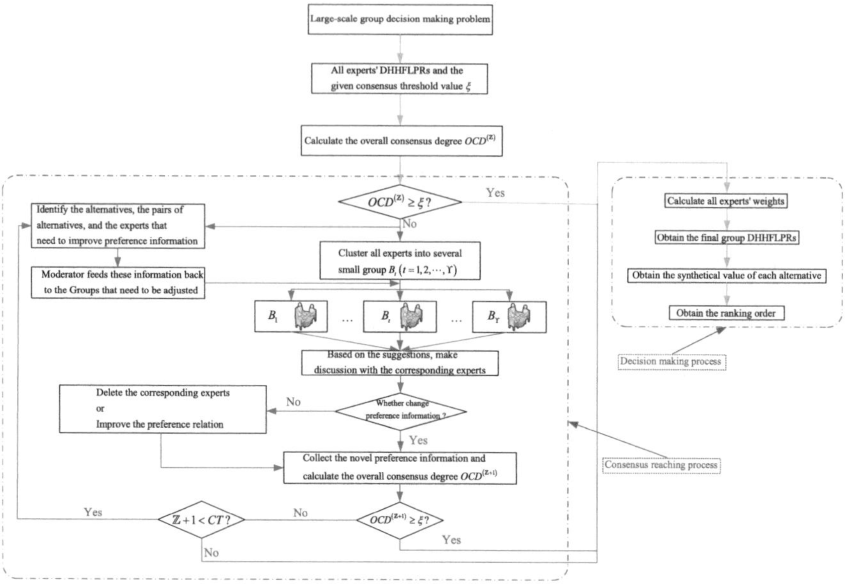

Fig. 5.3 LSGDM model with DHHFLPRs

Example 5.1 (Gou et al. 2018). In China, the state council's opinions on the implementation of the strictest water resources management system was promulgated. According to the practical situation, Sichuan province introduced the following implementation opinions to take the strict water resources management as a strategic move for accelerating the transformation of economic development mode:

(1) Establish a total water control system. This measure mainly contains implementing the total amount of water control, strengthening water resources development and utilization management, strict water intaking permits, strengthening the unified deployment of water resources, strict groundwater management and protection, and strengthening the collection and use of water resources expenditure, etc.
(2) Establish water efficiency control system. This measure mainly contains accelerating the development of water-saving society, enhancing water management, and strengthening the oversight and management of water saving, etc.
(3) Establish water functional area to restrict the pollution system. This measure mainly contains strict water function area management, strengthening the pollution discharge outlets of rivers management, strengthening water conservation, strengthening the protection of drinking water, and carrying out pilot and creation of water ecological civilization.
(4) Promote the comprehensive implementation of the most stringent water resources management system. This measure mainly contains strengthening the

leadership of water resources management, establishing water resources management responsibility and examination system, improving the investment mechanism of water resources management, enhancing the team construction, improving the system and strengthening supervision.

Obviously, each policy discussed above is an important measure and all of them can be used to take the strict water resources management more efficiently. Therefore, in order to evaluate the implementation status of the above policies, a review meeting is hold and 20 experts $E = \{e^1, e^2, \ldots, e^{20}\}$ are invited to provide their preference information about the evaluations of four important cities: Chengdu (A_1), Panzhihua (A_2), Liangshan (A_3), and Nanchong (A_4). Let $S_O = \{s_{t\langle o_k \rangle} | t = -4, \ldots, -1, 0, 1, \ldots, 4;\ k = -4, \ldots, -1, 0, 1, \ldots, 4\}$ be a DHLTS with $S = \{s_{-4} = \textit{extremely bad}, s_{-3} = \textit{very bad}, s_{-2} = \textit{bad}, s_{-1} = \textit{slightly bad}, s_0 = \textit{medium}, s_1 = \textit{slightly good}, s_2 = \textit{good}, s_3 = \textit{very good}, s_4 = \textit{extremely good}\}$ and $O = \{o_{-4} = \textit{far from}, o_{-3} = \textit{scarcely}, o_{-2} = \textit{only a little}, o_{-1} = \textit{a little}, o_0 = \textit{just right}, o_1 = \textit{much}, o_2 = \textit{very much}, o_3 = \textit{extremely much}, o_4 = \textit{entirely}\}$. Then the experts provide their evaluations with linguistic information, we collect these linguistic information and transform them into DHHFLEs, which can be contained in the following DHHFLPRs $\widetilde{H}^r_{S_O} (r = 1, 2, \ldots, 20)$:

$$\widetilde{H}^1_{S_O} = \begin{pmatrix} \{s_{0\langle o_0\rangle}\} & \{s_{0\langle o_1\rangle}\} & \{s_{1\langle o_{-2}\rangle}, s_{1\langle o_2\rangle}\} & \{s_{-1\langle o_{-1}\rangle}\} \\ \{s_{0\langle o_{-1}\rangle}\} & \{s_{0\langle o_0\rangle}\} & \{s_{0\langle o_0\rangle}\} & \{s_{-1\langle o_1\rangle}\} \\ \{s_{-1\langle o_2\rangle}, s_{-1\langle o_{-2}\rangle}\} & \{s_{0\langle o_0\rangle}\} & \{s_{0\langle o_0\rangle}\} & \{s_{2\langle o_{-3}\rangle}\} \\ \{s_{1\langle o_1\rangle}\} & \{s_{1\langle o_{-1}\rangle}\} & \{s_{-2\langle o_3\rangle}\} & \{s_{0\langle o_0\rangle}\} \end{pmatrix}$$

$$\widetilde{H}^2_{S_O} = \begin{pmatrix} \{s_{0\langle o_0\rangle}\} & \{s_{1\langle o_{-2}\rangle}, s_{2\langle o_{-2}\rangle}\} & \{s_{1\langle o_{-2}\rangle}\} & \{s_{0\langle o_2\rangle}, s_{1\langle o_{-1}\rangle}\} \\ \{s_{-1\langle o_2\rangle}, s_{-2\langle o_2\rangle}\} & \{s_{0\langle o_0\rangle}\} & \{s_{-1\langle o_1\rangle}, s_{0\langle o_{-1}\rangle}\} & \{s_{-2\langle o_{-1}\rangle}, s_{-1\langle o_2\rangle}\} \\ \{s_{-1\langle o_2\rangle}\} & \{s_{1\langle o_{-1}\rangle}, s_{0\langle o_1\rangle}\} & \{s_{0\langle o_0\rangle}\} & \{s_{0\langle o_{-3}\rangle}\} \\ \{s_{0\langle o_{-2}\rangle}, s_{-1\langle o_1\rangle}\} & \{s_{2\langle o_1\rangle}, s_{1\langle o_{-2}\rangle}\} & \{s_{0\langle o_3\rangle}\} & \{s_{0\langle o_0\rangle}\} \end{pmatrix}$$

$$\widetilde{H}^3_{S_O} = \begin{pmatrix} \{s_{0\langle o_0\rangle}\} & \{s_{0\langle o_{-1}\rangle}\} & \{s_{1\langle o_{-1}\rangle}\} & \{s_{-1\langle o_3\rangle}\} \\ \{s_{0\langle o_1\rangle}\} & \{s_{0\langle o_0\rangle}\} & \{s_{0\langle o_2\rangle}\} & \{s_{-2\langle o_1\rangle}, s_{-1\langle o_1\rangle}\} \\ \{s_{-1\langle o_1\rangle}\} & \{s_{0\langle o_{-2}\rangle}\} & \{s_{0\langle o_0\rangle}\} & \{s_{2\langle o_{-1}\rangle}, s_{3\langle o_{-1}\rangle}\} \\ \{s_{1\langle o_{-3}\rangle}\} & \{s_{2\langle o_{-1}\rangle}, s_{1\langle o_{-1}\rangle}\} & \{s_{-2\langle o_1\rangle}, s_{-3\langle o_1\rangle}\} & \{s_{0\langle o_0\rangle}\} \end{pmatrix}$$

$$\widetilde{H}^4_{S_O} = \begin{pmatrix} \{s_{0\langle o_0\rangle}\} & \{s_{1\langle o_1\rangle}\} & \{s_{2\langle o_2\rangle}, s_{3\langle o_{-1}\rangle}\} & \{s_{0\langle o_3\rangle}\} \\ \{s_{-1\langle o_{-1}\rangle}\} & \{s_{0\langle o_0\rangle}\} & \{s_{1\langle o_{-1}\rangle}, s_{2\langle o_1\rangle}\} & \{s_{-1\langle o_2\rangle}, s_{-1\langle o_3\rangle}\} \\ \{s_{-2\langle o_{-2}\rangle}, s_{-3\langle o_1\rangle}\} & \{s_{-1\langle o_1\rangle}, s_{-2\langle o_{-1}\rangle}\} & \{s_{0\langle o_0\rangle}\} & \{s_{-2\langle o_1\rangle}, s_{-1\langle o_3\rangle}\} \\ \{s_{0\langle o_{-3}\rangle}\} & \{s_{1\langle o_{-2}\rangle}, s_{1\langle o_{-3}\rangle}\} & \{s_{2\langle o_{-1}\rangle}, s_{1\langle o_{-3}\rangle}\} & \{s_{0\langle o_0\rangle}\} \end{pmatrix}$$

$$\widetilde{H}_{S_O}^{5} = \begin{pmatrix} \{s_{0\langle o_0\rangle}\} & \{s_{1\langle o_2\rangle}, s_{2\langle o_2\rangle}\} & \{s_{1\langle o_{-1}\rangle}\} & \{s_{0\langle o_1\rangle}, s_{1\langle o_1\rangle}\} \\ \{s_{-1\langle o_{-2}\rangle}, s_{-2\langle o_{-2}\rangle}\} & \{s_{0\langle o_0\rangle}\} & \{s_{-1\langle o_{-1}\rangle}, s_{0\langle o_{-1}\rangle}\} & \{s_{-2\langle o_{-1}\rangle}, s_{-1\langle o_2\rangle}\} \\ \{s_{-1\langle o_1\rangle}\} & \{s_{1\langle o_1\rangle}, s_{0\langle o_1\rangle}\} & \{s_{0\langle o_0\rangle}\} & \{s_{0\langle o_{-3}\rangle}\} \\ \{s_{0\langle o_{-1}\rangle}, s_{-1\langle o_{-1}\rangle}\} & \{s_{2\langle o_1\rangle}, s_{1\langle o_{-2}\rangle}\} & \{s_{0\langle o_3\rangle}\} & \{s_{0\langle o_0\rangle}\} \end{pmatrix}$$

$$\widetilde{H}_{S_O}^{6} = \begin{pmatrix} \{s_{0\langle o_0\rangle}\} & \{s_{1\langle o_{-2}\rangle}, s_{2\langle o_{-2}\rangle}\} & \{s_{0\langle o_2\rangle}\} & \{s_{1\langle o_{-1}\rangle}, s_{2\langle o_{-3}\rangle}\} \\ \{s_{-1\langle o_2\rangle}, s_{-2\langle o_2\rangle}\} & \{s_{0\langle o_0\rangle}\} & \{s_{-1\langle o_1\rangle}, s_{0\langle o_{-1}\rangle}\} & \{s_{-2\langle o_{-1}\rangle}, s_{-1\langle o_2\rangle}\} \\ \{s_{0\langle o_{-2}\rangle}\} & \{s_{1\langle o_{-1}\rangle}, s_{0\langle o_1\rangle}\} & \{s_{0\langle o_0\rangle}\} & \{s_{0\langle o_{-3}\rangle}\} \\ \{s_{-1\langle o_1\rangle}, s_{-2\langle o_3\rangle}\} & \{s_{2\langle o_1\rangle}, s_{1\langle o_{-2}\rangle}\} & \{s_{0\langle o_3\rangle}\} & \{s_{0\langle o_0\rangle}\} \end{pmatrix}$$

$$\widetilde{H}_{S_O}^{7} = \begin{pmatrix} \{s_{0\langle o_0\rangle}\} & \{s_{-1\langle o_1\rangle}, s_{0\langle o_2\rangle}\} & \{s_{0\langle o_2\rangle}, s_{1\langle o_3\rangle}\} & \{s_{-1\langle o_{-2}\rangle}\} \\ \{s_{1\langle o_{-1}\rangle}, s_{0\langle o_{-2}\rangle}\} & \{s_{0\langle o_0\rangle}\} & \{s_{0\langle o_{-1}\rangle}, s_{1\langle o_1\rangle}\} & \{s_{-2\langle o_1\rangle}, s_{-1\langle o_2\rangle}\} \\ \{s_{0\langle o_{-2}\rangle}, s_{-1\langle o_{-3}\rangle}\} & \{s_{0\langle o_1\rangle}, s_{-1\langle o_{-1}\rangle}\} & \{s_{0\langle o_0\rangle}\} & \{s_{1\langle o_1\rangle}, s_{2\langle o_3\rangle}\} \\ \{s_{1\langle o_2\rangle}\} & \{s_{2\langle o_{-1}\rangle}, s_{1\langle o_{-2}\rangle}\} & \{s_{-1\langle o_{-1}\rangle}, s_{-2\langle o_{-3}\rangle}\} & \{s_{0\langle o_0\rangle}\} \end{pmatrix}$$

$$\widetilde{H}_{S_O}^{8} = \begin{pmatrix} \{s_{0\langle o_0\rangle}\} & \{s_{1\langle o_3\rangle}\} & \{s_{-1\langle o_2\rangle}, s_{-1\langle o_3\rangle}\} & \{s_{-1\langle o_{-1}\rangle}, s_{-1\langle o_{-1}\rangle}\} \\ \{s_{1\langle o_3\rangle}\} & \{s_{0\langle o_0\rangle}\} & \{s_{-1\langle o_1\rangle}\} & \{s_{-1\langle o_1\rangle}\} \\ \{s_{1\langle o_{-2}\rangle}, s_{1\langle o_{-3}\rangle}\} & \{s_{-1\langle o_1\rangle}\} & \{s_{0\langle o_0\rangle}\} & \{s_{1\langle o_{-1}\rangle}\} \\ \{s_{1\langle o_1\rangle}, s_{1\langle o_1\rangle}\} & \{s_{-1\langle o_1\rangle}\} & \{s_{1\langle o_{-1}\rangle}\} & \{s_{0\langle o_0\rangle}\} \end{pmatrix}$$

$$\widetilde{H}_{S_O}^{9} = \begin{pmatrix} \{s_{0\langle o_0\rangle}\} & \{s_{1\langle o_{-1}\rangle}\} & \{s_{2\langle o_2\rangle}, s_{3\langle o_1\rangle}\} & \{s_{0\langle o_3\rangle}\} \\ \{s_{-1\langle o_1\rangle}\} & \{s_{0\langle o_0\rangle}\} & \{s_{1\langle o_{-1}\rangle}, s_{2\langle o_1\rangle}\} & \{s_{-1\langle o_3\rangle}, s_{0\langle o_2\rangle}\} \\ \{s_{-2\langle o_{-2}\rangle}, s_{-3\langle o_{=1}\rangle}\} & \{s_{-1\langle o_1\rangle}, s_{-2\langle o_{-1}\rangle}\} & \{s_{0\langle o_0\rangle}\} & \{s_{-2\langle o_1\rangle}, s_{-1\langle o_2\rangle}\} \\ \{s_{0\langle o_{-3}\rangle}\} & \{s_{1\langle o_{-3}\rangle}, s_{0\langle o_{-2}\rangle}\} & \{s_{2\langle o_{-1}\rangle}, s_{1\langle o_{-2}\rangle}\} & \{s_{0\langle o_0\rangle}\} \end{pmatrix}$$

$$\widetilde{H}_{S_O}^{10} = \begin{pmatrix} \{s_{0\langle o_0\rangle}\} & \{s_{1\langle o_{-2}\rangle}, s_{2\langle o_{-1}\rangle}\} & \{s_{1\langle o_{-2}\rangle}\} & \{s_{0\langle o_2\rangle}, s_{1\langle o_1\rangle}\} \\ \{s_{-1\langle o_2\rangle}, s_{-2\langle o_1\rangle}\} & \{s_{0\langle o_0\rangle}\} & \{s_{-1\langle o_1\rangle}, s_{0\langle o_{-1}\rangle}\} & \{s_{-2\langle o_{-2}\rangle}, s_{-1\langle o_2\rangle}\} \\ \{s_{-1\langle o_2\rangle}\} & \{s_{1\langle o_{-1}\rangle}, s_{0\langle o_1\rangle}\} & \{s_{0\langle o_0\rangle}\} & \{s_{0\langle o_{-2}\rangle}\} \\ \{s_{0\langle o_{-2}\rangle}, s_{-1\langle o_{-1}\rangle}\} & \{s_{2\langle o_2\rangle}, s_{1\langle o_{-2}\rangle}\} & \{s_{0\langle o_2\rangle}\} & \{s_{0\langle o_0\rangle}\} \end{pmatrix}$$

$$\widetilde{H}_{S_O}^{11} = \begin{pmatrix} \{s_{0\langle o_0\rangle}\} & \{s_{2\langle o_1\rangle}\} & \{s_{-1\langle o_2\rangle}, s_{-1\langle o_2\rangle}\} & \{s_{-2\langle o_{-1}\rangle}, s_{-1\langle o_{-2}\rangle}\} \\ \{s_{-2\langle o_{-1}\rangle}\} & \{s_{0\langle o_0\rangle}\} & \{s_{0\langle o_{-1}\rangle}, s_{0\langle o_1\rangle}\} & \{s_{-1\langle o_0\rangle}, s_{-1\langle o_1\rangle}\} \\ \{s_{1\langle o_{-2}\rangle}, s_{1\langle o_{-2}\rangle}\} & \{s_{0\langle o_1\rangle}, s_{0\langle o_{-1}\rangle}\} & \{s_{0\langle o_0\rangle}\} & \{s_{1\langle o_{-2}\rangle}, s_{2\langle o_3\rangle}\} \\ \{s_{2\langle o_1\rangle}, s_{1\langle o_2\rangle}\} & \{s_{1\langle o_0\rangle}, s_{1\langle o_{-1}\rangle}\} & \{s_{-1\langle o_2\rangle}, s_{-2\langle o_{-3}\rangle}\} & \{s_{0\langle o_0\rangle}\} \end{pmatrix}$$

$$\widetilde{H}_{S_O}^{12}=\begin{pmatrix} \{s_{0\langle o_0\rangle}\} & \{s_{2\langle o_{-2}\rangle}\} & \{s_{2\langle o_2\rangle},s_{3\langle o_{-1}\rangle}\} & \{s_{0\langle o_2\rangle}\} \\ \{s_{-2\langle o_2\rangle}\} & \{s_{0\langle o_0\rangle}\} & \{s_{1\langle o_{-1}\rangle},s_{2\langle o_{-1}\rangle}\} & \{s_{-1\langle o_3\rangle},s_{0\langle o_2\rangle}\} \\ \{s_{-2\langle o_{-2}\rangle},s_{-3\langle o_1\rangle}\} & \{s_{-1\langle o_1\rangle},s_{-2\langle o_1\rangle}\} & \{s_{0\langle o_0\rangle}\} & \{s_{-2\langle o_1\rangle},s_{-1\langle o_3\rangle}\} \\ \{s_{0\langle o_{-2}\rangle}\} & \{s_{1\langle o_{-3}\rangle},s_{0\langle o_{-2}\rangle}\} & \{s_{2\langle o_{-1}\rangle},s_{1\langle o_{-3}\rangle}\} & \{s_{0\langle o_0\rangle}\} \end{pmatrix}$$

$$\widetilde{H}_{S_O}^{13}=\begin{pmatrix} \{s_{0\langle o_0\rangle}\} & \{s_{1\langle o_{-2}\rangle},s_{2\langle o_{-2}\rangle}\} & \{s_{1\langle o_{-2}\rangle}\} & \{s_{0\langle o_2\rangle},s_{1\langle o_{-1}\rangle}\} \\ \{s_{-1\langle o_2\rangle},s_{-2\langle o_2\rangle}\} & \{s_{0\langle o_0\rangle}\} & \{s_{-1\langle o_{-1}\rangle},s_{0\langle o_1\rangle}\} & \{s_{-2\langle o_1\rangle},s_{-1\langle o_2\rangle}\} \\ \{s_{-1\langle o_2\rangle}\} & \{s_{1\langle o_1\rangle},s_{0\langle o_{-1}\rangle}\} & \{s_{0\langle o_0\rangle}\} & \{s_{0\langle o_{-3}\rangle}\} \\ \{s_{0\langle o_{-2}\rangle},s_{-1\langle o_1\rangle}\} & \{s_{2\langle o_{-1}\rangle},s_{1\langle o_{-2}\rangle}\} & \{s_{0\langle o_3\rangle}\} & \{s_{0\langle o_0\rangle}\} \end{pmatrix}$$

$$\widetilde{H}_{S_O}^{14}=\begin{pmatrix} \{s_{0\langle o_0\rangle}\} & \{s_{1\langle o_3\rangle}\} & \{s_{-1\langle o_2\rangle},s_{-1\langle o_3\rangle}\} & \{s_{-1\langle o_{-1}\rangle}\} \\ \{s_{-1\langle o_{-3}\rangle}\} & \{s_{0\langle o_0\rangle}\} & \{s_{-1\langle o_1\rangle}\} & \{s_{-1\langle o_{-1}\rangle},s_{-1\langle o_3\rangle}\} \\ \{s_{1\langle o_{-2}\rangle},s_{1\langle o_{-3}\rangle}\} & \{s_{1\langle o_{-1}\rangle}\} & \{s_{0\langle o_0\rangle}\} & \{s_{1\langle o_{-3}\rangle},s_{1\langle o_3\rangle}\} \\ \{s_{1\langle o_1\rangle}\} & \{s_{1\langle o_1\rangle},s_{1\langle o_{-3}\rangle}\} & \{s_{-1\langle o_3\rangle},s_{-1\langle o_{-3}\rangle}\} & \{s_{0\langle o_0\rangle}\} \end{pmatrix}$$

$$\widetilde{H}_{S_O}^{15}=\begin{pmatrix} \{s_{0\langle o_0\rangle}\} & \{s_{-1\langle o_{-1}\rangle}\} & \{s_{1\langle o_{-2}\rangle},s_{2\langle o_1\rangle}\} & \{s_{2\langle o_{-3}\rangle}\} \\ \{s_{1\langle o_1\rangle}\} & \{s_{0\langle o_0\rangle}\} & \{s_{2\langle o_1\rangle},s_{3\langle o_{-1}\rangle}\} & \{s_{3\langle o_3\rangle}\} \\ \{s_{-1\langle o_2\rangle},s_{-2\langle o_{-1}\rangle}\} & \{s_{-2\langle o_{-1}\rangle},s_{-3\langle o_1\rangle}\} & \{s_{0\langle o_0\rangle}\} & \{s_{-1\langle o_3\rangle}\} \\ \{s_{-2\langle o_3\rangle}\} & \{s_{-3\langle o_{-3}\rangle}\} & \{s_{1\langle o_{-3}\rangle}\} & \{s_{0\langle o_0\rangle}\} \end{pmatrix}$$

$$\widetilde{H}_{S_O}^{16}=\begin{pmatrix} \{s_{0\langle o_0\rangle}\} & \{s_{1\langle o_1\rangle}\} & \{s_{3\langle o_{-2}\rangle},s_{4\langle o_{-3}\rangle}\} & \{s_{0\langle o_2\rangle}\} \\ \{s_{-1\langle o_{-1}\rangle}\} & \{s_{0\langle o_0\rangle}\} & \{s_{1\langle o_0\rangle},s_{2\langle o_2\rangle}\} & \{s_{-1\langle o_3\rangle},s_{0\langle o_2\rangle}\} \\ \{s_{-3\langle o_2\rangle},s_{4\langle o_3\rangle}\} & \{s_{-1\langle o_0\rangle},s_{-2\langle o_{-2}\rangle}\} & \{s_{0\langle o_0\rangle}\} & \{s_{-2\langle o_2\rangle},s_{-1\langle o_3\rangle}\} \\ \{s_{0\langle o_{-2}\rangle}\} & \{s_{1\langle o_{-3}\rangle},s_{0\langle o_{-2}\rangle}\} & \{s_{2\langle o_{-2}\rangle},s_{1\langle o_{-3}\rangle}\} & \{s_{0\langle o_0\rangle}\} \end{pmatrix}$$

$$\widetilde{H}_{S_O}^{17}=\begin{pmatrix} \{s_{0\langle o_0\rangle}\} & \{s_{-1\langle o_2\rangle},s_{0\langle o_1\rangle}\} & \{s_{1\langle o_{-3}\rangle},s_{2\langle o_3\rangle}\} & \{s_{0\langle o_2\rangle}\} \\ \{s_{1\langle o_{-2}\rangle},s_{0\langle o_{-1}\rangle}\} & \{s_{0\langle o_0\rangle}\} & \{s_{0\langle o_{-1}\rangle},s_{1\langle o_1\rangle}\} & \{s_{-2\langle o_1\rangle},s_{-1\langle o_2\rangle}\} \\ \{s_{-1\langle o_3\rangle},s_{-2\langle o_{-3}\rangle}\} & \{s_{0\langle o_1\rangle},s_{-1\langle o_{-1}\rangle}\} & \{s_{0\langle o_0\rangle}\} & \{s_{2\langle o_{-2}\rangle},s_{3\langle o_2\rangle}\} \\ \{s_{0\langle o_{-2}\rangle}\} & \{s_{2\langle o_{-1}\rangle},s_{1\langle o_{-2}\rangle}\} & \{s_{-2\langle o_2\rangle},s_{-3\langle o_{-2}\rangle}\} & \{s_{0\langle o_0\rangle}\} \end{pmatrix}$$

$$\widetilde{H}_{S_O}^{18}=\begin{pmatrix} \{s_{0\langle o_0\rangle}\} & \{s_{1\langle o_1\rangle}\} & \{s_{2\langle o_2\rangle},s_{3\langle o_1\rangle}\} & \{s_{0\langle o_3\rangle}\} \\ \{s_{-1\langle o_{-1}\rangle}\} & \{s_{0\langle o_0\rangle}\} & \{s_{1\langle o_{-1}\rangle},s_{2\langle o_{-1}\rangle}\} & \{s_{-1\langle o_3\rangle},s_{0\langle o_1\rangle}\} \\ \{s_{-2\langle o_{-2}\rangle},s_{-3\langle o_{-1}\rangle}\} & \{s_{-1\langle o_1\rangle},s_{-2\langle o_1\rangle}\} & \{s_{0\langle o_0\rangle}\} & \{s_{-2\langle o_1\rangle},s_{-1\langle o_3\rangle}\} \\ \{s_{0\langle o_{-3}\rangle}\} & \{s_{1\langle o_{-3}\rangle},s_{0\langle o_{-1}\rangle}\} & \{s_{2\langle o_{-1}\rangle},s_{1\langle o_{-3}\rangle}\} & \{s_{0\langle o_0\rangle}\} \end{pmatrix}$$

$$\widetilde{H}_{S_O}^{19}=\begin{pmatrix} \{s_{0\langle o_0\rangle}\} & \{s_{-1\langle o_0\rangle}\} & \{s_{1\langle o_{-2}\rangle},s_{2\langle o_{-1}\rangle}\} & \{s_{2\langle o_{-3}\rangle}\} \\ \{s_{1\langle o_0\rangle}\} & \{s_{0\langle o_0\rangle}\} & \{s_{2\langle o_1\rangle},s_{3\langle o_1\rangle}\} & \{s_{3\langle o_3\rangle}\} \\ \{s_{-1\langle o_2\rangle},s_{-2\langle o_1\rangle}\} & \{s_{-2\langle o_{-1}\rangle},s_{-3\langle o_{-1}\rangle}\} & \{s_{0\langle o_0\rangle}\} & \{s_{0\langle o_{-2}\rangle}\} \\ \{s_{-2\langle o_3\rangle}\} & \{s_{-3\langle o_{-3}\rangle}\} & \{s_{0\langle o_2\rangle}\} & \{s_{0\langle o_0\rangle}\} \end{pmatrix}$$

$$\widetilde{H}^{20}_{S_O} = \begin{pmatrix} \{s_{0\langle o_0\rangle}\} & \{s_{2\langle o_2\rangle}\} & \{s_{-1\langle o_2\rangle}, s_{0\langle o_{-1}\rangle}\} & \{s_{-2\langle o_{-1}\rangle}, s_{-1\langle o_{-3}\rangle}\} \\ \{s_{-2\langle o_{-2}\rangle}\} & \{s_{0\langle o_0\rangle}\} & \{s_{0\langle o_{-1}\rangle}, s_{0\langle o_{-1}\rangle}\} & \{s_{-1\langle o_1\rangle}, s_{-1\langle o_2\rangle}\} \\ \{s_{1\langle o_{-2}\rangle}, s_{0\langle o_1\rangle}\} & \{s_{0\langle o_1\rangle}, s_{0\langle o_1\rangle}\} & \{s_{0\langle o_0\rangle}\} & \{s_{1\langle o_{-3}\rangle}, s_{2\langle o_3\rangle}\} \\ \{s_{2\langle o_1\rangle}, s_{1\langle o_3\rangle}\} & \{s_{1\langle o_{-1}\rangle}, s_{1\langle o_{-2}\rangle}\} & \{s_{-1\langle o_3\rangle}, s_{-2\langle o_{-3}\rangle}\} & \{s_{0\langle o_0\rangle}\} \end{pmatrix}$$

Next, we can utilize Algorithm 5.3 to deal with this LSGDM problem:

Step 1. Based on Eqs. (5.13)–(5.15), check whether all experts reach the given consensus threshold. The consensus degrees of the pair of alternatives $cpda^{(0)}$, the alternatives $cda^{(0)}$ and the overall consensus degree of preference relations $ocd^{(0)}$ can be obtained:

$$cpda^{(0)} = \begin{pmatrix} 1 & 0.8535 & 0.8356 & 0.8497 \\ 0.8535 & 1 & 0.8413 & 0.8275 \\ 0.8356 & 0.8413 & 1 & 0.8160 \\ 0.8497 & 0.8275 & 0.8160 & 1 \end{pmatrix}$$

$$cda^{(0)} = \{0.8463, 0.8408, 0.8310, 0.8311\}$$

$$ocd^{(0)} = 0.8310$$

In this LSGDM problem, the given consensus threshold is $\xi = 0.85$ and $ocd^{(0)} < \xi$. So all experts do not reach group consensus and go to Step 2.

Step 2. Based on Algorithm 5.1, we cluster all experts into several small groups. The clustering process can be shown as follows:

Firstly, based on Eq. (5.4), the overall similarity matrix $OSM = (osm^{ab})_{20\times 20}$ is established:

$$
\begin{pmatrix}
1.0000 & 0.8698 & 0.9219 & 0.8177 & 0.8646 & 0.8620 & 0.9505 & 0.9141 & 0.8099 & 0.8646 & 0.9036 & 0.8438 & 0.8750 & 0.9193 & 0.7318 & 0.8047 & 0.9010 & 0.8099 & 0.7318 & 0.8958 \\
 & 1.0000 & 0.8646 & 0.8802 & 0.9661 & 0.9922 & 0.8359 & 0.8776 & 0.8620 & 0.9844 & 0.8307 & 0.8177 & 0.9948 & 0.8724 & 0.7422 & 0.8620 & 0.8594 & 0.8724 & 0.7526 & 0.8281 \\
 & & 1.0000 & 0.8021 & 0.8411 & 0.8568 & 0.9557 & 0.8464 & 0.7943 & 0.8594 & 0.8568 & 0.8594 & 0.8646 & 0.8516 & 0.7370 & 0.7891 & 0.9583 & 0.7943 & 0.7370 & 0.8385 \\
 & & & 1.0000 & 0.8724 & 0.8724 & 0.7943 & 0.7943 & 0.9714 & 0.8750 & 0.7630 & 0.9115 & 0.8854 & 0.7891 & 0.7891 & 0.9714 & 0.8333 & 0.9818 & 0.7891 & 0.7604 \\
 & & & & 1.0000 & 0.9635 & 0.8125 & 0.8854 & 0.8438 & 0.8609 & 0.8385 & 0.8151 & 0.9609 & 0.8802 & 0.7240 & 0.8490 & 0.8359 & 0.8646 & 0.7344 & 0.8359 \\
 & & & & & 1.0000 & 0.8281 & 0.8698 & 0.8594 & 0.9870 & 0.8229 & 0.8099 & 0.9870 & 0.8646 & 0.7500 & 0.8542 & 0.8516 & 0.8698 & 0.7604 & 0.8203 \\
 & & & & & & 1.0000 & 0.8646 & 0.7865 & 0.8307 & 0.8854 & 0.8516 & 0.8411 & 0.8698 & 0.7240 & 0.7813 & 0.9401 & 0.7865 & 0.7240 & 0.8672 \\
 & & & & & & & 1.0000 & 0.7656 & 0.8776 & 0.9375 & 0.8099 & 0.8828 & 0.9948 & 0.6667 & 0.7813 & 0.8151 & 0.7865 & 0.6667 & 0.9401 \\
 & & & & & & & & 1.0000 & 0.8516 & 0.7344 & 0.9036 & 0.8672 & 0.7604 & 0.8021 & 0.9740 & 0.8255 & 0.9792 & 0.8021 & 0.7318 \\
 & & & & & & & & & 1.0000 & 0.8307 & 0.8177 & 0.9782 & 0.8724 & 0.7474 & 0.8516 & 0.8542 & 0.8672 & 0.7578 & 0.8281 \\
 & & & & & & & & & & 1.0000 & 0.8151 & 0.8359 & 0.9427 & 0.6354 & 0.7500 & 0.8255 & 0.7552 & 0.6354 & 0.9766 \\
 & & & & & & & & & & & 1.0000 & 0.8229 & 0.8151 & 0.7526 & 0.9193 & 0.8906 & 0.9193 & 0.7422 & 0.8073 \\
 & & & & & & & & & & & & 1.0000 & 0.8776 & 0.7474 & 0.8672 & 0.8646 & 0.8776 & 0.7578 & 0.8333 \\
 & & & & & & & & & & & & & 1.0000 & 0.6615 & 0.7760 & 0.8203 & 0.7813 & 0.6615 & 0.9453 \\
 & & & & & & & & & & & & & & 1.0000 & 0.7969 & 0.7578 & 0.7865 & 0.9792 & 0.6328 \\
 & & & & & & & & & & & & & & & 1.0000 & 0.8203 & 0.9792 & 0.7969 & 0.7474 \\
 & & & & & & & & & & & & & & & & 1.0000 & 0.8255 & 0.7474 & 0.8073 \\
 & & & & & & & & & & & & & & & & & 1.0000 & 0.7865 & 0.7526 \\
 & & & & & & & & & & & & & & & & & & 1.0000 & 0.6328 \\
 & & & & & & & & & & & & & & & & & & & 1.0000
\end{pmatrix}
$$

Table 5.1 The rate of threshold

C_1	C_2	C_3	C_4	C_5	C_6	C_7	C_8	C_9	C_{10}	C_{11}
0.0013	0.0026	0.0052	0.0026	0.0007	0.0013	0.0104	0.0039	0.0026	0.0052	0.0312

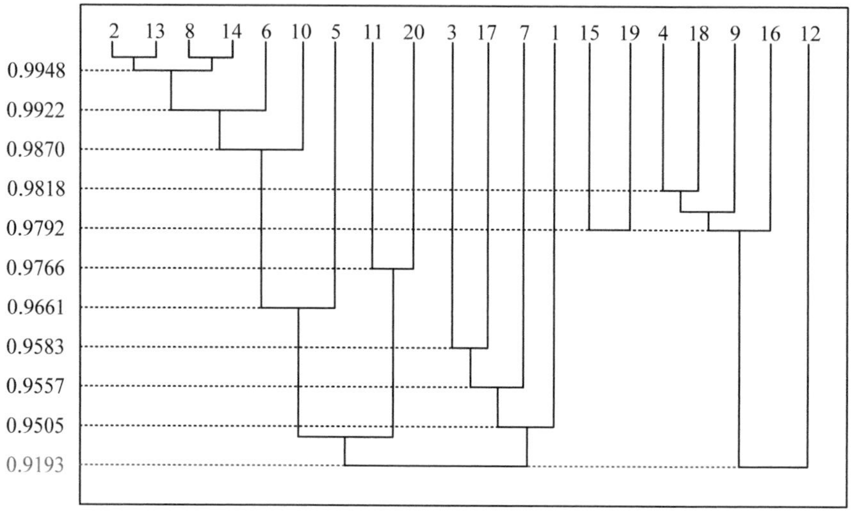

Fig. 5.4 The clustering process

Additionally, we rank all different elements of the upper triangular matrix of *OSM*, and then calculate the rate of threshold shown in Table 5.1.

Figure 5.4 can be drawn to describe the cluster process:

Therefore, all experts can be divided into three groups $\{B_1, B_2, B_3\}$:

$$B_1 = \{e^1, e^2, e^3, e^5, e^6, e^7, e^8, e^{10}, e^{11}, e^{13}, e^{14}, e^{17}, e^{20}\}$$
$$B_2 = \{e^{15}, e^{19}\}$$
$$B_3 = \{e^4, e^9, e^{12}, e^{16}, e^{18}\}$$

Then go to Step 3 and start the first round of consensus reaching process: **Round 1**.

Step 3[1]. Based on $cpda^{(0)}$, $cda^{(0)}$, $ocd^{(0)}$, and Eqs. (5.16)–(5.21), all alternatives need to be improved. In this round, we only discuss A_3 firstly. The experts and the parts of alternatives that need to repair their consensus degrees can be shown as follows:

(1) The experts e^9, e^{16} and e^{18} need to improve their preference information in pair of alternative (A_1, A_3);
(2) The expert e^{19} needs to improve their preference information in pair of alternatives (A_2, A_3);
(3) The expert e^{17} needs to improve their preference information in pair of alternatives (A_3, A_4).

And then we calculate all experts' weights based on Algorithm 5.2:

$$w^{(1)} = (0.0654, 0.0657, 0.0654, 0.0253, 0.0661, 0.0657, 0.0654, 0.0657, 0.0253, 0.0658, 0.0657, \\ 0.0251, 0.0657, 0.0657, 0.0101, 0.0253, 0.0655, 0.0253, 0.0101, 0.0658)^T$$

Furthermore, based on Eq. (3.24), the group DHHFLPR $(\widetilde{H}^c_{S_O})^{(1)} = ((h^c_{S_{O_{ij}}})_{4\times 4})^{(1)}$ can be obtained:

$$(\widetilde{H}^c_{S_O})^{(1)} = \begin{pmatrix} \{s_{0\langle o_0\rangle}\} & \{s_{0.95\langle o_{0.42}\rangle}\} & \{s_{0.67\langle o_{0.28}\rangle}\} & \{s_{-0.25\langle o_{0.14}\rangle}\} \\ \{s_{-0.95\langle o_{-0.42}\rangle}\} & \{s_{0\langle o_0\rangle}\} & \{s_{0.01\langle o_{0.12}\rangle}\} & \{s_{-1.13\langle o_{1.15}\rangle}\} \\ \{s_{-0.67\langle o_{-0.28}\rangle}\} & \{s_{-0.01\langle o_{-0.12}\rangle}\} & \{s_{0\langle o_0\rangle}\} & \{s_{0.76\langle o_{-0.82}\rangle}\} \\ \{s_{0.25\langle o_{-0.14}\rangle}\} & \{s_{1.13\langle o_{-1.15}\rangle}\} & \{s_{-0.76\langle o_{0.82}\rangle}\} & \{s_{0\langle o_0\rangle}\} \end{pmatrix}$$

Then, the moderator feeds all information obtained in this round back to the three groups as a reference. Each group discusses whether adjusts the corresponding expert's evaluation information and how to adjust them. Finally, all corresponding experts agree to change and the changed information is listed below:

(1) For the pair of alternatives (A_1, A_3), the expert e^9 decreases $\{s_{2\langle o_2\rangle}, s_{3\langle o_1\rangle}\}$ into $\{s_{0\langle o_2\rangle}\}$, e^{16} decreases $\{s_{3\langle o_{-2}\rangle}, s_{4\langle o_{-3}\rangle}\}$ into $\{s_{1\langle o_{-1}\rangle}\}$, and e^{18} decreases $\{s_{2\langle o_2\rangle}, s_{3\langle o_1\rangle}\}$ into $\{s_{2\langle o_{-2}\rangle}\}$;
(2) For the pair of alternatives (A_2, A_3), the expert e^{19} decreases $\{s_{2\langle o_1\rangle}, s_{3\langle o_1\rangle}\}$ into $\{s_{1\langle o_{-1}\rangle}\}$;
(3) For the pair of alternatives (A_3, A_4), the expert e^{17} decreases $\{s_{2\langle o_{-2}\rangle}, s_{3\langle o_2\rangle}\}$ into $\{s_{1\langle o_{-1}\rangle}\}$.

Step 4[1]. Collect all modified evaluation information, and go back to Step 1. Check whether all experts' reach the given consensus threshold again. The consensus degrees for the pair of alternatives $cpda^{(1)}$, the alternatives $cda^{(1)}$ and $cdpr^{(1)}$ are obtained:

$$cpda^{(1)} = \begin{pmatrix} 1 & 0.8535 & 0.8780 & 0.8497 \\ 0.8535 & 1 & 0.8594 & 0.8275 \\ 0.8780 & 0.8594 & 1 & 0.8315 \\ 0.8497 & 0.8275 & 0.8315 & 1 \end{pmatrix}$$

$$cda^{(1)} = \{0.8604, 0.8468, 0.8563, 0.8362\}$$

$$ocd^{(1)} = 0.8362$$

Obviously, all experts still do not reach the given consensus threshold. Then we need to go to Step 3 again and start the second round of consensus reaching process:

Round 2:

Step 3^2. Based on $cpda^{(1)}$, $cda^{(1)}$ and $ocd^{(1)}$, and Eqs. (5.16)–(5.21), we need to adjust the alternatives A_2 and A_4. In this round, we only discuss A_4. The experts and the parts of alternatives need to improve their consensus degrees, which can be shown as follows:

(1) The expert e^{20} needs to improve his/her preference information in pair of alternatives (A_1, A_4);
(2) The experts e^{15} and e^{19} need to improve their preference information in pair of alternatives (A_2, A_4);
(3) The expert e^3 needs to improve his/her preference information in pair of alternatives (A_3, A_4).

And then, we calculate all experts' weights again:

$$w^{(2)} = (0.0654, 0.0657, 0.0654, 0.0254, 0.0661, 0.0657, 0.0654, 0.0657, 0.0251, 0.0658, 0.0657, 0.0253, 0.0657, 0.0657, 0.0102, 0.0252, 0.0656, 0.0252, 0.0100, 0.0658)^T$$

Then we can obtain the group DHHFLPR $(\widetilde{H}_{S_O}^c)^{(2)} = ((h_{S_{O_{ij}}}^c)_{4\times4})^{(2)}$:

$$(\widetilde{H}_{S_O}^c)^{(2)} = \begin{pmatrix} \{s_{0\langle o_0\rangle}\} & \{s_{0.95\langle o_{0.42}\rangle}\} & \{s_{0.53\langle o_{0.24}\rangle}\} & \{s_{-0.25\langle o_{0.14}\rangle}\} \\ \{s_{-0.95<o_{-0.42}\rangle}\} & \{s_{0<o_0\rangle}\} & \{s_{-0.01\langle o_{0.09}\rangle}\} & \{s_{-1.13\langle o_{1.15}\rangle}\} \\ \{s_{-0.53\langle o_{-0.24}\rangle}\} & \{s_{0.01\langle o_{-0.09}\rangle}\} & \{s_{0\langle o_0\rangle}\} & \{s_{0.6\langle o_{-0.89}\rangle}\} \\ \{s_{0.25\langle o_{-0.14}\rangle}\} & \{s_{1.13\langle o_{-1.15}\rangle}\} & \{s_{-0.6\langle o_{0.89}\rangle}\} & \{s_{0\langle o_0\rangle}\} \end{pmatrix}$$

In this round, suppose that the expert e^{20} disagree to change, therefore, this group discuss and decide to improve his preference relation based on Eq. (5.23) randomly. The rest corresponding experts agree to change and the changed information is listed below:

(1) For the pair of alternatives (A_1, A_4), the expert e^{20} increases $\{s_{-2\langle o_{-1}\rangle}, s_{-1\langle o_{-3}\rangle}\}$ into $\{s_{-1\langle o_3\rangle}\}$;
(2) For the pair of alternatives (A_2, A_4), the expert e^{15} decreases $\{s_{3\langle o_3\rangle}\}$ into $\{s_{-1\langle o_3\rangle}\}$, and e^{19} decreases $\{s_{3\langle o_3\rangle}\}$ into $\{s_{2\langle o_1\rangle}\}$;
(3) For the pair of alternatives (A_3, A_4), the expert e^3 decreases $\{s_{2\langle o_{-1}\rangle}\}$ into $\{s_{1\langle o_2\rangle}\}$.

Step 4[2]. Collect all evaluation information, and go back to Step 1. Then the consensus degrees for the pair of alternatives $cpda^{(2)}$, the alternatives $cda^{(2)}$ and $ocd^{(2)}$ are obtained:

$$cpda^{(2)} = \begin{pmatrix} 1 & 0.8535 & 0.8780 & 0.8638 \\ 0.8535 & 1 & 0.8594 & 0.8505 \\ 0.8780 & 0.8594 & 1 & 0.8394 \\ 0.8638 & 0.8505 & 0.8394 & 1 \end{pmatrix}$$

$$cda^{(2)} = \{0.8651, 0.8545, 0.8589, 0.8512\}$$

$$ocd^{(2)} = 0.8512$$

Obviously, we obtain $ocd^{(2)} = 0.8512 > 0.85$. Therefore, all experts reach the given consensus threshold. Then go to Step 5.

Step 5. Calculate all experts' weights

$$w^* = (0.0654, 0.0657, 0.0653, 0.0254, 0.0661, 0.0657, 0.0654, 0.0657, 0.0251, 0.0658, 0.0657, 0.0253, 0.0657, 0.0657, 0.0102, 0.0252, 0.0656, 0.0252, 0.01, 0.0658)^T$$

and obtain the final group DHHFLPR:

$$*\widetilde{H}^c_{S_O} = \begin{pmatrix} \{s_{0\langle o_0\rangle}\} & \{s_{0.95\langle o_{0.42}\rangle}\} & \{s_{0.53\langle o_{0.24}\rangle}\} & \{s_{-0.22\langle o_{0.46}\rangle}\} \\ \{s_{-0.95\langle o_{-0.42}\rangle}\} & \{s_{0\langle o_0\rangle}\} & \{s_{0.01\langle o_{0.1}\rangle}\} & \{s_{-1.14\langle o_{1.11}\rangle}\} \\ \{s_{-0.53\langle o_{-0.24}\rangle}\} & \{s_{-0.01\langle o_{-0.1}\rangle}\} & \{s_{0\langle o_0\rangle}\} & \{s_{0.5\langle o_{-0.69}\rangle}\} \\ \{s_{0.22\langle o_{-0.46}\rangle}\} & \{s_{1.14\langle o_{-1.11}\rangle}\} & \{s_{-0.5\langle o_{0.69}\rangle}\} & \{s_{0\langle o_0\rangle}\} \end{pmatrix}$$

Then the synthetical value of each alternative is $SV(A) = \{2.1926, 1.7622, 1.9649, 2.0803\}$. Therefore, the ranking is $A_1 \succ A_4 \succ A_3 \succ A_2$. We can get the result that Chengdu is the optimal city in the process of the implementation status evaluations of the water resources management policies.

5.2 Large-Scale Group Consensus Decision-Making Method with DHHFLPRs by Managing Minority Opinions and Non-cooperative Behaviors

In the consensus reaching process of LSGDM, a noticeable drawback usually found in such large groups is the presence of experts and subgroups of experts who behave in a manner that does not contribute to reaching consensus (Yager 2001), since they are not going to adjust their preferences to reach the consensus. In large groups, there are often several subgroups or coalitions of experts with similar interests. Some of these subgroups are prone to modify their preferences to achieve an agreement, while some others do not modify their preferences or even do it in an opposite way to that of the remaining experts (Palomares et al. 2014a). These non-cooperating individuals and subgroups are called non-cooperative behaviors (Dong et al. 2016; Palomares et al. 2014a; Xu et al. 2015). In an actual LSGDM process, the arguments experts raise and the evidence they are based on are commonly rebutted, undercut or countered by the arguments of others, irrespective of the experts. However, the purpose of this subsection is only to identify and deal with these non-cooperative behaviors. Therefore, for convenience, we assume that the LSGDM process is unaffected by these factors and ignore these external influencing factors for the time being.

Additionally, in spite of different opinions or minority preferences often being cited as obstacles to decision-making, if they are appropriately processed, the decision result will be more reasonable and accurate (Xiong et al. 2008). experts who hold the minority opinions in a large-scale group mainly consist of four types (Xiong et al. 2008): (1) A leader, who is always able to give some unique points of views, and has enough rights to determine the final decision result. (2) An experienced expert, who often has a deep insight about the decision-making problem, and can propose constructive suggestions. (3) A young and aggressive expert, whose opinion is relatively extreme, and is rarely influenced by other experts' opinions. (4) A noteworthy and independent expert, whose view is usually a little out of the ordinary. As far as we know, the leader and the experienced expert, such as the CEO of a company and the experienced professor, are very powerful and experienced, so the preferences provided by them are positive in general. On the contrary, the other two kinds of experts are usually inexperienced or extreme, such as the new employee and the employee whose thinking is different from ordinary. Because these two kinds of experts' experiences are usually shallow or the thinking is difficult to be understood, so the preferences provided by them should be considered carefully. Therefore, the preferences provided by the first two should be given special attention, while the latter two should be considered prudently.

The existing research on the study of minority opinions and non-cooperative behaviors has some gaps. Firstly, some research only studied one part of the topic. Some only dealt with non-cooperative behaviors (Dong et al. 2016; Palomares et al. 2014a), and others only discussed the minority views (Xiong et al. 2008), and therefore results in incomplete information processing. Secondly, Xu et al. (2015)

developed a consensus model for multi-criteria large-group emergency decision-making by dealing with non-cooperative behaviors and minority opinions. However, the clustering method contains too many human factors and the normalization of individual decision matrices will lose lots of original information.

Therefore, this subsection will deal with non-cooperative behaviors and minority opinions simultaneously in the consensus reaching process of LSGDM with DHHFLPRs by proposing the novel clustering method, the novel weights-determining method and the novel consensus model.

5.2.1 Clustering, Weight-Determining Method and Consensus Measures

This subsection will develop a consensus model to manage minority opinions and non-cooperative behaviors in consensus reaching process of LSGDM with DHHFLPRs. Some basic contents are discussed first:

(1) Clustering. All experts can be classified into several small groups, which makes the consensus reaching process much simpler because the communication among small groups is smoother. Additionally, the minority opinions can be identified quickly and considered to be the group with the least number of experts.

(2) Weights-determining method. Weights of experts and all small groups are very important when aggregating preferences. Meanwhile, identifying and managing minority opinions also depends on the weight of each group obtained by the clustering.

(3) Consensus measures. By establishing some consensus measures, it is convenient to identify whether all experts reach the given consensus threshold result or not.

Different from the similarity degree-based clustering algorithm, Gou et al. (2020) developed a distance measure-based clustering algorithm.

Firstly, the distance measure between two DHHFLPRs can be defined as follows:

Definition 5.2 (Gou et al. 2020). Let $\tilde{H}_{S_O}^a = (h_{S_{O_{ij}}}^a)_{m \times m}$ and $\tilde{H}_{S_O}^b = (h_{S_{O_{ij}}}^b)_{m \times m}$ be two DHHFLPRs. Then we call

$$d\left(\tilde{H}_{S_O}^a, \tilde{H}_{S_O}^b\right) = \sqrt{\frac{2}{m(m-1)} \sum_{i=1}^{m} \sum_{i<j}^{m} \left(f\left(\left(le\left(h_{S_{O_{ij}}}^a\right)\right) - f\left(le\left(h_{S_{O_{ij}}}^a\right)\right)\right)\right)^2} \quad (5.30)$$

the distance measure between $\tilde{H}_{S_O}^a$ and $\tilde{H}_{S_O}^b$. Clearly, $d(\tilde{H}_{S_O}^a, \tilde{H}_{S_O}^b) \in [0, 1]$.

In general, the smaller the distance between two experts, the greater the possibility that they will be in a same group. According to this idea, a clustering method is developed:

Algorithm 5.4 (Gou et al. 2020). **Distance measure-based clustering algorithm**

Step 1. Establish the overall distance matrix. Based on Eq. (5.30), an overall distance matrix $ODM = \left(odm^{ab}\right)_{n\times n}$ associated with all pairs of experts is obtained, where

$$odm^{ab} = d(\widetilde{H}^{a}_{S_O}, \widetilde{H}^{b}_{S_O})(a, b = 1, 2, \ldots, n) \tag{5.31}$$

Step 2. Choose the classification threshold. Ranking all the different elements of the upper triangular matrix of $ODM = \left(odm^{ab}\right)_{n\times n}$ (except the diagonal elements) following the ascending order, denoted by $\Delta_1 < \Delta_2 < \cdots < \Delta_p < \cdots < \Delta_q$, where Δ_i is the ith smallest value and $q \leq n(n-1)/2$.

Step 3. Determine the optimal classification threshold $\Delta*$. Let TC_p be the rate of threshold change, obtained by

$$TC_p = \frac{\Delta_p - \Delta_{p-1}}{n_p - n_{p-1}} \tag{5.32}$$

where Δ_{p-1} and Δ_p are the numbers of experts in the $(p-1)-th$ and $p-th$ classifications, respectively. n_p and n_{p-1} be the numbers of classifications in the $(p-1)-th$ and $p-th$ classifications. When $n_p = n$, all experts are considered, so the calculation process is over and all TC_p are collected. If

$$TC_\mu = \max_p\{TC_p\} \tag{5.33}$$

then the pth classification threshold is called the optimal classification threshold, namely, $\Delta* = \Delta_\mu$.

Step 4. Determine the clustering result. Firstly, all pairs of experts (e^a, e^b) are classified into the overall groups as $\Im_1, \Im_2, \ldots, \Im_\zeta$, where $odm^{ab} \leq \Delta^*$ following the ascending order of odm^{ab}. If $\Im_{\zeta_i} \cap \Im_{\zeta_j} \neq \varnothing$ $(\zeta_i \neq \zeta_j;\ \zeta_i, \zeta_j = 1, 2, \ldots, \zeta)$, then these elements of the overall group are combined into a group. Finally, the clustering result $G_\phi(\phi = 1, 2, \ldots, \Phi)$ can be obtained when $\Im_{\zeta_i} \cap \Im_{\zeta_j} = \varnothing$. In other words, the final clustering result can be obtained when all groups do not intersect each other.

Step 5. End.

An example can be set up to show the clustering process of Algorithm 5.4:

Example 5.2 (Gou et al. 2020). Let $S_O = \{s_{t\langle o_k\rangle} | t = -4, \ldots, 4;\ k = -4, \ldots, 4\}$ be a DHLTS. Four experts propose their preferences and the overall distance matrix is calculated as:

$$ODM = \left(odm^{ab}\right)_{4\times 4} = \begin{pmatrix} 0 & 0.3687 & 0.1060 & 0.3409 \\ 0.3687 & 0 & 0.4050 & 0.3487 \\ 0.1060 & 0.4050 & 0 & 0.3407 \\ 0.3409 & 0.3487 & 0.3407 & 0 \end{pmatrix}$$

and then, all the different elements of the upper triangular matrix of ODM are ranked, and the optimal classification threshold is calculated as $\Delta* = \Delta_2 = 0.3407$. Therefore, the overall groups are obtained as $\Im_1 = (e^1, e^3)$ and $\Im_2 = (e^3, e^4)$. Based on Step 4, the final clustering result can be obtained: $G_1 = \{e^1, e^3, e^4\}$ and $G_2 = \{e^2\}$.

And then, when investigating the consensus reaching process in LSGDM based on the clustering result, but each group's weight is also an essential element. Suppose that all experts are classified into $\Phi(1 \leq \Phi \leq n)$ groups. Each group's weight at the beginning of the decision can be obtained by satisfying two hypotheses: (1) The experts in the same group can be given the same weight because of their preferences are very close and it can be considered that there is no difference among them. Especially the experienced experts should be given a larger weight, such as the leader and the experienced expert. The young or aggressive expert, however, should be assigned a smaller weight. (2) The group with a larger number of experts should be given a larger weight based on the majority principle.

Therefore, let n_ϕ be the number of experts included in a group G_ϕ $(\phi = 1, 2, \ldots, \Phi)$. Then, the weights of the experts e^a $(a = 1, 2, \ldots, n_\phi)$ in the group G_ϕ are obtained using

$$\omega_\phi^a = \frac{1}{n_\phi}, \quad \phi = 1, 2, \ldots, \Phi; a = 1, 2, \ldots, n_\phi \tag{5.34}$$

Furthermore, based on the number of experts in a group, the weight of each group G_ϕ is obtained as:

$$w_\phi = \frac{n_\phi}{\sum_{\phi=1}^{\Phi} n_\phi} \tag{5.35}$$

where $0 \leq w_\phi \leq 1$, and $\sum_{\phi=1}^{\Phi} w_\phi = 1$. Then, the weight of every expert in the overall group can be achieved using

$$\omega^a = \omega_\phi^a \times w_\phi, \quad \phi = 1, 2, \ldots, \Phi; a = 1, 2, \ldots, n_\phi \tag{5.36}$$

In LSGDM, the ideal result of the consensus reaching process is a stable state where each expert completely agrees with all other experts' preferences. However, it is very difficult and almost impossible to achieve when considering the differences among people. Therefore, setting a consensus threshold value is very reasonable and necessary, that is, the consensus reaching process can be considered to be over when their overall consensus degree reaches or exceeds the given threshold value. Let ξ be the given consensus threshold value, which can be used to decide whether the consensus reaching process is carried out. Similar to Subsection 5.1.3, the consensus threshold is usually set to be in the interval [0.7386,0.85]. Besides, the overall consensus degree can be calculated by the similarity measure among the experts' preferences.

Next, a consensus model for LSGDM with DHHFLPRs can be developed as follows:

Firstly, based on the clustering result and the DHHFLWA operator, the group preference matrix $\widetilde{H}_{S_O}^{G_\phi} = (h_{S_{O_{ij}}}^{G_\phi})_{m \times m}$ of a group G_ϕ $(\phi = 1, 2, \ldots, \Phi)$ and the overall preference matrix $\widetilde{H}_{S_O}^c = (h_{S_{O_{ij}}}^c)_{m \times m}$ of all groups are achieved, where $h_{S_{O_{ij}}}^c = \sum_{a=1}^n \omega^a \times h_{S_{O_{ij}}}^a$.

In addition, the consensus degree (CD) between a group preference matrix $\widetilde{H}_{S_O}^{G_\phi}$ and the overall preference matrix $\widetilde{H}_{S_O}^c$ is defined as:

$$\begin{aligned} CD\left(\tilde{H}_{S_O}^{G_\phi}\right) &= 1 - d\left(\tilde{H}_{S_O}^{G_\phi}, \tilde{H}_{S_O}^c\right) \\ &= 1 - \sqrt{\frac{2}{m(m-1)} \sum_{i=1}^{m} \sum_{i<j}^{m} \left(f\left(le\left(h_{S_{O_{ij}}}^{G_\phi} \right) \right) - f\left(le\left(h_{S_{O_{ij}}}^c \right) \right) \right)^2} \end{aligned} \tag{5.37}$$

where $d(\widetilde{H}_{S_O}^{G_\phi}, \widetilde{H}_{S_O}^c)$ is the distance between $\widetilde{H}_{S_O}^{G_\phi}$ and $\widetilde{H}_{S_O}^c$.

Then, based on $CD(\widetilde{H}_{S_O}^{G_\phi})(\phi = 1, 2, \ldots, \Phi)$, the overall consensus degree (OCD) is obtained using

$$\mathrm{OCD} = \frac{\sum_{\phi=1}^{\Phi} CD\left(\tilde{H}_{S_O}^{G_\phi}\right)}{\Phi} \tag{5.38}$$

Clearly, $0 \leq OCD \leq 1$, and the bigger the value of OCD is, the higher the consensus degree among all experts will be. If $OCD \geq \xi$, then the consensus degree of all experts is sufficiently high and the consensus reaching process is over. Otherwise, some changes about preferences or weights need to be made to improve the consensus degree and reach the given consensus threshold value. In next subsection, some methods are developed to improve the consensus degree by identifying and managing minority opinions and non-cooperative behaviors.

5.2.2 *Managing Minority Opinions and Non-cooperative Behaviors*

Gou et al. (2020) developed a method to determine some necessary parameters in the consensus reaching process, incorporate minority opinions and non-cooperative behaviors into the consensus model and proposed an algorithm to manage them in LSGDM with DHHFLPRs.

(a) **Determination of comprehensive adjustment coefficient**

In the consensus reaching process of an LSGDM, it is common for experts to be easily influenced by subjective and objective factors, so there may be uncertainty and subjectivity in the opinion adjustment coefficients provided by the experts (Xu et al. 2015). Therefore, some adjustment coefficients need to be developed to improve decision credibility. Firstly, subjective and objective adjustment coefficients are discussed. Then, the comprehensive adjustment coefficient can be obtained based on two rules:

(1) Subjective adjustment coefficient

Suppose that $\widetilde{H}_{S_O}^{G_\phi(\mathbb{Z})} = (h_{S_{O_{ij}}}^{G_\phi(\mathbb{Z})})_{m\times m}$ is the group preference matrix of the group G_ϕ and $\widetilde{H}_{S_O}^{c(\mathbb{Z})} = (h_{S_{O_{ij}}}^{c(\mathbb{Z})})_{m\times m}$ is the overall preference matrix in the $\mathbb{Z}$th iteration. If $OCD^{(\mathbb{Z})} < \xi$, then it means that the consensus has not been reached. Let G_{ϕ^*} be the group that has the largest difference out of all the groups, i.e., $CD(\widetilde{H}_{S_O}^{G_{\phi^*}(\mathbb{Z})}) = \min\{CD(\widetilde{H}_{S_O}^{G_\phi(\mathbb{Z})}) | \phi = 1, 2, \ldots, \Phi\}$. Considering the group consensus degree and practical situation, the group G_{ϕ^*} needs to discuss and provide an adjustment coefficient first, denoted by $\vartheta_{G_{\phi^*}}^{S(\mathbb{Z})}$ $(0 \le \vartheta_{G_{\phi^*}}^{S(\mathbb{Z})} \le 1)$, to modify its preference. This adjustment coefficient depends entirely on the group's judgment and it reflects the degree to which groups tend to modify their own preference information. The larger the value of $\vartheta_{G_{\phi^*}}^{S(\mathbb{Z})}$ is, the stronger the modification willingness of the group G_{ϕ^*} has. Because the adjustment coefficient provided by group G_{ϕ^*} reflects its subjective attitude towards the group consensus degree and the opinions of modifications, it can be called a subjective adjustment coefficient.

(2) Objective adjustment coefficient

In general, the larger the difference between group G_ϕ and the overall group, the more this group needs to be improved to reach the consensus threshold value ξ. That is, the lower the consensus degree of the group G_ϕ, the bigger the objective adjustment coefficient $\vartheta_{G_\phi}^{O(\mathbb{Z})}$ the group G_ϕ needs. Specifically, the objective adjustment coefficient is calculated using

$$\vartheta_{G_\phi}^{O(\mathbb{Z})} = 1 - \frac{1-\xi}{1 - CD(\widetilde{H}_{S_O}^{G_\phi(\mathbb{Z})})} \tag{5.39}$$

Clearly, $0 \le \vartheta_{G_\phi}^{O(\mathbb{Z})} \le 1$. From Eq. (5.39), it is clear and logical that the higher the given consensus threshold value ξ, the greater the effort the experts need to make.

(3) Comprehensive adjustment coefficient

Combining the subjective adjustment coefficient $\vartheta_{G_\phi}^{S(\mathbb{Z})}$ and the objective adjustment coefficient $\vartheta_{G_\phi}^{O(\mathbb{Z})}$, the comprehensive adjustment coefficient of group $G_\phi^{(\mathbb{Z})}$, denoted as $\vartheta_{G_\phi}^{(\mathbb{Z})}$, can be obtained based on the following rules:

(1) If $\vartheta_{G_\phi}^{S(\mathbb{Z})} \ge \vartheta_{G_\phi}^{O(\mathbb{Z})}$, then $\vartheta_{G_\phi}^{(\mathbb{Z})} \ge \vartheta_{G_\phi}^{S(\mathbb{Z})}$;
(2) If $\vartheta_{G_\phi}^{S(\mathbb{Z})} < \vartheta_{G_\phi}^{O(\mathbb{Z})}$, then $\vartheta_{G_\phi}^{(\mathbb{Z})} = \varepsilon \vartheta_{G_\phi}^{S(\mathbb{Z})} + (1-\varepsilon)\vartheta_{G_\phi}^{O(\mathbb{Z})}$, where $0 \le \vartheta_{G_\phi}^{(\mathbb{Z})} \le 1$, $\varepsilon \in [0,1]$ is a parameter which reflects the importance degree of the subjective adjustment coefficient.

Based on the comprehensive adjustment coefficient, the preferences of a group $G_\phi^{(\mathbb{Z})}$ is improved using

$$\widetilde{H}_{S_O}^{G_\phi(\mathbb{Z}+1)} = \vartheta_{G_\phi}^{(\mathbb{Z})} \widetilde{H}_{S_O}^{c(\mathbb{Z})} \oplus (1 + \vartheta_{G_\phi}^{(\mathbb{Z})}) \widetilde{H}_{S_O}^{G_\phi(\mathbb{Z})} \tag{5.40}$$

Motivated by Xu (2009), it is convenient to improve the group consensus degree and reach the given consensus threshold value by Eq. (5.40).

(b) **Managing Minority Opinions**

Gou et al. (2020) developed a method to deal with minority opinions, and it consists of three parts: identifying the minority opinions, making a discussion among the experts and adjusting the corresponding weight information.

Method 1 (Gou et al. 2020). **Identify and manage minority opinions**

Part 1. Identify minority opinions. A group can be identified as a minority subgroup if it satisfies two conditions:

(a) The consensus degree of the group should be the smallest;
(b) The group consists of only one or a few expert(s).

Let $E = \{e^1, e^2, \ldots, e^n\}$ be a set of experts, and all of them are classified into $\Phi(1 \le \Phi \le n)$ groups. Suppose that a group G_{ϕ^*} (n_{ϕ^*} is the number of experts in this group) has the biggest difference out of all the groups (smallest consensus degree). $\tilde{n} = [n/\Phi]$ ([] is a bracket function) is the threshold which is used to determine which group belongs to the minority opinion group. If $n_{\phi^*} \le \tilde{n}$, then G_{ϕ^*} is called a minority opinion group.

Part 2. Explain the rationality of the minority opinion and make a discussion among all groups.

First, the group with minority opinion explains the rationality of its opinion, then a discussion about the group with minority opinion is put into force among the remaining groups. Based on the principle that the minority opinion should be considered fully and treated reasonably, each group should hold a broad discussion and give its attitude and opinion.

If more than half of the remaining groups think that the opinion of group G_{ϕ^*} is worth consideration, namely, $\tilde{n} \geq n/2$, then it is necessary to increase the weight of this group to enhance its importance degree on overall groups. Meanwhile, the adjustment function should be close to the number of groups who support the minority opinion group. The more groups that support the minority opinion group, the higher the weight the group should be given.

Part 3. Improvement.

Based on the analyses above, a weight-improving method can be developed for the minority opinion group. Firstly, ranking the weight vector of all groups in ascending order, denoted as $w'^{(\mathbb{Z})} = (w_1'^{(\mathbb{Z})}, w_2'^{(\mathbb{Z})}, \ldots, w_\Phi'^{(\mathbb{Z})})^T$, where $w_\phi'^{(\mathbb{Z})} (\phi = 1, 2, \ldots, \Phi)$ is the ϕth smallest weight. Then the difference value, denoted as $dv_{\phi^*}^{M1(\mathbb{Z})}$, between the number of the groups who support the group G_{ϕ^*} (denoted as $n_{\phi^*}^{M1(\mathbb{Z})}$) and the half of the number of the remaining groups can be obtained by

$$dv_{\phi^*}^{M1(\mathbb{Z})} = \begin{cases} round\left(n_{\phi^*}^{M1(\mathbb{Z})} - (\Phi - 1)/2\right), & if\ \Phi\ is\ an\ even\ number \\ n_{\phi^*}^{M1(\mathbb{Z})} - (\Phi - 1)/2, & if\ \Phi\ is\ an\ odd\ number \end{cases} \tag{5.41}$$

where $round(\cdot)$ is the round operation.

Then, a weight improvement function is defined as follows:

Definition 5.3 (Gou et al. 2020). Let $w^{(\mathbb{Z})} = (w_1^{(\mathbb{Z})}, w_2^{(\mathbb{Z})}, \ldots, w_\Phi^{(\mathbb{Z})})^T$ be the weight vector of all groups in the $\mathbb{Z}$th iteration, and $w'^{(\mathbb{Z})} = (w_1'^{(\mathbb{Z})}, w_2'^{(\mathbb{Z})}, \ldots, w_\Phi'^{(\mathbb{Z})})^T$ be the weight vector in ascending order. Suppose that G_{ϕ^*} is the group with minority opinion and its weight is the υ^*th smallest weight, namely, $w_{\phi^*}^{(\mathbb{Z})} = w_{\upsilon^*}'^{(\mathbb{Z})}$, then the weight improvement function is developed as:

$$w_{\phi^*}^{M1(\mathbb{Z})} = \min\left\{\max\left\{w_\phi^{(\mathbb{Z})} | \phi = 1, 2, \ldots, \Phi\right\}, w'^{(\mathbb{Z})}_{\upsilon^* + dv_{\phi^*}^{M1(\mathbb{Z})}}\right\} \tag{5.42}$$

where $w_{\phi^*}^{M1(\mathbb{Z})}$ is the adjusted weight and the weight of the group G_{ϕ^*} becomes the $(\upsilon^* + dv_{\phi^*}^{M1(\mathbb{Z})})$th smallest weight in the new weight vector.

The consensus measure will then be repeated based on the method discussed above. However, if there are no more than half of the groups in favor of the minority opinion group, which means that most experts hold opposite opinions about the rationality of the opinion given by the minority opinion group, so both the weight improving process and the processing of minority opinions will end.

(c) **Handling Non-Cooperative Behaviors**

As we mentioned above, this subsection is committed to developing a method to handle the non-cooperative behaviors.

Method 2 (Gou et al. 2020). **Identify and manage non-cooperative behaviors**

Part 1. Identify the non-cooperative group(s)

According to the opinion of the group G_{ϕ^*}, the remaining groups $G_{\phi'}(\phi' = 1, 2, \ldots, \Phi; \phi' \neq \phi^*)$ provide their adjustment suggestions, denoted as $\vartheta^{(\mathbb{Z})}_{G_{\phi'}G_{\phi^*}}$ $(0 \leq \vartheta^{(\mathbb{Z})}_{G_{\phi'}G_{\phi^*}} \leq 1)$. Based on Eq. (5.39), the objective adjustment coefficient $\vartheta^{O(\mathbb{Z})}_{G_{\phi^*}}$ is obtained. Then, the expected adjustment suggestion interval is obtained and denoted as $\overline{\vartheta}^{(\mathbb{Z})}_{G_{\phi^*}} = [\vartheta^{(\mathbb{Z})L}_{G_{\phi^*}}, \vartheta^{(\mathbb{Z})U}_{G_{\phi^*}}] = [\min\{\vartheta^{(\mathbb{Z})}_{G_{\phi'}G_{\phi^*}}, \vartheta^{O(\mathbb{Z})}_{G_{\phi^*}}\}, \max\{\vartheta^{(\mathbb{Z})}_{G_{\phi'}G_{\phi^*}}, \vartheta^{O(\mathbb{Z})}_{G_{\phi^*}}\}]$. If the subjective adjustment coefficient $\vartheta^{S(\mathbb{Z})}_{G_{\phi^*}}$ of this group is included in or is smaller than the left boundary of the interval $\overline{\vartheta}^{(\mathbb{Z})}_{G_{\phi^*}}$, then the group G_{ϕ^*} belongs to a non-cooperative group.

Part 2. Measure the non-cooperative degree

To determine the degree of a group that is unwilling to repair its opinion, the non-cooperative degree should be defined:

Definition 5.4 (Gou et al. 2020). Let $\vartheta^{S(\mathbb{Z})}_{G_{\phi^*}}$ be the subjective adjustment coefficient provided by group G_{ϕ^*}, and so it can be written in interval form, i.e., $\overline{\vartheta}^{S(\mathbb{Z})}_{G_{\phi^*}} = \left[\vartheta^{S(\mathbb{Z})L}_{G_{\phi^*}}, \vartheta^{S(\mathbb{Z})U}_{G_{\phi^*}}\right]$ with $\vartheta^{S(\mathbb{Z})}_{G_{\phi^*}} = \vartheta^{S(\mathbb{Z})L}_{G_{\phi^*}} = \vartheta^{S(\mathbb{Z})U}_{G_{\phi^*}}$. Then based on the possibility degree proposed in (Da and Liu 1999), the non-cooperative degree of the group G_{ϕ^*} is obtained using

$$\begin{aligned} N^{(\mathbb{Z})}(G_{\phi^*}) &= 1 - p(\overline{\vartheta}^{S(\mathbb{Z})}_{G_{\phi^*}} \geq \overline{\vartheta}^{(\mathbb{Z})}_{G_{\phi^*}}) \\ &= 1 - \min\left\{\max\left\{\frac{\vartheta^{S(\mathbb{Z})U}_{G_{\phi^*}} - \vartheta^{(\mathbb{Z})L}_{G_{\phi^*}}}{(\vartheta^{S(\mathbb{Z})U}_{G_{\phi^*}} - \vartheta^{S(\mathbb{Z})L}_{G_{\phi^*}}) + (\vartheta^{(\mathbb{Z})U}_{G_{\phi^*}} - \vartheta^{(\mathbb{Z})L}_{G_{\phi^*}}}, 0\right\}, 0\right\} \end{aligned} \tag{5.43}$$

where $0 \leq N^{(\mathbb{Z})}(G_{\phi^*}) \leq 1$.

Part 3. Manage the non-cooperative behaviors

(a) If $N^{(\mathbb{Z})}(G_{\phi^*}) = 0$, then the group G_{ϕ^*} can be regarded as a completely cooperative group. Therefore, it is not necessary to change the weight of G_{ϕ^*}, and the comprehensive adjustment coefficient is only used to repair its preference directly.
(b) If $N^{(\mathbb{Z})}(G_{\phi^*}) = 1$, then the group G_{ϕ^*} can be regarded as a completely non-cooperative group. A lot of time will be wasted trying to improve this group, so the best choice is to remove it.
(c) If $0 \le N^{(\mathbb{Z})}(G_{\phi^*}) \le 1$, then the group G_{ϕ^*} can be regarded as a partly non-cooperative group. Therefore, it is necessary to adjust its weight to reduce its reflection, and then utilize the comprehensive adjustment coefficient to repair its preference.

Xu et al. (2015) developed a non-cooperative degree-based staircase weight adjustment function but it is not very precise. Therefore, a new weight adjustment function is developed:

$$w_{\phi^*}^{M2(\mathbb{Z})} = \begin{cases} w_{\phi^*}^{(\mathbb{Z})}, & N^{(\mathbb{Z})}(G_{\phi^*}) \in [0, 0.1) \\ w_{\phi^*}^{(\mathbb{Z})} \times 0.9, & N^{(\mathbb{Z})}(G_{\phi^*}) \in [0.1, 0.2) \\ w_{\phi^*}^{(\mathbb{Z})} \times 0.8, & N^{(\mathbb{Z})}(G_{\phi^*}) \in [0.2, 0.3) \\ w_{\phi^*}^{(\mathbb{Z})} \times 0.7, & N^{(\mathbb{Z})}(G_{\phi^*}) \in [0.3, 0.4) \\ w_{\phi^*}^{(\mathbb{Z})} \times 0.6, & N^{(\mathbb{Z})}(G_{\phi^*}) \in [0.4, 0.5) \\ w_{\phi^*}^{(\mathbb{Z})} \times 0.5, & N^{(\mathbb{Z})}(G_{\phi^*}) \in [0.5, 0.6) \\ w_{\phi^*}^{(\mathbb{Z})} \times 0.4, & N^{(\mathbb{Z})}(G_{\phi^*}) \in [0.6, 0.7) \\ w_{\phi^*}^{(\mathbb{Z})} \times 0.3, & N^{(\mathbb{Z})}(G_{\phi^*}) \in [0.7, 0.8) \\ w_{\phi^*}^{(\mathbb{Z})} \times 0.2, & N^{(\mathbb{Z})}(G_{\phi^*}) \in [0.8, 0.9) \\ w_{\phi^*}^{(\mathbb{Z})} \times 0.1, & N^{(\mathbb{Z})}(G_{\phi^*}) \in [0.9, 1) \\ 0, & N^{(\mathbb{Z})}(G_{\phi^*}) = 1 \end{cases} \tag{5.44}$$

where $w_{\phi^*}^{(\mathbb{Z})}$ is the weight of the group G_{ϕ^*} and $w_{\phi^*}^{M2(\mathbb{Z})}$ expresses the adjusted weight of the group G_{ϕ^*} in the $\mathbb{Z}$th iteration.

Based on the discussion above, an algorithm is established to handle LSGDM with DHHFLPRs by managing minority opinions and non-cooperative behaviors:

Algorithm 5.5 (Gou et al. 2020). **LSGDM consensus model with DHHFLPRs by managing minority opinions and non-cooperative behaviors**

Input: Preference matrices $\widetilde{H}_{S_O}^{a(\mathbb{Z})} = (h_{S_{O_{ij}}}^{a(\mathbb{Z})})_{m \times m}$ $(a = 1, 2, \ldots, n)$, iteration number $\mathbb{Z}$, and the given threshold value ξ.

Output: The final preference matrix of each group $\widetilde{H}_{S_O}^{G_\phi(\mathbb{Z}^*)}$, the final overall preference matrix $\widetilde{H}_{S_O}^{c(\mathbb{Z}^*)}$ and the rank of experts.

Step 1. Clustering experts into Φ groups G_ϕ $(\phi = 1, 2, \ldots, \Phi)$. and calculate the weight vector of all groups using Eq. (5.35). Then the group preference matrix $\widetilde{H}_{S_O}^{G_\phi} = (h_{S_{O_{ij}}}^{G_\phi})_{m\times m}$ $(\phi = 1, 2, \ldots, \Phi)$ of each group is calculated based on Eq. (3.24). Let $\mathbb{Z} = 0$ and go to Step 2.

Step 2. Aggregate all group preference matrices into the overall preference matrix $\widetilde{H}_{S_O}^{c} = (h_{S_{O_{ij}}}^{c})_{m\times m}$ based on Eq. (3.24).

Step 3. Calculate the consensus degree of each group preference matrix, i.e., $CD(\widetilde{H}_{S_O}^{G_\phi})$ based on Eq. (5.37), and obtain the overall consensus degree OCD based on Eq. (5.38). If $OCD \geq \xi$, then go to Step 5; otherwise, go to Step 4.

Step 4. Consensus improvement process

(1) Use Method 1 to identify and manage the group with minority opinion and determine whether the weight of this group needs to be repaired. If so, we use Eq. (5.41) to modify it, let $\mathbb{Z} = \mathbb{Z} + 1$ and go back to Step 2; otherwise, go to Step 4 (2).

(2) Use Method 2 to identify whether there is a group with non-cooperative behavior. If so, firstly we should reduce its weight based on Eq. (5.44), and then we need to calculate the comprehensive adjustment coefficient based on Eq. (5.40) and use it to repair preference information; Otherwise, we can only calculate the comprehensive adjustment coefficient and repair its preference. Let $\mathbb{Z} = \mathbb{Z} + 1$ and go back to Step 2.

Step 5. Let $\mathbb{Z}^* = \mathbb{Z}$. Output the group preference matrix $\widetilde{H}_{S_O}^{G_\phi(\mathbb{Z}^*)} = (h_{S_O}^{G_\phi(\mathbb{Z}^*)})_{m\times m}$ $(\phi = 1, 2, \ldots, \Phi)$ of each group and the overall preference matrix $\widetilde{H}_{S_O}^{c(\mathbb{Z}^*)} = (h_{S_O}^{c(\mathbb{Z}^*)})_{m\times m}$.

Step 6. Rank alternatives based on the expected values of alternatives: $SV(A_i) = \sum_{j=1}^{m} E(h_{S_O}^{c(\mathbb{Z}^*)})(i = 1, 2, \ldots, m)$.

Step 7. End.

A figure is drawn to explain this algorithm (Fig. 5.5).

5.2.3 *Application of the Algorithm 5.5*

This subsection applies Algorithm 5.5 to deal with a practical LSGDM problem that is to determine the main reason of haze pollution in Chengdu, Sichuan province. Firstly, we describe the background about the reasons of haze pollution, and then we utilize Algorithm 5.5 to deal with this LSGDM problem, finally we make some comparative analyses.

Example 5.3 (Gou et al. 2020). In recent years, haze remains an important issue in China's development process. Based on the urban environment and air quality

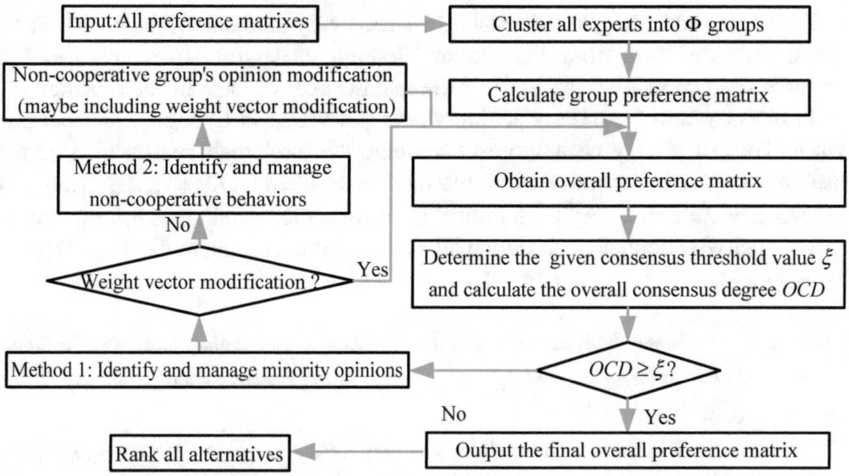

Fig. 5.5 The consensus reaching process in LSGDM based on Algorithm 5.5

monitoring data, the haze treatment of China has achieved initial results and the overall picture has improved. However, the pollution has not been effectively curbed, and local air pollution remains serious such as Henan province, Shandong province, and Shanxi province, etc. Therefore, the situation is not optimistic and China still faces with many problems and challenges.

Next, four main reasons can be summarized as follows:

(1) Economic restructuring is lagging behind. The formation and distribution of haze in China is closely related to industrial pollutant emission. Most of the areas with heavy haze are the Industrial cities or old industrial bases with relatively lagging economic structure.
(2) Energy consumption structure dominated by fossil energy. China is the largest energy producer and consumer, the largest coal consumer, the first big environment pollutants and greenhouse gases in the world. The production and living is highly dependent on fossil fuels such as coal and oil. Additionally, unreasonable energy production and consumption structure, as well as the emission of pollutants produced in the process of using are important reasons for haze weather.
(3) Environmental responsibilities in some areas are weakened. Environmental law enforcement is one of the important means to strengthen environmental management and environmental protection. However, some areas believe that haze control will affect local economic development. Then, the effects of environmental protection and management are greatly reduced.
(4) Regional coordination and governance mechanism still needs to be further deepened. The existing regional coordination and governance mechanism is very imperfect, so the cooperation mechanisms are also difficult to sustain.

Therefore, according to the existing issues, every area should be based on the actual situation, learn from domestic and foreign successful experience, and then promote haze control on the basis of the main causes of haze in each district.

In recent years, the haze is getting worse and worse in Chengdu, Sichuan province. Thus, it is very necessary to determine the most main reason of haze formation. Suppose that the above four reasons are the alternatives $\{A_1, A_2, A_3, A_4\}$, 20 experts $E = \{e^1, e^2, \ldots, e^{20}\}$ are invited to provide their preference information by making pairwise comparisons among all alternatives according to the given DHLTS $S_O = \{s_{t\langle o_k\rangle} | t = -4, \ldots, 4;\ k = -4, \ldots, 4\}$, where

$$\begin{aligned} S = \{&s_{-4} = \textit{extremely bad}, s_{-3} = \textit{very bad}, s_{-2} = \textit{bad}, s_{-1} = \textit{slightly bad}, s_0 = \textit{medium}, \\ &s_1 = \textit{slightly good}, s_2 = \textit{good}, s_3 = \textit{very good},\ s_4 = \textit{extremely good}\} \end{aligned}$$

$$\begin{aligned} O = \{&o_{-4} = \textit{far from}, o_{-3} = \textit{scarcely}, o_{-2} = \textit{only a little}, o_{-1} = \textit{a little}, o_0 = \textit{just right}, \\ &o_1 = \textit{much}, o_2 = \textit{very much}, o_3 = \textit{extremely much}, o_4 = \textit{entirely}\} \end{aligned}.$$

These preference information can be transformed into 20 DHHFLPRs $\widetilde{H}^r_{S_O} = (h^r_{S_{O_{ij}}})_{4\times4}$ $(r = 1, 2, \ldots, 20)$ as follows:

$$\widetilde{H}^1_{S_O} = \begin{pmatrix} \{s_{0\langle o_0\rangle}\} & \{s_{0\langle o_1\rangle}\} & \{s_{1\langle o_1\rangle}, s_{1\langle o_3\rangle}\} & \{s_{-2\langle o_{-1}\rangle}\} \\ \{s_{0\langle o_{-1}\rangle}\} & \{s_{0\langle o_0\rangle}\} & \{s_{1\langle o_0\rangle}\} & \{s_{-3\langle o_2\rangle}, s_{-2\langle o_2\rangle}, s_{-1\langle o_2\rangle}\} \\ \{s_{-1\langle o_{-1}\rangle}, s_{-1\langle o_{-3}\rangle}\} & \{s_{-1\langle o_0\rangle}\} & \{s_{0\langle o_0\rangle}\} & \{s_{2\langle o_{-3}\rangle}\} \\ \{s_{2\langle o_1\rangle}\} & \{s_{3\langle o_{-2}\rangle}, s_{2\langle o_{-2}\rangle}, s_{1\langle o_{-2}\rangle}\} & \{s_{-2\langle o_3\rangle}\} & \{s_{0\langle o_0\rangle}\} \end{pmatrix}$$

$$\widetilde{H}^2_{S_O} = \begin{pmatrix} \{s_{0\langle o_0\rangle}\} & \{s_{1\langle o_{-3}\rangle}, s_{1\langle o_{-1}\rangle}\} & \{s_{-1\langle o_{-2}\rangle}\} & \{s_{1\langle o_{-1}\rangle}\} \\ \{s_{-1\langle o_3\rangle}, s_{-1\langle o_1\rangle}\} & \{s_{0\langle o_0\rangle}\} & \{s_{-3\langle o_0\rangle}, s_{-3\langle o_2\rangle}\} & \{s_{-1\langle o_2\rangle}\} \\ \{s_{1\langle o_2\rangle}\} & \{s_{3\langle o_0\rangle}, s_{3\langle o_{-2}\rangle}\} & \{s_{0\langle o_0\rangle}\} & \{s_{-2\langle o_{-3}\rangle}\} \\ \{s_{-1\langle o_1\rangle}\} & \{s_{1\langle o_{-2}\rangle}\} & \{s_{2\langle o_3\rangle}\} & \{s_{0\langle o_0\rangle}\} \end{pmatrix}$$

$$\widetilde{H}^3_{S_O} = \begin{pmatrix} \{s_{0\langle o_0\rangle}\} & \{s_{0\langle o_{-1}\rangle}\} & \{s_{1\langle o_{-2}\rangle}, s_{1\langle o_0\rangle}\} & \{s_{-2\langle o_3\rangle}\} \\ \{s_{0\langle o_1\rangle}\} & \{s_{0\langle o_0\rangle}\} & \{s_{2\langle o_2\rangle}\} & \{s_{-3\langle o_2\rangle}, s_{-2\langle o_2\rangle}, s_{-1\langle o_2\rangle}\} \\ \{s_{-1\langle o_2\rangle}, s_{-1\langle o_0\rangle}\} & \{s_{-2\langle o_{-2}\rangle}\} & \{s_{0\langle o_0\rangle}\} & \{s_{2\langle o_{-1}\rangle}\} \\ \{s_{2\langle o_{-3}\rangle}\} & \{s_{3\langle o_{-2}\rangle}, s_{2\langle o_{-2}\rangle}, s_{1\langle o_{-2}\rangle}\} & \{s_{-2\langle o_1\rangle}\} & \{s_{0\langle o_0\rangle}\} \end{pmatrix}$$

$$\widetilde{H}^4_{S_O} = \begin{pmatrix} \{s_{0\langle o_0\rangle}\} & \{s_{0\langle o_1\rangle}\} & \{s_{3\langle o_{-2}\rangle}, s_{3\langle o_0\rangle}\} & \{s_{2\langle o_3\rangle}\} \\ \{s_{0\langle o_{-1}\rangle}\} & \{s_{0\langle o_0\rangle}\} & \{s_{1\langle o_{-1}\rangle}\} & \{s_{1\langle o_2\rangle}\} \\ \{s_{3\langle o_2\rangle}, s_{3\langle o_0\rangle}\} & \{s_{-1\langle o_1\rangle}\} & \{s_{0\langle o_0\rangle}\} & \{s_{-2\langle o_0\rangle}, s_{-2\langle o_2\rangle}\} \\ \{s_{-2\langle o_{-3}\rangle}\} & \{s_{-1\langle o_{-2}\rangle}\} & \{s_{2\langle o_0\rangle}, s_{2\langle o_{-2}\rangle}\} & \{s_{0\langle o_0\rangle}\} \end{pmatrix}$$

$$\widetilde{H}^5_{S_O} = \begin{pmatrix} \{s_{0\langle o_0\rangle}\} & \{s_{1\langle o_1\rangle}, s_{1\langle o_3\rangle}\} & \{s_{0\langle o_{-1}\rangle}\} & \{s_{1\langle o_1\rangle}\} \\ \{s_{-1\langle o_{-1}\rangle}, s_{-1\langle o_{-3}\rangle}\} & \{s_{0\langle o_0\rangle}\} & \{s_{-3\langle o_{-1}\rangle}, s_{-2\langle o_{-1}\rangle}, s_{-1\langle o_{-1}\rangle}\} & \{s_{-1\langle o_2\rangle}\} \\ \{s_{0\langle o_1\rangle}\} & \{s_{3\langle o_{-1}\rangle}, s_{2\langle o_{-1}\rangle}, s_{1\langle o_{-1}\rangle}\} & \{s_{0\langle o_0\rangle}\} & \{s_{-3\langle o_{-3}\rangle}\} \\ \{s_{-1\langle o_{-1}\rangle}\} & \{s_{1\langle o_{-2}\rangle}\} & \{s_{3\langle o_3\rangle}\} & \{s_{0\langle o_0\rangle}\} \end{pmatrix}$$

$$\widetilde{H}_{S_O}^{6} = \begin{pmatrix} \{s_{0\langle o_0\rangle}\} & \{s_{1\langle o_{-2}\rangle}\} & \{s_{-1\langle o_2\rangle}\} & \{s_{1\langle o_{-2}\rangle}, s_{1\langle o_0\rangle}\} \\ \{s_{-1\langle o_2\rangle}\} & \{s_{0\langle o_0\rangle}\} & \{s_{-2\langle o_{-1}\rangle}\} & \{s_{-3\langle o_{-1}\rangle}, s_{-2\langle o_{-1}\rangle}, s_{-1\langle o_{-1}\rangle}\} \\ \{s_{1\langle o_{-2}\rangle}\} & \{s_{2\langle o_1\rangle}\} & \{s_{0\langle o_0\rangle}\} & \{s_{-2\langle o_{-3}\rangle}\} \\ \{s_{-1\langle o_2\rangle}, s_{-1\langle o_0\rangle}\} & \{s_{3\langle o_1\rangle}, s_{2\langle o_1\rangle}, s_{1\langle o_1\rangle}\} & \{s_{2\langle o_3\rangle}\} & \{s_{0\langle o_0\rangle}\} \end{pmatrix}$$

$$\widetilde{H}_{S_O}^{7} = \begin{pmatrix} \{s_{0\langle o_0\rangle}\} & \{s_{-1\langle o_1\rangle}\} & \{s_{0\langle o_3\rangle}, s_{1\langle o_3\rangle}, s_{2\langle o_3\rangle}\} & \{s_{-1\langle o_{-1}\rangle}\} \\ \{s_{1\langle o_{-1}\rangle}\} & \{s_{0\langle o_0\rangle}\} & \{s_{0\langle o_{-1}\rangle}\} & \{s_{-1\langle o_1\rangle}, s_{-1\langle o_3\rangle}\} \\ \{s_{0\langle o_{-3}\rangle}, s_{-1\langle o_{-3}\rangle}, s_{-2\langle o_{-3}\rangle}\} & \{s_{0\langle o_1\rangle}\} & \{s_{0\langle o_0\rangle}\} & \{s_{1\langle o_1\rangle}\} \\ \{s_{1\langle o_1\rangle}\} & \{s_{1\langle o_{-1}\rangle}, s_{1\langle o_{-3}\rangle}\} & \{s_{-1\langle o_{-1}\rangle}\} & \{s_{0\langle o_0\rangle}\} \end{pmatrix}$$

$$\widetilde{H}_{S_O}^{8} = \begin{pmatrix} \{s_{0\langle o_0\rangle}\} & \{s_{1\langle o_3\rangle}\} & \{s_{-1\langle o_2\rangle}\} & \{s_{-1\langle o_{-1}\rangle}\} \\ \{s_{1\langle o_3\rangle}\} & \{s_{0\langle o_0\rangle}\} & \{s_{-2\langle o_1\rangle}, s_{-1\langle o_1\rangle}, s_{0\langle o_1\rangle}\} & \{s_{-1\langle o_1\rangle}\} \\ \{s_{1\langle o_{-2}\rangle}\} & \{s_{2\langle o_{-1}\rangle}, s_{1\langle o_{-1}\rangle}, s_{0\langle o_{-1}\rangle}\} & \{s_{0\langle o_0\rangle}\} & \{s_{1\langle o_{-1}\rangle}\} \\ \{s_{1\langle o_1\rangle}\} & \{s_{-1\langle o_1\rangle}\} & \{s_{1\langle o_{-1}\rangle}\} & \{s_{0\langle o_0\rangle}\} \end{pmatrix}$$

$$\widetilde{H}_{S_O}^{9} = \begin{pmatrix} \{s_{0\langle o_0\rangle}\} & \{s_{-2\langle o_{-1}\rangle}, s_{-1\langle o_{-1}\rangle}, s_{0\langle o_{-1}\rangle}\} & \{s_{2\langle o_2\rangle}\} & \{s_{3\langle o_3\rangle}\} \\ \{s_{2\langle o_1\rangle}, s_{1\langle o_1\rangle}, s_{0\langle o_1\rangle}\} & \{s_{0\langle o_0\rangle}\} & \{s_{1\langle o_{-2}\rangle}, s_{1\langle o_0\rangle}\} & \{s_{1\langle o_3\rangle}\} \\ \{s_{-2\langle o_{-2}\rangle}\} & \{s_{-1\langle o_2\rangle}, s_{-1\langle o_0\rangle}\} & \{s_{0\langle o_0\rangle}\} & \{s_{-2\langle o_1\rangle}\} \\ \{s_{-3\langle o_{-3}\rangle}\} & \{s_{-1\langle o_{-3}\rangle}\} & \{s_{2\langle o_{-1}\rangle}\} & \{s_{0\langle o_0\rangle}\} \end{pmatrix}$$

$$\widetilde{H}_{S_O}^{10} = \begin{pmatrix} \{s_{0\langle o_0\rangle}\} & \{s_{2\langle o_{-1}\rangle}\} & \{s_{0\langle o_{-2}\rangle}\} & \{s_{0\langle o_1\rangle}, s_{1\langle o_1\rangle}, s_{2\langle o_1\rangle}\} \\ \{s_{-2\langle o_1\rangle}\} & \{s_{0\langle o_0\rangle}\} & \{s_{-2\langle o_{-1}\rangle}\} & \{s_{-1\langle o_1\rangle}, s_{-1\langle o_3\rangle}\} \\ \{s_{0\langle o_2\rangle}\} & \{s_{2\langle o_1\rangle}\} & \{s_{0\langle o_0\rangle}\} & \{s_{-3\langle o_{-2}\rangle}\} \\ \{s_{0\langle o_{-1}\rangle}, s_{-1\langle o_{-1}\rangle}, s_{-2\langle o_{-1}\rangle}\} & \{s_{1\langle o_{-1}\rangle}, s_{1\langle o_{-3}\rangle}\} & \{s_{3\langle o_2\rangle}\} & \{s_{0\langle o_0\rangle}\} \end{pmatrix}$$

$$\widetilde{H}_{S_O}^{11} = \begin{pmatrix} \{s_{0\langle o_0\rangle}\} & \{s_{1\langle o_1\rangle}, s_{2\langle o_1\rangle}, s_{3\langle o_1\rangle}\} & \{s_{-1\langle o_2\rangle}\} & \{s_{-2\langle o_{-1}\rangle}\} \\ \{s_{-1\langle o_{-1}\rangle}, s_{-2\langle o_{-1}\rangle}, s_{-3\langle o_{-1}\rangle}\} & \{s_{0\langle o_0\rangle}\} & \{s_{0\langle o_{-2}\rangle}, s_{0\langle o_0\rangle}\} & \{s_{-1\langle o_0\rangle}\} \\ \{s_{1\langle o_{-2}\rangle}\} & \{s_{0\langle o_2\rangle}, s_{0\langle o_0\rangle}\} & \{s_{0\langle o_0\rangle}\} & \{s_{2\langle o_3\rangle}\} \\ \{s_{2\langle o_1\rangle}\} & \{s_{1\langle o_0\rangle}\} & \{s_{-2\langle o_{-3}\rangle}\} & \{s_{0\langle o_0\rangle}\} \end{pmatrix}$$

$$\widetilde{H}_{S_O}^{12} = \begin{pmatrix} \{s_{0\langle o_0\rangle}\} & \{s_{0\langle o_{-2}\rangle}\} & \{s_{3\langle o_{-1}\rangle}\} & \{s_{1\langle o_2\rangle}, s_{2\langle o_2\rangle}, s_{3\langle o_2\rangle}\} \\ \{s_{0\langle o_2\rangle}\} & \{s_{0\langle o_0\rangle}\} & \{s_{1\langle o_{-1}\rangle}\} & \{s_{2\langle o_1\rangle}, s_{2\langle o_3\rangle}\} \\ \{s_{-3\langle o_1\rangle}\} & \{s_{-1\langle o_1\rangle}\} & \{s_{0\langle o_0\rangle}\} & \{s_{-2\langle o_1\rangle}\} \\ \{s_{-1\langle o_{-2}\rangle}, s_{-2\langle o_{-2}\rangle}, s_{-3\langle o_{-2}\rangle}\} & \{s_{-2\langle o_{-1}\rangle}, s_{-2\langle o_{-3}\rangle}\} & \{s_{2\langle o_{-1}\rangle}\} & \{s_{0\langle o_0\rangle}\} \end{pmatrix}$$

$$\widetilde{H}_{S_O}^{13} = \begin{pmatrix} \{s_{0\langle o_0\rangle}\} & \{s_{1\langle o_{-2}\rangle}, s_{2\langle o_{-2}\rangle}, s_{3\langle o_{-2}\rangle}\} & \{s_{-1\langle o_{-3}\rangle}, s_{-1\langle o_{-1}\rangle}\} & \{s_{0\langle o_2\rangle}\} \\ \{s_{-1\langle o_2\rangle}, s_{-2\langle o_2\rangle}, s_{-3\langle o_2\rangle}\} & \{s_{0\langle o_0\rangle}\} & \{s_{-3\langle o_1\rangle}\} & \{s_{-1\langle o_2\rangle}\} \\ \{s_{1\langle o_3\rangle}, s_{1\langle o_1\rangle}\} & \{s_{3\langle o_{-1}\rangle}\} & \{s_{0\langle o_0\rangle}\} & \{s_{-2\langle o_{-3}\rangle}\} \\ \{s_{0\langle o_{-2}\rangle}\} & \{s_{1\langle o_{-2}\rangle}\} & \{s_{2\langle o_3\rangle}\} & \{s_{0\langle o_0\rangle}\} \end{pmatrix}$$

$$\widetilde{H}_{S_O}^{14} = \begin{pmatrix} \{s_{0\langle o_0\rangle}\} & \{s_{0\langle o_3\rangle}, s_{1\langle o_3\rangle}, s_{2\langle o_3\rangle}\} & \{s_{-1\langle o_3\rangle}\} & \{s_{-1\langle o_{-1}\rangle}\} \\ \{s_{0\langle o_{-3}\rangle}, s_{-1\langle o_{-3}\rangle}, s_{-2\langle o_{-3}\rangle}\} & \{s_{0\langle o_0\rangle}\} & \{s_{-1\langle o_1\rangle}\} & \{s_{-1\langle o_{-1}\rangle}\} \\ \{s_{1\langle o_{-3}\rangle}\} & \{s_{1\langle o_{-1}\rangle}\} & \{s_{0\langle o_0\rangle}\} & \{s_{1\langle o_3\rangle}\} \\ \{s_{1\langle o_1\rangle}\} & \{s_{1\langle o_1\rangle}\} & \{s_{-1\langle o_{-3}\rangle}\} & \{s_{0\langle o_0\rangle}\} \end{pmatrix}$$

$$\widetilde{H}^{15}_{S_O}=\begin{pmatrix}\{s_{0\langle o_0\rangle}\} & \{s_{-1\langle o_{-1}\rangle}\} & \{s_{1\langle o_{-3}\rangle},s_{1\langle o_{-1}\rangle}\} & \{s_{1\langle o_{-3}\rangle},s_{2\langle o_{-3}\rangle},s_{3\langle o_{-3}\rangle}\}\\ \{s_{1\langle o_{1}\rangle}\} & \{s_{0\langle o_0\rangle}\} & \{s_{3\langle o_{-1}\rangle}\} & \{s_{3\langle o_{3}\rangle}\}\\ \{s_{-1\langle o_{3}\rangle},s_{-1\langle o_{1}\rangle}\} & \{s_{-3\langle o_{1}\rangle}\} & \{s_{0\langle o_0\rangle}\} & \{s_{-1\langle o_{3}\rangle}\}\\ \{s_{-1\langle o_{3}\rangle},s_{-2\langle o_{3}\rangle},s_{-3\langle o_{3}\rangle}\} & \{s_{-3\langle o_{-3}\rangle}\} & \{s_{1\langle o_{-3}\rangle}\} & \{s_{0\langle o_0\rangle}\}\end{pmatrix}$$

$$\widetilde{H}^{16}_{S_O}=\begin{pmatrix}\{s_{0\langle o_0\rangle}\} & \{s_{-1\langle o_{1}\rangle}\} & \{s_{3\langle o_{-2}\rangle}\} & \{s_{3\langle o_{1}\rangle},s_{3\langle o_{3}\rangle}\}\\ \{s_{1\langle o_{-1}\rangle}\} & \{s_{0\langle o_0\rangle}\} & \{s_{2\langle o_{2}\rangle}\} & \{s_{0\langle o_{3}\rangle},s_{1\langle o_{3}\rangle},s_{2\langle o_{3}\rangle}\}\\ \{s_{-3\langle o_{2}\rangle}\} & \{s_{-2\langle o_{-2}\rangle}\} & \{s_{0\langle o_0\rangle}\} & \{s_{-1\langle o_{3}\rangle}\}\\ \{s_{-3\langle o_{-1}\rangle},s_{-3\langle o_{-3}\rangle}\} & \{s_{0\langle o_{-3}\rangle},s_{-1\langle o_{-3}\rangle},s_{-2\langle o_{-3}\rangle}\} & \{s_{1\langle o_{-3}\rangle}\} & \{s_{0\langle o_0\rangle}\}\end{pmatrix}$$

$$\widetilde{H}^{17}_{S_O}=\begin{pmatrix}\{s_{0\langle o_0\rangle}\} & \{s_{0\langle o_{1}\rangle}\} & \{s_{1\langle o_{1}\rangle},s_{1\langle o_{3}\rangle}\} & \{s_{-2\langle o_{-2}\rangle}\}\\ \{s_{0\langle o_{-1}\rangle}\} & \{s_{0\langle o_0\rangle}\} & \{s_{2\langle o_{0}\rangle}\} & \{s_{-3\langle o_{1}\rangle},s_{-2\langle o_{1}\rangle},s_{-1\langle o_{1}\rangle}\}\\ \{s_{-1\langle o_{-1}\rangle},s_{-1\langle o_{-3}\rangle}\} & \{s_{-2\langle o_{0}\rangle}\} & \{s_{0\langle o_0\rangle}\} & \{s_{3\langle o_{-3}\rangle}\}\\ \{s_{2\langle o_{2}\rangle}\} & \{s_{3\langle o_{-1}\rangle},s_{2\langle o_{-1}\rangle},s_{1\langle o_{-1}\rangle}\} & \{s_{-3\langle o_{3}\rangle}\} & \{s_{0\langle o_0\rangle}\}\end{pmatrix}$$

$$\widetilde{H}^{18}_{S_O}=\begin{pmatrix}\{s_{0\langle o_0\rangle}\} & \{s_{-1\langle o_{1}\rangle}\} & \{s_{3\langle o_{0}\rangle},s_{3\langle o_{2}\rangle}\} & \{s_{2\langle o_{3}\rangle}\}\\ \{s_{1\langle o_{-1}\rangle}\} & \{s_{0\langle o_0\rangle}\} & \{s_{1\langle o_{-1}\rangle}\} & \{s_{1\langle o_{1}\rangle},s_{2\langle o_{1}\rangle},s_{3\langle o_{1}\rangle}\}\\ \{s_{-3\langle o_{0}\rangle},s_{-3\langle o_{-2}\rangle}\} & \{s_{-1\langle o_{1}\rangle}\} & \{s_{0\langle o_0\rangle}\} & \{s_{-2\langle o_{1}\rangle}\}\\ \{s_{-2\langle o_{-3}\rangle}\} & \{s_{-1\langle o_{-1}\rangle},s_{-2\langle o_{-1}\rangle},s_{-3\langle o_{-1}\rangle}\} & \{s_{2\langle o_{-1}\rangle}\} & \{s_{0\langle o_0\rangle}\}\end{pmatrix}$$

$$\widetilde{H}^{19}_{S_O}=\begin{pmatrix}\{s_{0\langle o_0\rangle}\} & \{s_{-1\langle o_{0}\rangle}\} & \{s_{1\langle o_{-1}\rangle},s_{2\langle o_{-1}\rangle},s_{3\langle o_{-1}\rangle}\} & \{s_{2\langle o_{-3}\rangle}\}\\ \{s_{1\langle o_{0}\rangle}\} & \{s_{0\langle o_0\rangle}\} & \{s_{2\langle o_{0}\rangle},s_{2\langle o_{2}\rangle}\} & \{s_{3\langle o_{3}\rangle}\}\\ \{s_{-1\langle o_{1}\rangle},s_{-2\langle o_{1}\rangle},s_{-3\langle o_{1}\rangle}\} & \{s_{-2\langle o_{0}\rangle},s_{-2\langle o_{-2}\rangle}\} & \{s_{0\langle o_0\rangle}\} & \{s_{0\langle o_{-2}\rangle}\}\\ \{s_{-2\langle o_{3}\rangle}\} & \{s_{-3\langle o_{-3}\rangle}\} & \{s_{0\langle o_{2}\rangle}\} & \{s_{0\langle o_0\rangle}\}\end{pmatrix}$$

$$\widetilde{H}^{20}_{S_O}=\begin{pmatrix}\{s_{0\langle o_0\rangle}\} & \{s_{1\langle o_{2}\rangle},s_{2\langle o_{2}\rangle},s_{3\langle o_{2}\rangle}\} & \{s_{-1\langle o_{2}\rangle}\} & \{s_{-1\langle o_{-3}\rangle}\}\\ \{s_{-1\langle o_{-2}\rangle},s_{-2\langle o_{-2}\rangle},s_{-3\langle o_{-2}\rangle}\} & \{s_{0\langle o_0\rangle}\} & \{s_{0\langle o_{-2}\rangle},s_{0\langle o_{0}\rangle}\} & \{s_{-1\langle o_{0}\rangle},s_{-1\langle o_{2}\rangle}\}\\ \{s_{1\langle o_{-2}\rangle}\} & \{s_{0\langle o_{2}\rangle},s_{0\langle o_{0}\rangle}\} & \{s_{0\langle o_0\rangle}\} & \{s_{2\langle o_{3}\rangle}\}\\ \{s_{1\langle o_{3}\rangle}\} & \{s_{1\langle o_{0}\rangle},s_{1\langle o_{-2}\rangle}\} & \{s_{-2\langle o_{-3}\rangle}\} & \{s_{0\langle o_0\rangle}\}\end{pmatrix}$$

Next, we start the consensus reaching process according to the proposed Algorithm 5.5.

Step 1. Cluster all students into five groups G_ϕ $(\phi = 1, 2, \ldots, 5)$ and calculate the weights of all groups, the results are shown in Fig. 5.6 and Table 5.2.

Step 2. Aggregate all group preference matrices $\widetilde{H}^{G_\phi}_{S_O} = (h^{G_\phi}_{S_{O_{ij}}})_{4\times 4}(\phi = 1, 2, \ldots, 5)$ into the overall preference matrix $\widetilde{H}^{c}_{S_O} = (h^{c}_{S_{O_{ij}}})_{4\times 4}$.

Step 3. The consensus threshold value is given as $\xi = 0.85$. Then the consensus degree of each group $CD(\widetilde{H}^{G_\phi(0)}_{S_O})$ $(\phi = 1, 2, \ldots, 5)$ and the overall consensus degree $OCD^{(0)}$ are obtained and shown in Table 5.3.

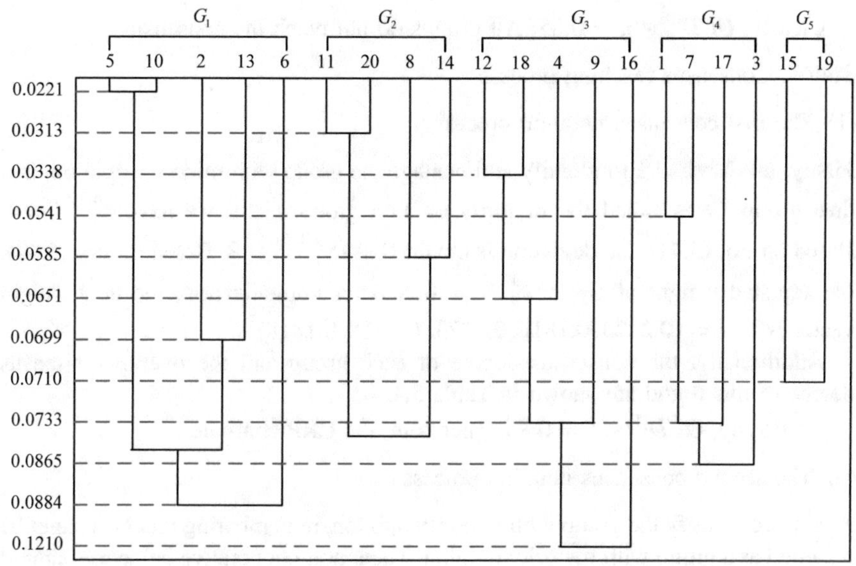

Fig. 5.6 The clustering result

Table 5.2 The clustering result and the weight information of each group

Group	The experts in each group	The weight vector of each group	The weight of each group
G_1	$\{e^2, e^5, e^6, e^{10}, e^{13}\}$	$(0.2, 0.2, 0.2, 0.2, 0.2)^T$	$w_1^{(0)} = 0.25$
G_2	$\{e^8, e^{11}, e^{14}, e^{20}\}$	$(0.25, 0.25, 0.25, 0.25)^T$	$w_2^{(0)} = 0.2$
G_3	$\{e^4, e^9, e^{12}, e^{16}, e^{18}\}$	$(0.2, 0.2, 0.2, 0.2, 0.2)^T$	$w_3^{(0)} = 0.25$
G_4	$\{e^1, e^3, e^7, e^{17}\}$	$(0.25, 0.25, 0.25, 0.25)^T$	$w_4^{(0)} = 0.2$
G_5	$\{e^{15}, e^{19}\}$	$(0.5, 0.5)^T$	$w_5^{(0)} = 0.1$

Table 5.3 The consensus degrees of groups and overall consensus degree

	$G_1^{(0)}$	$G_2^{(0)}$	$G_3^{(0)}$	$G_4^{(0)}$	$G_5^{(0)}$
$CD(\widetilde{H}_{S_O}^{G_\phi^{(0)}})$	0.7789	0.797	0.7871	0.7980	0.7598
$OCD^{(0)}$			0.7843		

Clearly, $OCD^{(0)} < \xi = 0.85$. All groups do not reach the consensus.

Step 4. Consensus reaching process.

(1) The first consensus iteration process

Firstly, use Method 1 to identify and manage the group with minority opinion, and then group G_5 is called the minority opinion group. Next, we have $n_5^{M1(0)} = 4$. Based on Eq. (5.41), the deviation is obtained as $dv_5^{M1(0)} = 2$. Based on Eq. (5.42), the adjusted weight of G_5 is $w_5^{M1(0)} = 0.2$. After normalization, the new weight vector is $w^{(1)} = (0.2273, 0.1818, 0.2273, 0.1818, 0.1818)^T$.

Additionally, the consensus degree of each group and the overall consensus degree of this round are shown in Table 5.4.

Obviously, $OCD^{(1)} < \xi = 0.85$. Therefore, the CRP continues.

(2) The second consensus iteration process

Again, we identify the group with minority opinion, remembering that G_1 cannot be regarded as a group with minority opinion. Then, non-cooperative behaviors should be taken into consideration. The remaining groups provide their adjustment suggestions for the opinions of G_1 as $\vartheta_{G_2G_1}^{(1)} = 0.56$, $\vartheta_{G_3G_1}^{(1)} = 0.6$, $\vartheta_{G_4G_1}^{(1)} = 0.65$ and $\vartheta_{G_5G_1}^{(1)} = 0.43$. Using Eq. (5.39), we get $\vartheta_{G_1}^{O(1)} = 0.6658$. Then the expected adjustment suggestion interval is $\overline{\vartheta}_{G_1}^{(1)} = [0.43, 0.6658]$. The students in G_1 provide their subjective adjustment coefficient $\vartheta_{G_1}^{S(1)} = 0.8$. With Eq. (5.43), $N^{(1)}(G_1) = 0$, which means that G_1 can be regarded as a completely cooperative group. Therefore, it is unnecessary to change the weight of G_1. Considering $\vartheta_{G_1}^{S(1)} > \vartheta_{G_1}^{O(1)}$, the comprehensive adjustment coefficient is $\vartheta_{G_1}^{(1)} = \vartheta_{G_1}^{S(1)} = 0.8$, and it is utilized to repair the preference of G_1 on the basis of Eq. (5.40).

Then the consensus degree of each group and the overall consensus degree of this round are shown in Table 5.5.

Obviously, $OCD^{(2)} < \xi$. Therefore, the CRP continues.

(3) The third consensus iteration process

Firstly, group G_2 can be regarded as a group with minority opinion. However, there is only one group that supports the opinion of group G_2. Thus, we need to deal with non-cooperative behaviors. Based on the discussion results and the overall

Table 5.4 The consensus degrees of groups and overall consensus degree

	$G_1^{(1)}$	$G_2^{(1)}$	$G_3^{(1)}$	$G_4^{(1)}$	$G_5^{(1)}$
$CD(\widetilde{H}_{S_O}^{G_\phi(1)})$	0.7655	0.7856	0.8015	0.7929	0.7816
$OCD^{(1)}$			0.7854		

Table 5.5 The consensus degrees of groups and overall consensus degree

	$G_1^{(2)}$	$G_2^{(2)}$	$G_3^{(2)}$	$G_4^{(2)}$	$G_5^{(2)}$
$CD(\widetilde{H}_{So}^{G_\phi(2)})$	0.9105	0.7801	0.8079	0.8089	0.8062
$OCD^{(2)}$			0.8227		

Table 5.6 The consensus degrees of groups and overall consensus degree

	$G_1^{(3)}$	$G_2^{(3)}$	$G_3^{(3)}$	$G_4^{(3)}$	$G_5^{(3)}$
$CD(\widetilde{H}_{So}^{G_\phi(3)})$	0.8960	0.9305	0.8399	0.7847	0.8318
$OCD^{(3)}$			0.8566		

consensus degree, the remaining groups provide their adjustment suggestions for the opinions of G_2 as $\vartheta_{G_1G_2}^{(2)} = 0.85$, $\vartheta_{G_3G_2}^{(2)} = 0.7$, $\vartheta_{G_4G_2}^{(2)} = 0.67$, and $\vartheta_{G_5G_2}^{(2)} = 0.9$. Using Eq. (5.39), we get $\vartheta_{G_1}^{O(2)} = 0.6682$. Then the expected adjustment suggestion interval is $\overline{\vartheta}_{G_2}^{(2)} = [0.6682, 0.9]$. Suppose that the group G_2 provides their subjective adjustment coefficient $\vartheta_{G_1}^{S(2)} = 0.85$. Based on Eq. (5.43), $N^{(1)}(G_1) = 2157$, which means that G_2 is regarded as a partly non-cooperative group. Therefore, G_2's weight needs to be adjusted to reduce its reflection. Based on Eq. (5.44), we get $w_2^{M2(2)} = 0.1454$. After normalization, the new weight vector is $w^{(3)} = (0.2359, 0.1509, 0.2359, 0.1887, 0.1887)^T$.

Considering that $\vartheta_{G_2}^{S(2)} > \vartheta_{G_2}^{O(2)}$, the comprehensive adjustment coefficient is $\vartheta_{G_2}^{(2)} = \vartheta_{G_2}^{S(2)}, = 0.85$, which can be utilized to repair the preference of G_2 on the basis of Eq. (5.39).

Then the consensus degree of each group and the overall consensus degree of this round are shown in Table 5.6.

Obviously, $OCD^{(3)} > \xi = 0.85$. Therefore, the CRP is over.

Step 5. Let $\mathbb{Z}^* = 3$. Output the final group preference matrices $\widetilde{H}_{So}^{G_\phi(3)} = (h_{So}^{G_\phi(3)})_{4\times4}$ $(\phi = 1, 2, \ldots, 5)$ and the final overall preference matrix $\widetilde{H}_{So}^{c(3)} = (h_{So}^{c(3)})_{4\times4}$.

Step 6. Calculate the expected values of all alternatives and obtain $SV(A_1) = 2.2324$, $SV(A_2) = 2.2840$, $SV(A_3) = 1.6628$ and $SV(A_4) = 1.8232$. Then the ranking of all alternatives is $A_2 \succ A_1 \succ A_4 \succ A_3$, which means that "the energy consumption structure dominated by fossil energy" is the main reason.

References

Brandt F, Conitzer V, Endriss U, Lang J, Procaccia AD (2016) Handbook of computational social choice. Cambridge University Press

Chevaleyre Y, Endriss U, Lang J, Maudet N (2007) A short introduction to computational social choice. In: van Leeuwen J, Italiano GF, van der Hoek W, Meinel C, Sack H, Plášil F (2007) SOFSEM 2007: Theory and practice of computer science. Lecture notes in computer science, vol 4362, Springer, Berlin, Heidelberg, pp 51–69

Da QL, Liu XW (1999) Interval number linear programming and the satisfactory solution. Syst Eng Theor Pract 19:3–7

Dong YC, Zhang HJ, Herrera-Viedma E (2016) Integrating experts' weights generated dynamically into the consensus reaching process and its applications in managing non-cooperative behaviors. Decis Support Syst 84:1–15

Gou XJ, Xu ZS, Herrera F (2018) Consensus reaching process for large-scale group decision making with double hierarchy hesitant fuzzy linguistic preference relations. Knowl-Based Syst 157:20–33

Gou XJ, Xu ZS, Liao HC, Herrera F (2020) A consensus model to manage minority opinions and noncooperative behaviors in large-scale GDM with double hierarchy linguistic preference relations. IEEE Trans Cybern. https://doi.org/10.1109/TCYB.2020.2985069

Hao ZN, Xu ZS, Zhao H, Fujita H (2018) A dynamic weight determination approach based on the intuitionistic fuzzy Bayesian network and its application to emergency decision making. IEEE Trans Fuzzy Syst 26(4):1893–1907

Hsu HM, Chen CT (1996) Aggregation of fuzzy opinions under group decision making. Fuzzy Sets Syst 79(3):279–285

Labella Á, Liu Y, Rodríguez RM, Martínez L (2018) Analyzing the performance of classical consensus models in largescale group decision making: a comparative study. Appl Soft Comput 67:677–690

Lee LS (2000) Optimal consensus of fuzzy opinions under group decision making environment. Fuzzy Sets Syst 132(3):303–315

Linstone H, Turoff M (1975) The Delphi method: techniques and applications. Addison-Wesley Publishing Company, London

Liu R, Chen XH (2006) Improved clustering algorithm and its application in complex huge group decision-making. J Syst Eng Electron 28:1695–1699

Liu BS, Huo TF, Liao PC, Gong J, Xue B (2015a) A group decision-making aggregation model for contractor selection in large scale construction projects based on two-stage partial least squares (PLS) path modeling. Group Decis Negot 24:855–883

Liu BS, Shen YH, Chen Y, Chen XH, Chen Y, Wang XQ (2015b) A two-layer weight determination method for complex multi-attribute large-group decision making experts in a linguistic environment. Inf Fusion 23:156–165

Liu BS, Shen YH, Chen XH, Chen Y, Wang XQ (2014a) A partial binary tree DEA-DA cyclic classification model for decision makers in complex multi-attribute large-group interval-valued intuitionistic fuzzy decision making problems. Inf Fusion 18(1):119–130

Liu BS, Shen YH, Chen XH, Sun H, Chen Y (2014b) A complex multi-attribute large-group PLS decision-making method in the interval-valued intuitionistic fuzzy environment. Appl Math Model 38:4512–4527

Liu BS, Shen YH, Zhang W, Chen XH, Wang XQ (2015c) An interval-valued intuitionistic fuzzy principal component analysis model-based method for complex multi-attribute large-group decision-making. Eur J Oper Res 245(1):209–225

Liu Y, Fan ZP, Zhang X (2016) A method for large group decision making based on evaluation information provided by participators from multiple groups. Inf Fusion 29(C):132–141

Palomares I, Martínez L, Herrera F (2014a) A consensus model to detect and manage noncooperative behaviors in large-scale group decision making. IEEE Trans Fuzzy Syst 22 (3):516–530

Palomares I, Martínez L, Herrera F (2014b) MENTOR: A graphical monitoring tool of preferences evolution in large-scale group decision making. Knowl-Based Syst 58:66–74
Pang Q, Wang H, Xu ZS (2016) Probabilistic linguistic term sets in multi-attribute group decision making. Inf Sci 369:128–143
Quesada FJ, Palomares I, Martínez L (2015) Managing experts behavior in large-scale consensus reaching processes with uninorm aggregation operators. Appl Soft Comput 35:873–887
Ramanathan R, Ganesh LS (1994) Group preference aggregation methods employed in AHP: an evaluation and an intrinsic process for deriving members' weightages. Eur J Oper Res 79 (2):249–265
Wu T, Liu XW (2016) An interval type-2 fuzzy clustering solution for large-scale multiple-criteria group decision-making problems. Knowl-Based Syst 114:118–127
Wu ZB, Xu JP (2018) A consensus model for large-scale group decision making with hesitant fuzzy information and changeable clusters. Inf Fusion 41:217–231
Xia MM, Xu ZS (2011) Hesitant fuzzy information aggregation in decision making. Int J Approx Reason 52:395–407
Xiong CQ, Li DH, Jin LH (2008) Group consistency analysis for protecting the minority view. Syst Eng Theory Pract 10:102–107
Xu ZS (2009) An automatic approach to reaching consensus in multiple attribute group decision making. Comput Ind Eng 56(4):1369–1374
Xu XH, Zhang LY, Wan QF (2012) A variation coefficient similarity measure and its application in emergency group decision-making. Syst Eng Procedia 5:119–124
Xu XH, Du ZJ, Chen XH (2015) Consensus model for multi-criteria large-group emergency decision making considering non-cooperative behaviors and minority opinions. Decis Support Syst 79:150–160
Yue ZL (2011) An extended TOPSIS for determining weights of decision experts with interval numbers. Knowl-Based Syst 24(1):146–153
Yue ZL (2012) Approach to group decision making based on determining the weights of experts by using projection method. Appl Math Model 36(7):2900–2910
Yager RR (2001) Penalizing strategic preference manipulation in multi-agent decision making. IEEE Trans Fuzzy Syst 9(3):393–403
Zhang XL (2018) A novel probabilistic linguistic approach for large-scale group decision making with incomplete weight information. Int J Fuzzy Syst 20:2245–2256
Zhang HJ, Dong YC, Herrera-Viedma E (2018) Consensus building for the heterogeneous LSGDM with the individual concerns and satisfactions. IEEE Trans Fuzzy Syst 26(2):884–898
Zhu J, Zhang ST, Chen Y, Zhang LL (2016) A hierarchical clustering approach based on three-dimensional gray relational analysis for clustering a large group of decision makers with double information. Group Decis Negot 25(2):325–354

Printed in the United States
by Baker & Taylor Publisher Services